Lectures on Elementary Particles and Quantum Field Theory
Volume 2

Edited by
Stanley Deser, Marc Grisaru, and Hugh Pendleton

Lectures by Rudolph Haag, Maurice Jacob, Henry Primakoff, Michael C. Reed, and Bruno Zumino

Lectures on Elementary Particles and Quantum Field Theory

1970 Brandeis University Summer Institute in Theoretical Physics, Volume 2

THE M.I.T. PRESS
Cambridge, Massachusetts, and London, England

ISBN 0 262 04032 8 (hardcover)
ISBN 0 262 54015 0 (paperback)

Library of Congress catalog card number: 70-138840

CONTENTS

FOREWORD

The notes contained in this volume, the second of two, are based on a lecture series delivered by the authors at the 1970 Brandeis Summer Institute in Theoretical Physics. In order to permit rapid publication, they are printed directly from typewritten copy. We are grateful to the lecturers for their assistance in the preparation of the typescripts, and to Betty Griffin, Sylvia Pendleton, and Mary Sider for their typing. We thank Geraldine Prentice, secretary of the Brandeis Summer Institute, for her continuing helpfulness.

The 1970 Brandeis Summer Institute in Theoretical Physics was made possible by the generous support of the National Science Foundation and of the North Atlantic Treaty Organization.

The first volume of notes contains lectures by Steven Adler on perturbation theory anomalies, by Stanley Mandelstam on the Veneziano formula and its ramifications, by Steven Weinberg on soft-meson theorems stemming from dynamical symmetries, and by Wolfhart Zimmermann on local operator products in renormalizable field theories.

Stanley Deser
Marc Grisaru
Hugh Pendleton
—Editors

Observables and Fields

Rudolf Haag
II. Institut für Theoretische Physik
der Universität Hamburg
Hamburg, Germany

CONTENTS

Introduction

The following story was reported to me. A few years ago Klaus Hepp gave some lectures in the Brandeis summer school. At some stage he praised the beauty of axiomatic field theory. Next day he found the note on the blackboard:

"Axiom 1: Axiomatic Field Theory is beautiful in an empty sort of way."

Presumably this note expresses also pretty accurately the feelings of the majority of today's audience and indeed there is an element of truth in it. Specifically, after about 18 years of hard efforts, the principal objective of this enterprise has not yet been achieved. This objective was to find out whether an adequate framework for the description of elementary particle physics could be developed within the conceptual structure provided by the principles of quantum physics, special relativity theory and locality (= "Nahwirkungsprinzip"). Of course, from the point of view of the development of basic physical theory, this question has been one of the central themes for the past 35 years and is by no means a monopoly of axiomatic field theorists. What distinguishes various groups is not the question itself, but rather the attitude towards it. There are three major ideologies:

1) The answer to the above question will certainly be no. We need a radical change of our concepts, some brilliant new idea. It is just as futile to approach elementary particle physics with the conceptual structure of 1930 as it was to attack atomic physics within the frame of classical mechanics. Therefore we should look for daring new approaches. Some examples: modification of geometry by assumption of a fundamental length, elimination of concepts which are far removed from experimental possibilities or, on the formalistic level, non associative algebra, indefinite metric in Hilbert spaces, etc.

2) The answer may be yes if we are sufficiently careful. It is worthwhile to develop a framework which incorporates the old principles, formulating them precisely, separating the essential and the peripheral features of traditional Quantum Field Theory, recognizing the numerous mathematical pitfalls. One should then demonstrate that this framework is internally consistent and study whether it leads to any consequences which are in disagreement with experience.

3) The time is not ripe for any assessment of the fundamental principles. The most fruitful task for the theoretician at present is to analyze experiments, looking for regularities and for phenomenological models which describe the essential features.

It is unfortunately in the nature of ideologies that they tend to crystallize. One has to make a determined effort to keep the channels of communication between the different camps open. I think that the organizers of this summer school had this need in mind when they asked me to lecture here and therefore I do not feel apologetic for exposing you to some ideas and problems in axiomatic field theory.

Our first concern will be with the "axioms" themselves, the formulation of the input assumptions. In the course of the years there has been some development both in the direction of simplification by recognizing the essential elements and in the direction of enriching the structure. Let me give a brief sketch of this development.

I. Axiomatic Quantum Field Theory in Various Formulations

A. The Simplest Kind of Field Theory

In the years 1953-56 the motivation was provided by the divergence difficulties of standard Lagrangian field theory models and the wish to see whether the renormalization procedures could be welded into a mathematically well defined scheme independent of a perturbation expansion. For this purpose it seemed adequate to consider the simplest type of field theory, namely that of a single neutral scalar field A describing a single type of particle (of course with interaction). The generalization to several fields with more complicated transformation properties seemed obvious and straight forward. In laying out the framework one was guided then by the experience gained from the perturbation treatment of a Lagrangian field theory keeping those structural features of the renormalized perturbation solution which could be precisely expressed mathematically. This led to the following principal assumptions:

1) Principles of Quantum Physics. Essentially the mathemathical and conceptual structure outlined in the books by Dirac and von Neumann. It may suffice here to say that the mathematics deals with a Hilbert space $\mathcal{H}$ whose vectors correspond to physical states; observables are represented by self adjoint operators acting in $\mathcal{H}$ and there are the well known rules for calculating probabilities for the results of observations.

2) Poincaré invariance

The Poincaré group (inhomogeneous Lorentz group) consists of translations in space time and homogeneous Lorentz transformations. A general element is denoted by (a, Λ) where a is a 4-vector of translation, Λ a homogeneous Lorentz transformation and the notation suggests that Λ is applied first, a later. We assume that the Poincaré group is represented by unitary operators in $\mathcal{H}$

$$(a,\Lambda) \rightarrow U(a,\Lambda).$$

The action of the unitary operator $U(a,\Lambda)$ on a vector in $\mathcal{H}$ shall have the obvious physical interpretation; i.e., the image vector corresponds to the state which is prepared by the same intrinsic apparatus as the original state but shifted in its space-time placement and motion by (a,Λ). We shall write $U(a)$ instead of $U(a,1)$ (pure translation) and $U(\Lambda)$ instead of $U(0,\Lambda)$ (Lorentz transformation).

Actually the situation is slightly more complicated. One does not need (and in general does not have) a true representation of the Poincaré group but rather a representation of its "covering group." This corresponds to the well known replacement of a Lorentz-matrix Λ by a complex 2 x 2 matrix with determinant 1. The correspondence of Λ to such a matrix $\alpha(\Lambda)$ is determined only up to a sign, see [2],[3].

3) <u>The vacuum state and stability</u>

Writing

$$U(a) = e^{-iP_\mu a^\mu} \tag{1.1}$$

the infinitesimal generators P_μ of the translations may be interpreted as observables. They correspond to the total energy and linear momentum of the system. We assume:

a) There is a vector Ω in $\mathcal{H}$ which is invariant under all $U(a, \Lambda)$. It is the only discrete eigenvector of any $U(a)$. It corresponds to the physical vacuum state. Clearly it has zero energy and momentum.

b) Ω is the ground state of the system. Frequently one makes stronger assumptions on the energy-momentum spectrum in order to introduce particles and to exclude the additional complications associated with the occurrence of zero mass particles.

For instance instead of b)

b') The simultaneous spectrum of the operators P_μ is as in Fig. 1. It consists of the single point $P_\mu = 0$ corresponding to the vacuum state, then the hyperboloid $p^2 = -m^2$, $p^0 > 0$ corresponding to the states of a single particle (mass m) and then the continuum above the 2-particle threshold.

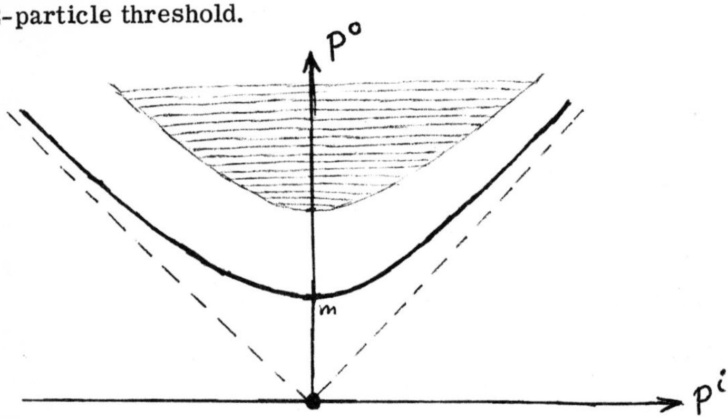

Figure 1.

4) The Field

It had been recognized very early in the development of Quantum Electrodynamics that the field at a point cannot be a proper observable (e.g., the analysis of idealized measurements of electromagnetic field strengths by Bohr and Rosenfeld,[1]). One has to consider averages of the field (denoted in our case by A) over space-time regions, such as $\frac{1}{VT}\int_0^T dt \int_V d^3x\, A(x)$ or, in general, weighted averages with smooth weight functions $f(x)$,

$$A(f) = \int d^4x\, A(x)\, f(x)\,. \tag{1.2}$$

In a handwaving way the mathematical nature of the field can be understood if we have the physical picture that $A(x)$ shall represent an operation on the physical system at the point x. One may anticipate then that such an operation must transfer an unlimited amount of energy-momentum to the system.

To express this expectation more precisely we define some notation (corresponding to a direct integral decomposition of the Hilbert space with respect to energy-momentum). Let $|p,n\rangle$ be a (improper = continuous) basis of state vectors, where p denotes the total energy-momentum of the state and the discrete index n is used to distinguish the states with the same p. A general state may be written

$$\Psi = \sum_n \int d\mu(p)\, \psi_n(p)\, |p,n\rangle\,, \tag{1.3}$$

where the "spectral measure" $d\mu(p)$ may be chosen without loss of generality as

$$d\mu(p) = d^4p\left\{\delta^4(p) + \theta(p^0)\delta(p^2+m^2) + \theta(p^0)\theta(-p^2-4m^2)\right\}, \tag{1.4}$$

the first term corresponding to the vacuum state, the second to the single particle states, and the last to the states above the 2-particle threshold. The normalization of Ψ is given by

$$(\Psi, \Psi) = \sum_n \int d\mu(p)\, |\psi_n(p)|^2 \tag{1.5}$$

corresponding to orthogonality relations

$$\langle p', n' | p, n\rangle\, d\mu(p) = \delta_{n'n}\, \delta^4(p'-p)\,. \tag{1.6}$$

The integrand in (1.3), i.e., the object

$$\Psi_p = \sum_n \psi_n(p)\, |p,n\rangle \tag{1.7}$$

for fixed p can be considered as a vector in a Hilbert space $\mathcal{H}_p$, the metric in $\mathcal{H}_p$ being given by

$$(\Psi_p, \Psi_p)_p = \sum_n |\psi_n(p)|^2\,. \tag{1.7'}$$

We expect now that the matrix elements $\langle p', n' | A(x) | p, n\rangle$ are finite. In fact, we may even assume that the restriction of $A(x)$ which maps from $\mathcal{H}_p$ to $\mathcal{H}_{p'}$ is a bounded operator with a norm $N_{p'p}$. But even for fixed p the norm $N_{p'p}$ will not decrease sufficiently for large p' to make $N^2_{p'p}$ integrable with respect to $d\mu(p')$. In fact as a consequence of covariance and locality of A (assumptions 5 and 6) one would guess that for large p' and fixed p

$$N^2_{p'p} \rightarrow \| A(x)\Omega \|^2_{p'} \equiv \rho(p'^2) \qquad (1.8)$$

because all states with bounded energy-momentum should be equivalent in their aspect around a single point x (since such a state cannot give a singular preferential emphasis to any point).

If we define $\mathcal{D}$ as the set of states with fast enough decrease of $\|\Psi_p\|_p$ so that

$$\int d\mu(p')\, d\mu(p) \; \|\Psi_{p'}\|_{p'} \, N_{p'p} \, \|\Psi_p\| < \infty ,$$

then the matrix elements

$$\langle \Phi | A(x) | \Psi \rangle$$

with both Φ and Ψ from $\mathcal{D}$ will be finite. Thus the field at a point may be regarded as a bilinear form over the domain $\mathcal{D}$. Alternatively, if instead of $A(x)$ we take $A(f)$ we get due to (1.1) and the covariance of A under translations

$$\langle p', n' | A(f) | p, n \rangle = \langle p', n' | A(0) | p, n \rangle \tilde{f}(p'-p) \qquad (1.9)$$

with $\tilde{f}(q) = \int d^4x \, e^{-iq\cdot x} f(x)$.

If $f(x)$ is very smooth then $\tilde{f}(q)$ will have a fast decrease for large q. In particular, if $f(x)$ is infinitely often differentiable with respect to the four coordinates x_μ then $\tilde{f}(q)$ will decrease faster than any power in any direction in q-space. Hence for such f and for $\Psi \in D$ the vector $A(f)\,\Psi$ will exist, provided that the growth of $N_{p'p}$ is bounded by a polynomial. Under these circumstances the domain D will also be stable under the application of $A(f)$.

To summarize: The field A at a point may be regarded as a bilinear form on the domain D but not as an operator. Alternatively, the field averaged with a sufficiently smooth weight function is an operator on D.

Consequences: a) In traditional field theories the dynamics is given by field equations which involve nonlinear functions of the field at a point. The above remarks indicate that it is not at all clear how such nonlinear functions of $A(x)$ can be defined or, in fact, whether they can be defined at all. The answer is known for free fields (where we do not need it) see e.g. [4], [5], and it has been studied for the renormalized perturbation series in some models [6], [7]. It has, however, so far not been possible to incorporate a formulation of a specific dynamical law into the framework of axiomatic field theory.

b) The kinematics of traditional field theories, i.e. the specification of the "degrees of freedom", is given by canonical commutation relations between the field at a fixed time but at different points in space. For the formulation of such relations we do not need the field at a point but only at a sharp time. We must ask therefore whether

$$A(\sigma, t) = \int d^3x \; \sigma(\underset{\sim}{x}) \, A(\underset{\sim}{x}, t)$$

is an operator, when σ is a sufficiently smooth function in 3-dimensional space. The averaging with σ provides a cut-off in spatial momentum transfer but not in energy. Looking at the expected behavior (1.8) of $N_{p'p}$ one sees that such a cut-off will be sufficient if $\rho(\varkappa^2)$ decreases sufficiently fast for large $\varkappa^2$ and thus furnishes a cut-off in the mass transfer. This is the case for a free field, where $\rho(\varkappa^2) = \delta(\varkappa^2 - m^2)$ but probably not in any model with non-vanishing interaction. The expectation that $\rho(\varkappa^2)$ does not have a fast decrease comes originally from the experience with renormalized perturbation expansions. In recent years, however, some progress has been made towards an understanding of the general reasons (independent of consideration of specific models) [8], [9]. If we accept the fact that $\rho(\varkappa^2)$ does not decrease fast enough to make $A(\sigma, t)$ into a well defined operator then also the formulation of kinematical relations becomes a difficult problem. In fact it is questionable whether the separation between kinematics and dynamics makes any sense at all.

5) Covariance of the Field

The field shall have a simple transformation property under the Poincaré group. In the example considered here it is

$$U(a, \Lambda)\, A(x)\, U^{-1}(a, \Lambda) = A(\Lambda x + a) \quad (1.10)$$

6) Locality: Einstein's Causality Principle

Whatever the complete physical interpretation of $A(f)$ may be, one wants to assert that it corresponds to an observable whose measurement involves only the part of space-time in which the function $f(x)$ does not vanish. This space-time region is called the "Support of f." If we accept Einstein's postulate that no physical effect can propagate faster than light then the measurement of $A(f)$ cannot perturb that of $A(g)$ whenever the supports of f and g lie space-like to each other. Under these circumstances $A(f)$ and $A(g)$ are compatible observables; the operators $A(f)$ and $A(g)$ should commute:

$$[A(f), A(g)] = 0 \text{ if supp } f \text{ is spacelike to supp } g, \quad (1.11)$$

or symbolically

$$[A(x), A(y)] = 0 \text{ for } (x-y)^2 > 0 . \quad (1.11')$$

7) Primitive Causality

Einstein's principle covers only one aspect of causality, the one which is added when we pass from a non-relativistic to a relativistic theory. Common to both is the requirement that the knowledge about the state of a system which is obtainable by measurements at one time suffices to determine the state. In view of the comments made under 4) about the field at a sharp time, we should allow finite time intervals for the measurement. The requirement of primitive causality is then the following: The set of operators $A(f)$ for all functions f with support in a time interval $t-\epsilon < x^0 < t+\epsilon$ should generate a complete system of observables. This requirement may be mathematically expressed in another way: If an operator Q commutes with the $A(f)$ for all functions f with support in a time interval, then Q is a multiple of the identity operator.

B. Adaptation to More Realistic Situations

From a formalistic point of view this is a straight forward matter. The guiding ideas came originally from the study of free fields (linear field equations).

There one has a very direct connection between the nature of the field and the types of particles described by the theory. This connection is at the root of the famous particle-wave dualism. Its overemphasis has been responsible for widespread and long lasting misunderstandings about the role of Quantum Fields. According to this view each field should be associated with a certain type of particle. The transformation properties of the field are related to the spin of the particle; the equal time commutation relations of the field are determined by the statistics of the particle. Thus one generalizes the assumptions 5) to

5B) The theory deals with several fields ϕ^i_ρ, the upper index distinguishing the different fields, the lower index the components of one field which go over into each other under Lorentz transformations. For each i one has a finite dimensional representation of the homogeneous Lorentz group by matrices $D^i(\Lambda)$. The transformation law of the field is

$$U(a,\Lambda)\,\phi^i_\rho(x)\,U^{-1}(a,\Lambda) = \sum_\sigma D^i_{\rho\sigma}(\Lambda^{-1})\,\phi^i_\sigma(\Lambda x + a). \tag{1.12}$$

To take care of the occurrence of Fermi statistics the causality principle 6) is generalized to

6B) If ϕ^i is associated with a Bose particle then one has

$$[\phi^i(x), \phi^i(y)] = 0 \quad \text{for } x-y \text{ spacelike} \tag{1.13}$$

if ϕ^i is associated with a Fermi particle then one has instead

$$[\phi^i(x), \phi^i(y)]_+ \equiv \phi^i(x)\phi^i(y) + \phi^i(y)\phi^i(x) = 0$$

$$\text{for } x-y \text{ spacelike} . \qquad (1.14)$$

The requirement (1.14) can be put in accordance with Einstein's causality principle if one assumes that every observable contains only even polynomials in the Fermi Fields. The commutation relations between different fields at spacelike distances are usually assumed to be: all Fermi fields anticommute with each other, all Bose fields commute, any Bose field commutes with any Fermi field.

Comments: Replacing 5) and 6) by 5B), 6B) gives a framework which appears flexible enough to accommodate elementary particle phenomenology. The predictive power derived from the bare framework has so far not been spectacular but not entirely void either. The three celebrated successes (all till now in agreement with experience) are the PCT - theorem, the connection between spin and statistics and a few quantitative statements about scattering amplitudes which follow from analytic properties, for instance dispersion relations for π-N scattering. The first two of these successes are of a qualitative nature, and it is somewhat disconcerting that with our present understanding of their roots they depend absolutely crucially on the detailed assumptions listed above. Thus these two conclusions disappear if in 6B) strict locality is replaced by macroscopic locality. They disappear even if in 5B) one allows fields with an infinite number of components. These questions will be one of the principal concerns of this series of lectures. In particular we shall focus on the problem of statistics.

The framework sketched under B is usually called the Wightman frame not only because Wightman gave the most precise mathematical characterization of (most of) the assumptions, but also because he pointed out one of the most important methods in analyzing the consequences, namely the study of the vacuum expectation values of products of fields [10].

C. Local Quantum Theory (Field Theory without Fields)

The point of view described under B has some unsatisfactory features. First, there is the association of fields with particle types which, though not necessarily tied to this framework, is nevertheless heuristically in the background. It is quite clear that if one insists on such an association and wants to go beyond a purely phenomenological description then one is forced to make a sharp distinction between elementary and composite particles. No satisfactory criterion (of more than approximative value) for such a distinction has been given. The opposite point of view, paraphrased by Chew as "the democracy of particles" has grown since the early thirties. It was in particular the realization that - in spite of the existence of β-decay - the neutron should not be considered as a composite of a proton and an electron which supported the idea that the era of atomistic thinking in physics was over, that the division of structures into elementary building blocks could be regarded only as an approximative model, very successful in nonrelativistic situations but with no place in high energy physics. The most emphatic and consistent advocate of this attitude was Heisenberg who worked hard to develop a theory involving a minimal number of basic fields. These fields should then be associated with the fundamental quantum numbers or conservation laws rather than with observed particle types. Although in recent years the possibility of a quark model of elementary particles has given the atomistic idea a new lease on life this does not necessarily contradict the trend towards "democracy." If the quarks turn out to be real and heavy and the models successful then one has an unexpected extension of the regime in which nonrelativistic approximations are meaningful but still one is led to a theory with few fields associated with fundamental quantum numbers rather than particles.

We may look at the relation between fields and particles from another angle: the collision theory of particles in the framework of local field theory. We adopt for the moment the field theoretic framework sketched above leaving the detailed physical interpretation of the fields open and keeping only the generic statements

(i) the operator $\phi^i(f)$ shall represent a physical operation on the system which can be performed in the space-time region given by the support of the weight function f,

(ii) the Hilbert space $\mathcal{H}$ shall contain (the state vectors of) all the different single particle states; the fields shall be complete in this Hilbert space, i.e. there shall be no operator in $\mathcal{H}$ apart from multiples of the identity operator which commutes with the $\phi^i(f)$ for all i and f.

One finds then that the identification of those vectors in $\mathcal{H}$ which correspond to arbitrary configurations of incoming or outgoing particles is already implied and hence the S-matrix elements for all possible processes in such a model are uniquely determined. [11], [12] See also the exposition in [13]. In other words it is neither necessary nor possible to add any further independent assumption concerning the physical significance of the fields. Their only role is to fix for each space-time region $\mathcal{O}$ the set of operators $\mathcal{F}(\mathcal{O})$ which correspond to physical operations performable in the region. For reasons which will be discussed a little later we shall call $\mathcal{F}(\mathcal{O})$ the "field algebra of the region $\mathcal{O}$." Once the correspondence

$$\mathcal{O} \rightarrow \mathcal{F}(\mathcal{O}) \tag{1.15}$$

is established the theory is defined and its physical content fixed. We may consider (1.15) as the intrinsic definition of the theory and the description by a set of fields as a special way to parametrize (1.15). Loosely speaking, if we start from the frame B then $\mathcal{F}(\mathcal{O})$ will consist of all functions of the operators $\phi^i(f)$ for all weight functions f which have their support in $\mathcal{O}$.

Let me try to explain what I mean by the word "intrinsic" in this context by an analogy. In geometry we would consider as intrinsic objects the space and its points. Now we may introduce coordinates, thus assigning to each point an n-tuple of numbers. These coordinates are not intrinsic because for the same geometry the introduction of a coordinate system can be done in many different ways; any choice involves some convention. In a similar way, suppose we are given the collection (net) of field algebras $\mathfrak{F}(\mathcal{O})$ for the various regions $\mathcal{O}$. Then the theory is fixed. We may try to find a system of local fields $\phi^i(f)$ which generates the net of field algebras $\mathfrak{F}(\mathcal{O})$. If such a system ϕ^i can be found it may be regarded as a coordinate system for the net $\mathfrak{F}$. As in the above analogy the system ϕ^i is not uniquely determined by the net $\mathfrak{F}$. There are many other field systems $\phi^{k'}, \phi^{l''}, \ldots$ which lead to the same net and hence to the same physics.

This non-uniqueness of the choice of a set of fields within one physical theory has been stressed by H. J. Borchers [14]. I shall call therefore the collection of all local fields associated with the same net $\mathfrak{F}$ a "Borchers class." It is true that the definition of such a class of "relatively local" fields used by Borchers is slightly different from the one I use here but this is a difference in technicalities and not in spirit. So let us not worry about it at this stage. To fix the ideas I want to illustrate the nature of a Borchers class in an example well known to all of you, the theory of a free scalar field A. In that case we may define other "local functions" of A by the Wick-ordered powers:

$$A_n(x) = :A(x)^n:$$

We can also obtain local vector fields, e.g.

$$B_\mu(x) = :\partial_\mu A(x)\,\partial^\nu A(x)\,\partial_\nu A(x):$$

or tensor fields

$$B_{\mu\nu}(x) = :\partial_\mu A(x)\,\partial_\nu A(x):$$

etc.

From any one of these fields one generates either the same local net $\mathfrak{F}$ as from the original field A or a subnet of it.

Let us now look a little closer at the nature of $\mathfrak{F}(\mathcal{O})$ and the necessary structural relationships within a net $\mathfrak{F}$. We have called $\mathfrak{F}(\mathcal{O})$ the "field algebra" of $\mathcal{O}$, implying thereby that $\mathfrak{F}(\mathcal{O})$ is an algebra. In fact it is a *-algebra, i.e. an algebra permitting an adjoint operation. This means that if F_1 and F_2 belong to $\mathfrak{F}(\mathcal{O})$ and if α_1, α_2 are any complex numbers, then the operators

$$F_1 F_2 \;, \quad \alpha_1 F_1 + \alpha_2 F_2 \;, \quad \text{and} \quad F_1^*$$

again belong to $\mathfrak{F}(\mathcal{O})$. If we have in mind the specific construction of $\mathfrak{F}(\mathcal{O})$ by functions of a set of basic fields then this statement is obvious. If, on the other hand we want to start from the physical significance of $\mathfrak{F}(\mathcal{O})$ as the set of all "operations performable in the region $\mathcal{O}$," then the statement that $\mathfrak{F}(\mathcal{O})$ must be a *-algebra is not quite so evident. It is tied to peculiarities of the general quantum theoretic description buried in our assumption 1). I am not too happy about the status of our understanding of this formalism from operational principles but it would carry us too far astray if we tried to discuss such questions here.

Some further more technical specifications about the nature of $\mathfrak{F}(\mathcal{O})$ have to be added. For many purposes it is convenient to consider only bounded operators. This is no restriction from the physical point of view. Suppose q is an unbounded self adjoint operator and ρ a function from the real numbers to the real numbers which maps the whole real line on a finite interval so that $|\rho(\xi)| \leq a$ for all ξ. Then the operator $\rho(q)$ is bounded and has a norm $\|\rho(q)\| \leq a$. If the mapping $\xi \to \rho(\xi)$ has a unique inverse then the measurement of the observable q and that of the observable $\rho(q)$ are the same physical operations. Only the scale of the measured values has been regauged from ξ to $\rho(\xi)$.

Thus we want to assert that each $\mathfrak{F}(\mathcal{O})$ shall be a *-algebra of bounded operators. The set of all bounded operators in a Hilbert space $\mathcal{H}$ is denoted by $\mathfrak{B}(\mathcal{H})$. There are several important topologies in the set $\mathfrak{B}(\mathcal{H})$ which have been studied extensively in the mathematical literature. It turns out that those topologies which appear natural in the mathematical context also have a rather direct physical significance in our discussion. Let me say some words at this stage about the relevant mathematical concepts (topologies, von Neumann rings, C*-algebras, etc.) and the most important theorems relating to them. I shall collect this material in an appendix so that the reader who is familiar with these mathematical matters can ignore the digression.

For the moment we need only note that $\mathfrak{F}(\mathcal{O})$ will be taken to be a von Neumann algebra and that this specification is a matter of convenience and choice rather than a restrictive assumption concerning the physics.

The adaptation of 5B and 6B to this language is trivial. Instead of 5B we have the simple transformation law

$$U(a,\Lambda)\ \mathfrak{F}(\mathcal{O})\ U^{-1}(a,\Lambda) = \mathfrak{F}(\Lambda\mathcal{O}+a) \qquad (1.16)$$

meaning that the field algebra of one region is mapped by the Poincaré operators $U(a,\Lambda)$ exactly onto the field algebra of the image region. In the case of 6B we still

have the somewhat artificial sounding assumption:
Each $F \in \mathfrak{F}(\mathcal{O})$ can be uniquely decomposed into

$$F = F_B + F_F \tag{1.17}$$

with F_B and F_F from $\mathfrak{F}(\mathcal{O})$. (The indices B and F stand for "Bose part" and "Fermi part"). If $F^{(1)}$ and $F^{(2)}$ belong to algebras of two regions which are space-like separated, then

$$[F_B^{(1)}, F_B^{(2)}] = [F_B^{(1)}, F_F^{(2)}] = 0 \tag{1.18}$$

$$[F_F^{(1)}, F_F^{(2)}]_+ = 0 \tag{1.19}$$

Again, if F is an observable then $F_F = 0$ and, of course, the product of two Bose type or two Fermi type operators is of Bose type, the product of a Bose-type with a Fermi-type operator is of Fermi type.

The unsatisfactory feature of this formulation of the locality principle does not only lie in its complicated structure. Let us denote by $R(\mathcal{O})$ the subalgebra of $\mathfrak{F}(\mathcal{O})$ generated by the observables. Then for $\mathcal{O}_1$ spacelike to $\mathcal{O}_2$ and $A_1 \in R(\mathcal{O}_1)$, $A_2 \in R(\mathcal{O}_2)$ we must have

$$[A_1, A_2] = 0 . \tag{1.20}$$

The commutativity for observables is the well known condition for compatibility of their measurement and the compatibility is demanded by Einstein's causality principle. Let me emphasize again that the assumption (1.20) is not made in view of our knowledge about Bose-statistics but is an expression of the causality principle. This suggests a number of questions:

1) Why should we consider a "field algebra" $\mathfrak{F}$ which is essentially larger than the "observable algebra" R?

2) While it is clear that for nonobservable elements of $\mathfrak{F}$ we need not have space-like commutativity in order to satisfy the causality principle (as illustrated by the example of Fermi-type operators) it is to be expected that the causality principle puts some restrictions on the commutation relations of the "local operations" which are not observables. The relations (1.18), (1.19) specify one possible commutation structure. But is this the only possibility? Does causality alone imply that there exists only the Bose-Fermi alternative?

We may answer the first question for the moment in a preliminary, so to speak phenomenological, way. Recall that we insisted on incorporating in $\mathcal{H}$ the state vectors of all types of particles. From collision theory and the causality requirement (1.20) for observables it follows that the observable algebra cannot connect the vacuum state with a state of a single Fermi particle. Hence if we restrict our attention to the observable algebra and if Fermi particles occur then $\mathcal{H}$ must be decomposable into subspaces $\mathcal{H}_\sigma$

so that $\mathcal{R}$ transforms each $\mathcal{H}_\sigma$ into itself. In other words, we must have "superselection rules" [15]; the matrix elements of any observable between vectors from different subspaces $\mathcal{H}_\sigma$ vanish. This brings us again close to the second question; we should understand why superselection rules appear and whether we can say anything about their structure.

To investigate these questions one can start from two different angles. The first approach would be to realize that the concept of an "operation" on the system is more general than that of an "observable" and that correspondingly we get less stringent requirements for causally disjoint "operations" than for causally disjoint "observables." Since we shall not follow this approach I shall not try to make the notion of a "physical operation" more precise at this stage but refer to [16] in which this concept is partly used. The second approach, and this is the one we shall take, starts from the observation that all the physical information of the theory must already be contained in the net of observable algebras $\mathcal{R}$ and in fact even in the restriction of $\mathcal{R}$ to one of the invariant subspaces $\mathcal{H}_\sigma$, say to the space $\mathcal{H}_0$ which results by the application of $\mathcal{R}$ on the vacuum state vector. In $\mathcal{H}_0$ we have an irreducible but still faithful representation of the observable algebra. A justification of these claims will be given a little later. First I want now to formulate the framework as it results if we focus only on the observable algebra and its irreducible representation in the "vacuum sector" $\mathcal{H}_0$. The assumptions 1), 2), 3) concerning quantum physics, Poincaré invariance and P_μ -spectrum are not changed. We may, however, anticipate that in the spectrum of P_μ we do not necessarily find the mass-hyperboloids of all relevant particles (because some of these states may lie in other superselection sectors).

4C. To each double cone K in space-time one has a von Neumann ring $R(K)$, the algebra of observables localized in the region K.

Note: For reasons becoming apparent later we do not at this stage consider arbitrary regions of space-time but only the simplest set of Poincaré covariant, finitely extended regions, namely double cones.

Instead of 5B we have

5C $$U(a,\Lambda)\,R(K)\,U^{-1}(a,\Lambda) = R(\Lambda K + a)\,. \qquad (1.21)$$

In order to formulate the primitive causality we need consider also the algebras of regions which are not double cones. Let us denote by $\bigvee_i R_i$ the von Neumann ring generated by all the rings R_i, so that one has

$$V_i R_i = \{\cap R_i'\}' \tag{1.22}$$

If $\mathcal{O}$ is any space-time region then the following definition suggests itself.

Definition.

$$R(\mathcal{O}) = V_i R(K_i) \text{ over all } K_i \subset \mathcal{O} . \tag{1.23}$$

Note that this introduces no essential additional restriction on the relation of the $R(K)$ because, if $\mathcal{O}$ is itself a double cone then on the right hand sight K also appears so that in this case the only consequence is the monotony requirement

$$R(K_1) \subset R(K_2) \text{ if } K_1 \subset K_2 . \tag{1.24}$$

We can now formulate a (strengthened) version of the primitive causality assumption.

7C. Let $\{K_i\}$ be an arbitrary covering of the base of a double cone K,

then $$V_i R(K_i) \supset R(K) . \tag{1.25}$$

Note that this assumption implies first that (at least for the simple types of regions considered here) the algebra of the region is generated by the algebras of an arbitrary set of covering subregions. This property, called "additivity" is suggested by the field theoretic background. For a field ϕ we can decompose $\phi(f)$ into a sum of $\phi(f_i)$ where the f_i have their respective support in the subregions corresponding to an arbitrarily chosen covering of the support of f. This decomposition of the smeared out fields corresponds to the additivity property of the rings. Secondly, (1.25) demands the hyperbolic propagation character of the equations of motion. Loosely speaking the Cauchy data on the base of a double cone determine the quantities everywhere in the double cone.

Another requirement, mentioned before, is

8C. Irreducibility

$$\bigvee_{\text{all } K} R(K) = \mathcal{B}(\mathcal{H}), \text{ or } \bigcap \{R(K)'\} = \{\lambda 1\}. \tag{1.26}$$

Finally there is the Einstein causality which we shall incorporate in a somewhat stronger assumption:

6C. Duality (for double cones)

$$R(K)' = R(K'), \tag{1.27}$$

where K' is the causal complement of K, i.e. the set of points which are space-like to K.

Clearly this assumption is a strengthened version of the locality postulate since the latter could also have been written

$$R(K') \subset R(K)'. \tag{1.28}$$

Beyond the locality requirement the assumption of duality implies that the rings $R(K)$ are maximal. If we have a net of von Neumann rings $R(K)$ satisfying the requirements 1), 2), 3), 4C, 5C, and 1.28 we may ask whether we can find a richer net $\bar{R}(K) \supset R(K)$ still satisfying the mentioned requirements. Now if Q belongs to $\bar{R}(K)$ it has to commute at least with $R(K')$. Hence, if R satisfies (1.27) then $\bar{R}(K) = R(K)$, i.e. then R is already maximal. One may ask whether conversely maximality also implies duality. Suppose we want to enrich the net R by adjoining one more element Q to $R(K)$ for one special K. Because of the covariance we have to adjoin the element $U(a, \Lambda)\, Q\, U^{-1}(a, \Lambda)$ to $R(\Lambda K + a)$. There will be a subset S_K of Poincaré transformations for which $\Lambda K + a$ is totally spacelike to K. Therefore Q can be adjoined to $R(K)$ if and only if the following two requirements are met

(i) $Q \in \{R(K')\}'$

(ii) $[Q, U(a, \Lambda)\, Q\, U^{-1}(a, \Lambda)] = 0$ for $(a, \Lambda) \in S_K$.

In the case of fields (framework B) Borchers [14] has derived the interesting result that if $\phi^i, (i = 1, \cdots, N)$, is a complete set of local fields and if ϕ^0 is another field (not assumed local) satisfying

$$[\phi^0(x), \phi^i(y)] = 0 \quad \text{for } (x-y) \text{ space-like}$$

then ϕ^0 itself is local. This so called "transitivity of locality" suggests that perhaps the requirement (ii) on Q may follow from (i) under rather general circumstances. If this could be shown then we could always enrich the net R by adjoining elements from $R(K')'$ until the resulting net satisfies duality and is then maximal. For the moment, however, it is not known under what circumstances (ii) can be inferred from (i) and hence it is not clear whether the duality assumption is an extra restriction (beyond locality and maximality). For the net of von Neumann rings arising in free field theories Araki has shown that duality holds [17].

D. Algebraic Approach

In the early fifties a mathematical fact of seemingly great importance to quantum field theory was noticed and emphasized independently by several authors. Studying the canonical commutation relations for an infinite system of degrees of freedom $(q_k, p_k; k = 1, \cdots \infty)$ it was found that there are many inequivalent irreducible representations of the q_k, p_k by operators in a (separable) Hilbert space. The number of such inequivalent representations is not denumerable and a complete classification in any constructive sense seems impossible.*) This is in contrast to the situation of a finite number of degrees of freedom where there is essentially one unique equivalence class of irreducible representations. It was also realized that the many inequivalent representations could not be dismissed as pathological. In fact, if a Quantum Field Theory could be defined at all by field equations and kinematical commutation relations then the selection of the representation space of the kinematical relations so that it fits with the assumptions 2) and 3) is a problem determined by the dynamics. In his talk at the Lille Conference 1956 [18] I. E. Segal confronted a rather critical and disbelieving audience with

---

* For the canonical anticommutation relations this phenomenon had already been pointed out by J. von Neumann, Comp. Math. 6, 1 (1938) but for some reason its relevance for quantum field theory had not been realized till much later.

the claim that the representation problem was irrelevant, that one did not need operators in a Hilbert space but only an abstract C*-algebra. In a very interesting earlier paper [19] Segal had studied the mathematical structure of quantum mechanics and pointed out there that many questions of physical interest (e.g. the determination of spectral values) could be answered without reference to a Hilbert space if one chooses the algebra of observables to be a C* algebra. The skepticism of the physicists about the possibility of a purely algebraic approach to field theory was due to the lack of a convincing idea as to how a typical scattering experiment could be discussed in this frame. Also the physical significance of the myriads of inequivalent representation was not understood. But seven years later we realized that Segal's claim had been essentially correct.

I give a brief description of the algebraic version of general quantum physics, i.e., the mathematical and conceptual structure which constitutes assumption 1) in this language:

1 D) The central mathematical object is a C* algebra $\mathfrak{A}$. A state ω is mathematically described by a positive linear form over $\mathfrak{A}$. Every complex valued function on $\mathfrak{A}$ satisfying the two properties (with $A, B \in \mathfrak{A}$ and α, β complex numbers)

$$\omega(\alpha A + \beta B) = \alpha\,\omega(A) + \beta\,\omega(B) \qquad \text{(linearity)}$$

$$\omega(A^* A) \geq 0 \qquad \text{(positivity)}$$

is a state.

We may add the normalization convention

$$\omega(1) = 1$$

One can immediately distinguish pure states and mixtures (see appendix). Every element $C \in \mathfrak{A}$ induces a linear transformation of states $\omega \rightarrow \omega_C$ defined by

$$\omega_C(A) = \omega(C^* A C) \; . \qquad (1.29)$$

Such a transformation maps the pure states into pure states. The norm is changed by the factor $\dfrac{\omega(C^* C)}{\omega(1)}$.

The correspondence of these mathematical objects to physics is the following. A "state" represents a statistical ensemble of physical systems. The norm $\|\omega\| = \omega(1)$ may be regarded as a measure for the total number of systems in the ensemble (in arbitrary units, hence usually put equal to 1). The mixing of states is linearly represented here. For instance $\omega' = \lambda_1 \omega_1 + \lambda_2 \omega_2$ (with $\lambda_1, \lambda_2 > 0$) is the mixture of ω_1 and ω_2 with weights λ_1 , λ_2 . (In the Hilbert space version this corresponds to the mixing of density matrices, not the superposition of wave functions!).

The physical significance of the elements of the algebra is a double one. On the one hand, if $A \in \mathfrak{A}$ is self adjoint, then A may be regarded as an "observable" in the standard sense of this term. If A is regarded in this role then $\omega(A)$ is the expectation value of the observable A in the state ω. Secondly, any element of $\mathfrak{A}$ with norm less or equal to one (whether self adjoint or not) represents an "operation." By this we mean the change of state produced if an apparatus acts during a finite amount of time on the systems constituting the ensemble ω. The "operation" may include a selection process by which a certain fraction of the systems in the original ensemble is rejected by the apparatus (prototype of an "operation" is an arrangement of Nicol prisms and quarter wave plates). The fraction of the original systems which is transmitted will be called the transition probability for the state through the apparatus. If $C \in \mathfrak{A}$ is regarded as an operation then the change of state is given by (1.29) and the transition probability by $\dfrac{\omega(C^*C)}{\omega(1)}$.

Of course the measurement of an "observable" also implies an "operation" in the above sense if the measured systems are available for subsequent further observations. Such an operation transforms, however, in general pure states into mixtures and does not coincide with (1.29) except in the special case when C is a projection (i.e. $C = C^* = C^2$). In general, we have the double role of the algebraic elements and it is useful to keep both roles in mind.

These remarks may suffice to indicate how an experiment may be described in the algebraic frame, without recourse to Hilbert space. A typical experiment may be schematized by a source which prepares the initial state and an analyzing apparatus involving a selection process. The result of the experiment is then the transition probability of the initial state through the apparatus. The description of the source and of the initial state will be done by a combination of two methods both corresponding to actual experimental practice. The first is filtering, the second monitoring. In the first, one uses an "operation" with as small a range as possible on an entirely unknown original state. In the second, one obtains information measuring the transition probabilities through a certain number of monitoring apparatus.

We may now compare with the Hilbert space formulation. There one uses ordinarily an irreducible representation of the observable algebra $\mathfrak{A}$ by operators in a Hilbert space $\mathcal{H}$. Let us denote the operator representing the algebraic element A by $\pi(A)$. Picking up any vector $\Psi \in \mathcal{H}$ we obtain a state on $\mathfrak{A}$ by

$$\omega_{\Psi}(A) = \langle \Psi, \pi(A) \Psi \rangle . \qquad (1.30)$$

If we let Ψ run through $\mathcal{H}$ we obtain precisely all those states which result from one of them by transformations of the form (1.29) i.e. by application of some operation from $\mathfrak{A}$. We call one such family of states a "superselection sector" because, if Ψ_1 and Ψ_2 are state vectors belonging to two unitarily inequivalent irreducible representations π_1 resp. π_2 they are so to speak incomparable. A linear superposition of such vectors is meaningless. It may, of course, be formally defined in the representation $\pi_1 \oplus \pi_2$ but then it only corresponds to the mixing of the states, not to a coherent superposition. Thus, existence of unitarily inequivalent irreducible representations of $\mathfrak{A}$ is synonymous with existence of superselection sectors. It also means that there are "pure operations" (i.e. linear transformations of states mapping pure states into pure states) which are not induced by elements of the observable algebra (not of the form (1.29)).

In the example of the algebra generated by an infinite system of canonical quantities we remarked that there is a tremendously large multitude of inequivalent irreducible representations and, correspondingly, an overwhelmingly rich supply of states. This phenomenon is typical for the nets of local algebras encountered in field theory. One may ask whether really all of these representations should be considered or whether some additional physical restriction has to be imposed to limit the number of superselection sectors which are regarded as physically relevant. The answer to this question depends somewhat on the point of view and on a reasonable balance between considerations of principle and those of practical nature. From the point of view of principle one can say that actually it is impossible to know precisely which state is prepared by a given source. The available information is always such that it determines not a state but a so called "weak neighborhood" in state space. This is a set of states whose common feature is the validity of some finite set of inequalities

$$|\omega(A_i) - a_i| < \epsilon_i \;;\; i = 1, 2, \cdots, N \;;\; \epsilon_i > 0. \quad (1.31)$$

For instance in monitoring a state one is only able to make a finite number of measurements (say $A_1, \cdots, A_N$) with a limited accuracy ($\epsilon_1, \cdots, \epsilon_N$). Any choice of measurements A_i, mean values a_i and error limits ϵ_i defines one weak neighborhood in state space.

Theorem [20].

Let π and π_0 be two irreducible representations and ω_0 some vector state of π_0. If π is faithful then every weak neighborhood of ω_0 contains also a vector state of π.

This means that it is impossible to prepare an ensemble by a realistic source in such a way that we know in which superselection sector the corresponding state lies. The distinction based on unitary inequivalence of representations is much too fine to be physically measurable. We might say that all faithful representations are physically equivalent. We might select any one of them (or use no representation at all) in order to discuss a specific experimental set-up.

On the other hand it is usually very convenient to simplify the description by an idealization which restricts the set of states considered (adding some information about the occurring states which is neither needed in principle nor available in reality). We shall do this also subsequently and thereby reduce the number of relevant superselection sectors to a manageable size and the distinction between them to physically important quantum numbers. This discussion will be exemplified below.

Thus far we have described how the assumption 1), the principles of Quantum Physics are expressed in the algebraic approach. The expression of the other assumptions is to a large part a straightforward adaptation of the formulations under C .

For instance, instead of the von Neumann rings $R(K)$, we consider now C*-algebras $\mathfrak{A}(K)$ with the obvious monotony property that $K_1 \supset K_2$ implies $\mathfrak{A}(K_1) \supset \mathfrak{A}(K_2)$. The total algebra of observables is defined as

$$\mathfrak{A} = \overline{\bigcup_K \mathfrak{A}(K)}\,, \tag{1.32}$$

where the bar means the completion in the norm topology.

One feature of physical importance, tied to the definition (1.32), should be stressed. Every element of $\mathfrak{A}$ can be approximated by an element from some finite region K uniformly with respect to all states. Hence $\mathfrak{A}$ contains only elements which correspond to still essentially local quantities, observables which are "quasilocal." Truly "global" quantities, like (bounded functions of) the total energy are excluded as unmeasurable, and this is satisfactory because their measurement would require an infinitely extended apparatus. In this respect the difference between the norm topology and the strong or weak operator topologies is crucial. Take the example of a Poincaré operator $U(a,\Lambda)$. It is a global operation because its effect does not become weaker in faraway regions of space. The norm of the difference between $U(a,\Lambda)$ and any element from any $\mathfrak{A}(K)$ is always greater than 1 because, no matter how large K is we can always find states which are essentially different from the vacuum with respect to measurements outside of K. On such states the effect of $U(a,\Lambda)$ is very different from that of any $A \in \mathfrak{A}(K)$. If, on the other hand, we first pick a state vector Ψ in the Hilbert space $\mathcal{H}_0$ then we can find a region K large enough so that with respect to observations in the causal complement K' this state and the vacuum state are almost identical. Then one can find an $A \in \mathfrak{A}(K)$ so that $\|(U(a,\Lambda) - A)\Psi\| < \epsilon$. This means that $U(a,\Lambda)$ can be approximated by local quantities in the strong operator topology in the vacuum representation.

We have seen that $U(a,\Lambda)$ cannot belong to $\mathfrak{A}$. Hence, in a strictly algebraic formulation the Poincaré invariance cannot be expressed in terms of unitary operators. Instead we have to each Poincaré transformation (a,Λ) a corresponding <u>automorphism</u> $\alpha_{a,\Lambda}$ of the algebra $\mathfrak{A}$ (see appendix for an explanation of the term "automorphism"). Again we must have the product relation

$$\alpha_{a_1\Lambda_1}\alpha_{a_2\Lambda_2} = \alpha_{a_1+\Lambda_1 a_2,\Lambda_1\Lambda_2},$$

and, in fact, now we have this relation really applying to the Poincaré group itself, not to the covering group because the arbitrariness of phase of a state vector Ψ does not enter here. The elements of the algebra and the states considered as positive forms over the algebra are free from this ambiguity. The transformation law reads then

$$\alpha_{a,\Lambda}\left(\mathfrak{A}(K)\right) = \mathfrak{A}(\Lambda K + a). \tag{1.33}$$

From (1.32) we see one reason for the occurrence of many inequivalent irreducible representations of $\mathfrak{A}$. A state ω over $\mathfrak{A}$ defines a state over each subalgebra $\mathfrak{A}(K)$. This is called the restriction of $\mathfrak{A}$ to $\mathfrak{A}(K)$ or, for brevity, in more physical terms the "partial state" of the region K. It will be denoted by $\omega|_K$. Now consider a sequence of mutually space-like regions K_n, moving to infinity as $n \to \infty$. We may prescribe arbitrarily a partial state $\omega|_{K_n}$ for each of these regions and there exists always a total state ω which is the simultaneous extension of this collection of partial states to the algebra $\mathfrak{A}$. (Compare [21]). All vector states occurring in the Hilbert space of one irreducible representation of $\mathfrak{A}$ have a common asymptotic behavior for their partial states $\omega|_{K_n}$ as $n \to \infty$. In other words for any two such normalized states we have $\| \omega^{(1)}|_{K_n} - \omega^{(2)}|_{K_n} \| \to 0$. Omitting fine points the proof uses the following facts: The two state vectors in question, being associated with the same irreducible representation π, are related by

$$\Psi^{(2)} = \pi(C)\,\Psi^{(1)} \quad \text{with } C \in \mathfrak{A}.$$

Since C is quasilocal it cannot change the partial state far away unless the state $\Psi^{(1)}$ already has correlations between its partial states in regions which are infinitely far apart. This is however not possible for a pure state.

We may formalize this in the following way. Let K_ℓ be a sequence of increasing double cones which exhaust space-time:

$$K_{\ell+1} \supset K_\ell \quad ; \quad \cup K_\ell = \text{all space-time} . \qquad (1.34)$$

Take any sequence of elements $A_\ell \in \mathfrak{A}(K_\ell')$ with $\|A_\ell\| \le 1$. Then for every fixed $C \in \mathfrak{A}$ we have due to locality

$$\| [C, A_\ell] \| \to 0 .$$

Since A_ℓ tends to commute with all elements of the algebra, the representatives $\pi(A_\ell)$ in an irreducible representation approach multiples of the identity operator (by a slight generalization of Schur's lemma). One has therefore the "cluster property"

$$(\Psi, \pi(A_\ell C)\Psi) \to (\Psi, \pi(A_\ell)\Psi)(\Psi, \pi(C)\Psi)$$

for normalized Ψ, and the first factor is, in fact, even independent of the direction of the unit vector Ψ in the representation space. This includes the statement that all vector states in an irreducible representation have the same asymptotic tail of partial states.

We see therefore how to construct a great variety of inequivalent representations of $\mathfrak{A}$. We just have to choose at random sequences of partial states for space-like separated regions moving to infinity and then extend each such sequence to a total state on $\mathfrak{A}$. If the asymptotic tails of two such sequences do not coincide we obtain states belonging to inequivalent representations. Simple examples of such states are those which describe a non-vanishing density of particles extending to infinity in space. For the purpose of elementary particle physics it is, however, both legitimate and convenient to idealize the "cosmology" by the claim that all states of interest to us coincide with the vacuum state asymptotically for observations in far away regions.

For the remainder of these lectures we shall limit our attention to states ω for which the difference of the partial states $(\omega - \omega_0)|_{K'_\ell}$ in the causal complement K'_ℓ of the double cone K_ℓ goes to zero in norm as $\ell \to \infty$ for a sequence of double cones K_ℓ of the type (1.34). Thus, for any "state of interest" ω and any positive ϵ there is a double cone K_ϵ so that

$$\| (\omega - \omega_0)|_{K'_\epsilon} \| < \epsilon . \qquad (1.35)$$

This restriction on the "states of interest" is a very strong one. Let me point out that it is, in fact, too strong to be reasonable in Quantum Electrodynamics because there Gauss's law asserts that an electric charge located in a finite region can be determined by means of field strength measurements on the surface of an arbitrarily large sphere. Hence, no matter how large we choose K there is always some element of $\mathfrak{A}(K')$ for which a state with a localized charge and the vacuum give markedly different expectation values. We should really require the asymptotic coincidence of states not for their restrictions to the complements K_ℓ' of a sequence of increasing finite regions K_ℓ, but the partial states of finitely extended regions moving out to infinity. Still, excluding the much more difficult case of long range forces from our consideration, the restriction (1.35) appears reasonable.

Another condition on the states we wish to consider, a condition which is not unrelated to (1.35) but not quite a consequence of it is the following: all states considered shall lead (via the GNS-construction) to representations in which the Poincaré automorphism group $\alpha_{a,\Lambda}$ can be (continuously) implemented by unitary operators $U(a,\Lambda)$ and furthermore the resulting spectrum shall be contained in the forward cone (no negative energies).

With these limitations on the "states of interest" and some well supported properties of the vacuum representation*) one finds that all the representatives we want to consider are "strongly locally equivalent." This means that for any two

---

* For instance that $\pi_0(\mathfrak{A}(K))''$ is a factor of Type III.

such representations their restrictions to a subalgebra $\mathfrak{A}(K')$ (or to $\mathfrak{A}(K)$ for that matter) are unitarily equivalent. The set of partial states of a region $\mathcal{O}$ occurring in any of these representations are identical as long as $\mathcal{O}$ has a non-void space-like complement $\mathcal{O}'$. The distinction between the surviving superselection sectors appears only when we consider the total algebra $\mathfrak{A}$. In particular we have now not only the asymptotic coincidence of states as demanded by (1.35) but in each superselection sector of interest and for each region K we can find states which coincide exactly with ω_0 on $\mathfrak{A}(K')$. Such states are called "exactly localized" in K. If such a state lies in a different superselection sector than the vacuum, say it lies in the sector σ, then we may visualize the situation by the physical picture that the state has a "charge quantum number σ" but that this charge is strictly localized within K. It can be strictly localized in any region, no matter how small, but it has to sit somewhere. Again we see that this is not realistic if the charge is necessarily accompanied by an extended, observable field as in Electrodynamics. Our limitations probably exclude the most interesting case of "gauge invariances of the second kind" and refer only to charge quantum numbers which are not the sources of fields, to cases in which only the analogue of gauge transformations of the first kind exist.

Let me briefly indicate the essential point in the argument which leads from our limitations on the states of interest to the strong local equivalence of the corresponding representations. If $\pi^{(1)}$ and $\pi^{(2)}$ are two disjoint representations of a C^*-algebra $\mathcal{C}$ (meaning that no subrepresentation of $\pi^{(1)}$ is unitarily equivalent to any subrepresentation of $\pi^{(2)}$) and if $\omega^{(1)}$ and $\omega^{(2)}$ are normalized vector states in the respective representations, then $\| \omega^{(1)} - \omega^{(2)} \| = 2$. Thus (1.35) demands that no representation of interest, when restricted to a sufficiently far out region K_ℓ', can be disjoined from the vacuum representation of $\mathfrak{A}(K_\ell')$. For factors of Type III non-disjointness implies unitary equivalence. The assumed translational covariance allows us to extend this unitary equivalence to the algebras of other regions.

With these limitations on the "states of interest" we can quickly summarize the assumptions on which the remainder of these lectures will be based. We can now without loss of generality identify the local algebras $\mathfrak{A}(K)$ with the Von Neumann algebras $R(K)$ in the vacuum representation (since all "states of interest" are normal states on this net of Von Neumann algebras; see Appendix for definition of the term "normal state"). We can then take over all the structural assumptions concerning the net $R(K)$ as described under C. The question to be asked is: do there exist unitarily inequivalent irreducible representations π_σ of the total algebra $\mathfrak{A} = \overline{\cup \mathfrak{A}(K)}$ which are strongly locally equivalent, i.e. for which $\pi_\sigma(\mathfrak{A}(K'))$ is unitarily equivalent to $\pi_0(\mathfrak{A}(K'))$ for every K. If so, these representations will be the relevant superselection sectors for us. We wish to classify them.*

II. Structure of Superselection Rules; Charge Quantum Numbers; Statistics

The question to be studied here was described at the end of the last section. Consider a representation π_σ in a Hilbert space $\mathcal{H}_\sigma$ and pick some double cone K. Since the representations π_σ and π_0 when restricted to $\mathfrak{A}(K')$ are unitarily equivalent we can find a unitary mapping from $\mathcal{H}_0$ to $\mathcal{H}_\sigma$, denoted by V, such that

$$V\pi_0(A)\Psi = \pi_\sigma(A)V\Psi \quad \text{for all } A\in\mathfrak{A}(K') \text{ and } \Psi\in\mathcal{H}_0. \tag{2.1}$$

We may omit the symbol V if we identify the two spaces $\mathcal{H}_0$ and $\mathcal{H}_\sigma$ (identifying the vector Ψ with its image $V\Psi$). This identification also makes $\pi_\sigma(A)$ into an operator from $\mathfrak{B}(\mathcal{H}_0)$ which can be expressed in terms of the $\pi_0(\mathfrak{A})$. We shall also omit the symbol π_0 because we take the operator algebras $\pi_0(\mathfrak{A}(K))$ in the vacuum sector as identical by definition to the C^*-algebras $\mathfrak{A}(K)$. Then (2.1) reads

* This analysis was initiated by Borchers [22]; the following section is based on [23,24,25].

$$\pi_\sigma(A) = A \qquad \text{for all} \quad A \in \mathfrak{A}(K'). \tag{2.2}$$

On the other hand, if $B \in \mathfrak{A}(K_1)$ and $A \in \mathfrak{A}(K_1')$, then as soon as $K_1 \supset K$ we have

$$[\pi_\sigma(A), \pi_\sigma(B)] = [A, \pi_\sigma(B)] = 0 .$$

Hence by duality

$$\pi_\sigma(\mathfrak{A}(K_1)) \subset \mathfrak{A}(K_1) \quad \text{for } K_1 \supset K . \tag{2.3}$$

This means first of all that

$$\pi_\sigma(\mathfrak{A}) \subset \mathfrak{A} . \tag{2.4}$$

Also, π_σ is a faithful representation and therefore preserves the algebraic structure and the norm. The transition from the vacuum sector to the sector σ can therefore be described by a "localized endomorphism" ρ_σ. In detail this means that the representative $\pi_\sigma(A)$ may be considered as the image $\rho_\sigma(A)$ of a norm preserving mapping of the algebra <u>into</u> itself and such that ρ_σ acts like the identity mapping on $\mathfrak{A}(K')$ and only reshuffles the elements of the algebra of the finite region K. If the mapping ρ_σ is <u>onto</u> the whole of $\mathfrak{A}$ we call it an <u>automorphism</u>. In general, however, the image of ρ may be smaller than $\mathfrak{A}$. In either case the endomorphism is called <u>localized</u> in K because it does nothing to the elements belonging to the causal complement region K'.

Let me summarize this discussion and the resulting notation. Due to the unitary equivalence of all representations $\pi_\sigma(\mathfrak{A}(K'))$ we can identify all representation spaces with $\mathcal{H}_0$ and as a consequence all operators $\pi_\sigma(A)$ with operators from $\pi_0(\mathfrak{A})$. Since we consider now π_0 as the defining representation of $\mathfrak{A}(K)$ we omit the symbol π_0. So any $\pi_\sigma(A)$ will be identified with some element of $\mathfrak{A}$, namely with $\rho_\sigma(A)$, the image of A by the endomorphism ρ_σ. Of course the identification of $\mathcal{H}_\sigma$ with $\mathcal{H}_0$ and the corresponding identification of $\pi_\sigma(A)$ with $\rho_\sigma(A)$ is not canonical (=natural) and highly non-unique. It depends on the arbitrary choice of a reference region K and even beyond that on the choice of the V in (2.1) which is determined only up to a unitary operator from $\mathfrak{A}(K)$. Still, any localized endomorphism ρ will lead from a representation $\pi^{(1)}$ to a strongly locally equivalent representation, denoted by $\pi^{(2)} = \pi^{(1)} \circ \rho$ (in detail $\pi^{(2)}(A) = \pi^{(1)}(\rho(A))$) and any representation of interest can be obtained from the defining representation $\pi_0(\mathfrak{A}) = \mathfrak{A}$ by an endomorphism localized in an arbitrarily chosen region K:

$$\pi = \pi_0 \circ \rho$$

with $\rho(A) = A$ for $A \in \mathfrak{A}(K')$.

There is one class of localized automorphisms which does not lead to a change of sectors: the "inner" automorphisms. Let $\mathcal{A}(K)$ be the set of all unitary elements of $\mathfrak{A}(K)$, then

$$\sigma_U(A) = UAU^{-1} \quad \text{with } U \in \mathcal{A}(K)\,;\; A \in \mathfrak{A} \qquad (2.5)$$

defines an inner automorphism localized in K. The unitary U is determined by σ_U up to a phase factor. One has the

<u>Lemma 2.1.</u> The representations $\pi_0 \circ \rho_1$ and $\pi_0 \circ \rho_2$ are unitarily equivalent if and only if $\rho_2 = \sigma_U \rho_1$ where σ_U is an inner automorphism.

<u>Proof</u>: Omitting as explained before the symbol π_0, unitary equivalence means that there exists a unitary $V \in \mathcal{B}(\mathcal{H}_0)$ such that

$$\rho_2(A) = V \rho_1(A) V^{-1}.$$

We only have to show that V in fact belongs to $\mathfrak{A}$. Since ρ_2 and ρ_1 are localized in some finite regions we can choose a K large enough so that it encloses both localization regions. Then

$$\rho_2(A) = \rho_1(A) = A \qquad \text{for } A \in \mathfrak{A}(K').$$

Hence $V \in \{\mathfrak{A}(K')\}' = \mathfrak{A}(K)$, (by duality).

Next we note

Lemma 2.2 If ρ_1 and ρ_2 are endomorphisms localized respectively in the space-like separated double cones K_1 and K_2 then

$$\rho_1 \rho_2 = \rho_2 \rho_1 \qquad (2.6)$$

Proof: Consider the 7 regions drawn in Figure 2 (for simplicity we draw only the base of the double cones)

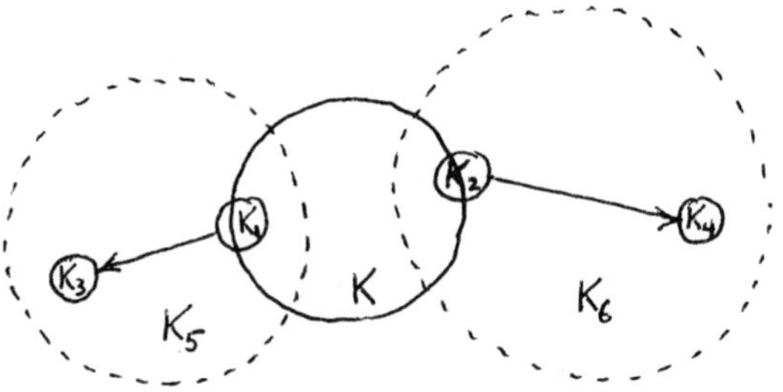

Figure 2.

K_1 and K_2 are the two given, space-like situated regions. K is arbitrary. We want to test the action of $\rho_1 \rho_2$ on $\mathfrak{A}(K)$. We shift K_2 by translation to K_4 and K_1 by opposite translation to K_3 till K_4 and K_3 are space-like to K . In addition the cone K_5 enclosing K_1 and K_3 will be space-like to K_6 , the cone enclosing K_2 and K_4 . Let ρ_3 and ρ_4

denote the endomorphisms localized respectively in K_3, K_4 which result by shifting ρ_1, ρ_2 by the relevant translations. Due to our covariance assumption the representation $\pi_0 \circ \rho_1$ is equivalent to $\pi_0 \circ \rho_3$ and $\pi_0 \circ \rho_2$ to $\pi_0 \circ \rho_4$ Hence by Lemma 2.1 we have

$$\rho_3 = \sigma_{U_5} \rho_1 \quad ; \rho_4 = \sigma_{U_6} \rho_2 ;$$

where σ_{U_5}, σ_{U_6} are inner automorphisms implemented by unitary operators U_5, U_6 which (by duality) belong to $\mathfrak{A}(K_5)$ respectively $\mathfrak{A}(K_6)$. Since K_3 and K_4 are space-like to K we have

$$\rho_3 \rho_4 (A) = A = \rho_4 \rho_3 (A) \qquad \text{for } A \in \mathfrak{A}(K).$$

Thus

$$\sigma_{U_5} \rho_1 \sigma_{U_6} \rho_2 (A) = \sigma_{U_6} \rho_2 \sigma_{U_5} \rho_1 (A) .$$

The right hand side is rewritten as

$$\sigma_{U_6} \rho_2 \left(U_5 \rho_1 (A) U_5^{-1}\right) = U_6 \rho_2 (U_5) \rho_2 \rho_1 (A) \rho_2 (U_5^{-1}) U_6^{-1}$$

$$= U_6 U_5 \rho_2 \rho_1 (A) U_5^{-1} U_6^{-1}$$

since $\rho_2(U_5) = U_5$ due to the support properties.

By the same manipulations the left hand side becomes

$$U_5 U_6 \rho_1 \rho_2 (A) U_6^{-1} U_5^{-1} .$$

But U_5 and U_6 commute too. Hence we have (2.6). We shall now specialize the discussion temporarily to the case of those sectors which can be reached from the vacuum sector by localized automorphisms. This allows a simpler analysis than the more general endomorphic case. We shall call such sectors therefore "simple sectors." We find then first of all a classification of simple sectors into Bose-type and Fermi type according to the following lemma:

Lemma 2.3 Let γ_1 and γ_2 be two automorphisms, leading to the same sector and based on space-like separated double cones K_1, K_2. By lemma (2.1) they are related by an inner automorphism, i.e. $\gamma_2 = \sigma_U \gamma_1$. One has then

$$\gamma_1 (U) = \pm U \tag{2.7}$$

where the sign depends only on the sector, not on the choice of γ_1, γ_2.

Sectors for which the + sign holds are called Bose-type; those for which the - sign holds Fermi type.

Proof: Since γ_1 and γ_2 commute, so do γ_1 and σ_U. But

$$\gamma_1 \sigma_U (A) = \gamma_1 (U A U^{-1}) = \gamma_1 (U) \gamma_1 (A) \gamma_1 (U^{-1}) = \sigma_{\gamma_1 (U)} \gamma_1 (A).$$

Thus $\gamma_1 \sigma_U = \sigma_{\gamma_1(U)} \gamma_1$. By commutativity

$$\sigma_U = \sigma_{\gamma_1(U)} \qquad \text{or} \qquad \gamma_1(U) = \epsilon_{\gamma_1,\gamma_2} U \tag{2.8}$$

where ϵ is a numerical phase factor depending possibly on γ_1 and γ_2. Interchanging the role of γ_1 and γ_2 and correspondingly replacing U by U^{-1}, we have

$$\gamma_2(U^{-1}) = \epsilon_{\gamma_2,\gamma_1} U^{-1} . \tag{2.9}$$

Multiplying (2.8) and (2.9) we get

$$\gamma_1(U)\gamma_2(U^{-1}) = \gamma_1(U\gamma_1^{-1}\gamma_2(U^{-1})) = \epsilon_{\gamma_1,\gamma_2}\epsilon_{\gamma_2,\gamma_1} . \tag{2.10}$$

By the definition of σ_U and commutativity we have

$$\gamma_1^{-1}\gamma_2 = \sigma_U .$$

Thus the left hand side of (2.10) is 1 and we have

$$\epsilon_{\gamma_1,\gamma_2} = \epsilon_{\gamma_2,\gamma_1}^{-1} . \tag{2.11}$$

Let us take now a third region K_3, space-like to both K_1 and K_2, and choose an automorphism γ_3 localized in K_3 and leading to the same sector. Then

$$\gamma_3 = \sigma_W \gamma_2 = \sigma_V \gamma_1 \ ; \quad \gamma_2 = \sigma_U \gamma_1$$

with $W \in \mathfrak{A}(K_2 \cup K_3)$; $V \in \mathfrak{A}(K_1 \cup K_3)$; $U \in \mathfrak{A}(K_1 \cup K_2)$.

Also $\sigma_W \sigma_U = \sigma_V$ and hence we can choose $V = WU$.

According to the earlier discussion we have

$$\gamma_1(V) = \epsilon_{\gamma_1,\gamma_3} V .$$

But

$$\gamma_1(V) = \gamma_1(WU) = \gamma_1(W)\gamma_1(U) = W\gamma_1(U) = \epsilon_{\gamma_1,\gamma_2} WU = \epsilon_{\gamma_1,\gamma_2} V$$

due to the support properties. Hence

$$\epsilon_{\gamma_1,\gamma_3} = \epsilon_{\gamma_1,\gamma_2} .$$

Repeating this process we see that $\epsilon_{\gamma_1,\gamma_2}$ is independent of its second argument and therefore by (2.11) also of its first argument. It is a fixed number depending only on the sector. Due to (2.11) its value can only be ± 1.

To see that the sign appearing in (2.7) has something to do with Bose - or Fermi - statistics we sketch the relation of this description with the conventional formulation by means of a field algebra. First it is evident that the set Γ of localized automorphisms forms a group, since $\gamma_1\gamma_2$ and γ^{-1} can be performed within Γ .*

The localized inner automorphisms form an invariant subgroup $\mathcal{J}$ since

$$\gamma \sigma_U \gamma^{-1} = \sigma_{\gamma(U)} .$$

* Actually we should restrict attention to those automorphisms which lead to sectors in which the Poincaré group is implementable. This subset of Γ has been called Γ_c in Ref. [24]. But it is shown in [24] that Γ_c is a group too and possibly we have $\Gamma_c = \Gamma$ so that we do not bother here to make the distinction.

The quotient group

$$\hat{\mathfrak{g}} = \Gamma / \mathfrak{J} \qquad (2.12)$$

corresponds by lemma (2.1) as a set to the collection of all simple sectors. We see that this set has a group structure and moreover this group is Abelian because if γ_1 and γ_2 are in Γ we can always find a γ_2' in the same equivalence class as γ_2 which has its support space-like to that of γ_1 (just shifting it by a sufficiently large translation). Then γ_1 and γ_2' commute due to lemma 2.2. Using an additive notation for the multiplication in the Abelian group $\hat{\mathfrak{g}}$ we see that the sectors can be labeled by generalized charge quantum numbers σ such that along with σ_1 and σ_2 also $\sigma_1 + \sigma_2$ and $-\sigma_1$ occur. Of course it is not implied that all such linear combinations belong actually to different sectors; e.g. we might have a relation like $-\sigma = \sigma$ in which case the chain would really consist only of two different sectors (a case usually described by a multiplicative quantum number). Still we have the typical chains: if the simple sector σ contains a single particle state, then the sector $n\sigma$ (with $n > 1$) contains the states with n such particles and $-\sigma$ contains the antiparticle.

Let us now turn to the construction of a field algebra for the set of simple sectors. I shall do this here only for the case of a single additive charge; i.e. the case where $\hat{\mathcal{G}}$ is the (additive) group of all positive and negative integers. For the case of a general $\hat{\mathcal{G}}$ I refer to [24]. The sectors are now labeled by an integer n running from $-\infty$ to ∞. We wish to consider all sectors simultaneously and take therefore the direct sum of the representations π_n. This "universal" representation π acts in the Hilbert space

$$\mathcal{H} = \sum_{\oplus} \mathcal{H}_n \quad . \qquad (2.13)$$

In $\mathcal{H}$ we wish to implement the automorphisms from Γ by unitary operators ψ so that

$$\pi(\gamma_\psi(A)) = \psi \, \pi(A) \, \psi^{-1} . \qquad (2.14)$$

If Ψ is a general vector in $\mathcal{H}$ we denote its projection on $\mathcal{H}_n$ by $\Psi(n)$. As before we may consider all $\mathcal{H}_n$ to be copies of $\mathcal{H}_0$ and regard $\Psi(n)$ as a vector of $\mathcal{H}_0$; then $\Psi \in \mathcal{H}$ is described by the set of vectors $\{\Psi(n)\}$ of $\mathcal{H}_0$. To describe the representation $\pi(\mathfrak{A})$ we pick an arbitrary reference automorphism γ_1 localized say in the region K_0 transferring unit charge so that we can reach all sectors by applying powers γ_1^n on the vacuum representation. Then π is taken to be

$$\left(\pi(A)\Psi\right)(n) = \pi_n(A)\Psi(n) = \gamma_1^n(A)\Psi(n). \quad (2.15)$$

To understand the notation keep in mind that $\pi_n(A) = \gamma_1^n(A)$ is regarded as an operator in $\mathcal{H}_0$ and each $\Psi(n)$ as a vector in $\mathcal{H}_0$. We implement the automorphism γ_1 by a unitary operator V_1 acting in $\mathcal{H}$ and defined by

$$(V_1\Psi)(n) = \Psi(n+1). \quad (2.16)$$

Correspondingly γ_1^n is implemented by the unitary V_1^n. One checks that this definition indeed gives

$$\pi(\gamma_1^m(A)) = V_1^m \pi(A) V_1^{-m}$$

because according to (2.16) and (2.15) we get

$$\left(V_1^m \pi(A) V_1^{-m}\Psi\right)(n) = \left(\pi(A)V_1^{-m}\Psi\right)(n+m) =$$

$$= \gamma_1^{n+m}(A)\Psi(n) = \left(\pi(\gamma_1^m(A))\Psi\right)(n).$$

An arbitrary $\gamma \in \Gamma$ is in some equivalence class modulo $\mathcal{J}$, which class contains also one of the powers γ_1^m. Hence any γ can be written as

$$\gamma = \sigma_U \gamma_1^m \quad (2.17)$$

and implemented by

$$\Psi = \pi(U)V_1^m. \quad (2.18)$$

The choice of the unitary Ψ implementing γ is again fixed up to the arbitrary phase in the choice of U (in the determination of U from σ_U).

Consider now the commutation relations between two such operators ψ' and ψ'' which are based on space-like separated regions and lead to the same sector (the corresponding automorphisms γ', γ'' being equivalent modulo $\mathfrak{J}$ to the same power of γ_1, say to γ_1^m. Then we have $\gamma'' = \sigma_W \gamma'$ where W now can be chosen to be $W = \psi'' \psi'^{-1}$. Lemma (2.3) gives

$$\gamma'(W) = \epsilon_m W \quad ; \quad \epsilon_m = \pm 1 .$$

Thus

$$\psi' \psi'' \psi'^{-1} \psi'^{-1} = \epsilon_m \psi'' \psi'^{-1}$$

or

$$\psi' \psi'' = \epsilon_m \psi'' \psi' \quad ; \qquad (2.19)$$

also, one immediately sees that

$$\epsilon_m = \epsilon_1^m . \qquad (2.20)$$

Thus the commutativity of space-like based automorphisms leads to the alternative between commutativity or anticommutativity for the corresponding implementing operators. If we use these operators in $\mathcal{H}$ to generate states $\psi'\Omega$, $\psi''\Omega$ then the state $\psi'\psi''\Omega$ may be interpreted as the "product state": we have the partial state equal to $\psi'\Omega$ in the one region and the partial state equal to $\psi''\Omega$ in the other region and Ω in the space-like complements of both regions. The sign ϵ_m determines whether linear combinations of such product state vectors behave like vectors in the symmetrized or antisymmetrized direct product space of the starting sectors. In other words it determines whether localized states in the sector $\psi'\Omega$ are of the Bose type or of Fermi type.

We still have to define the field algebra $\mathcal{F}(K)$ associated with the region K, the gauge group $\mathcal{G}$ and its representation $\mathcal{U}(\mathcal{G})$ in $\mathcal{H}$ and show

Lemma 2.4 The gauge invariant part of the field algebra of a region K is precisely the observable algebra of the region; i.e.

$$\pi(\mathfrak{A}(K)) = \mathcal{F}(K) \cap \mathcal{U}(\mathcal{G})' . \qquad (2.21)$$

The field algebra $\mathcal{F}(K)$ of the region K is of course defined as the von Neumann ring generated by all the ψ implementing automorphisms γ which are localized in K. The gauge group, abstractly, is defined as the dual group of $\hat{\mathcal{G}}$. Its elements are the characters of $\hat{\mathcal{G}}$; i.e. functions from $\hat{\mathcal{G}}$ to the complex numbers which furnish a one-dimensional representation of $\hat{\mathcal{G}}$. In our example, where $\hat{\mathcal{G}}$ is the additive group of integers, $\mathcal{G}$ is the group of the unit circle in the complex plane. Writing an element $g \in \mathcal{G}$ as $e^{i\delta}$ we have the character $g(n) = e^{in\delta}$. The unitary representative $\mathcal{U}(g)$ in the Hilbert space $\mathcal{H}$ is obviously given by

$$(\mathcal{U}(g)\Psi)(n) = g(n)\,\Psi(n) . \qquad (2.22)$$

For the proof of lemma (2.4) and the discussion of the general case of simple sectors (when the "charge group" $\hat{\mathfrak{g}}$ is generated by several, possibly not independent, elements) see ref. [24]. If there are several charges then the construction of the field algebra is not unique, the commutation properties of operators transferring different types of charge are not intrinsically fixed. One finds, however, that one can always achieve the "normal" commutation relations described in (1.17), (1.18), (1.19).

Non simple sectors. Endomorphic case.

We have seen that for simple sectors one always has the Bose-Fermi alternative and an Abelian gauge group. A more complicated structure results if $\mathfrak{A}$ admits localized endomorphisms (for which the image $\rho(\mathfrak{A})$ is strictly smaller than $\mathfrak{A}$). We denote the set of localized endomorphisms (including the automorphisms Γ) by Δ (respectively $\Delta(K)$ if the localization region is specified to be K). To avoid misunderstandings I should perhaps repeat that the term "endomorphism" is used here for a one-to-one mapping of $\mathfrak{A}$ into itself, conserving all the C^*-algebraic structure. Perhaps a better term might be "isomorphic injection."

We begin with some remarks on intertwining operators. Given ρ, $\rho' \in \Delta$ then the operator $S \in \mathcal{B}(\mathcal{H}_0)$ is called an intertwining operator from ρ to ρ' if

$$\rho(A)S = S\rho'(A) \quad \text{for all } A \in \mathfrak{A}. \tag{2.23}$$

Such intertwining operators exist if the representations $\rho(\mathfrak{A})$ and $\rho'(\mathfrak{A})$ are not disjoint. Actually in that case S belongs to $\mathfrak{A}(K)$ where K is the union of the support regions of ρ and ρ'. Let us write

$$\tilde{S} = (\rho; S; \rho'). \tag{2.24}$$

If we have two such triples $\tilde{S}_1 = (\rho_1; S_1; \rho_1')$ and $\tilde{S}_2 = (\rho_2; S_2; \rho_2')$ we can immediately construct an intertwining operator from $\rho_1\rho_2$ to $\rho_1'\rho_2'$. One checks that the operator $\rho_1(S_2)S_1 = S_1\rho_1'(S_2)$ performs this function. Thus one has a cross product of the triples, defined by

$$\tilde{S}_1 \times \tilde{S}_2 = (\rho_1\rho_2; \rho_1(S_2)S_1; \rho_1'\rho_2'). \tag{2.25}$$

This cross product is associative but in general not commutative

$$\tilde{S}_1 \times (\tilde{S}_2 \times \tilde{S}_3) = (\tilde{S}_1 \times \tilde{S}_2) \times \tilde{S}_3 = \tilde{S}_1 \times \tilde{S}_2 \times \tilde{S}_3. \tag{2.26}$$

If all the ρ_i' have their supports mutually spacelike and if the same holds for the ρ_i then the cross product (2.26) is commutative. The proof is analogous to that of lemma 2.2.

Consider now n endomorphisms $\rho_1, \rho_2, \cdots, \rho_n$ and a permutation P of the numbers $1, 2, \cdots, n$. The products $\rho_1 \rho_2 \cdots \rho_n$ and $\rho_{P(1)} \rho_{P(2)} \cdots \rho_{P(n)}$ lead to the same sector and hence there is a unitary intertwining operator $\epsilon_P(\rho_1, \cdots, \rho_n)$ from $\rho_1 \cdots \rho_n$ to $\rho_{P(1)} \cdots \rho_{P(n)}$:

$$\rho_1 \cdots \rho_n (A) \, \epsilon_P(\rho_1, \cdots, \rho_n) = \epsilon_P(\rho_1, \cdots, \rho_n) \, \rho_{P(1)} \cdots \rho_{P(n)}(A). \tag{2.27}$$

The relation (2.27) does not yet define $\epsilon_P(\rho_1, \cdots, \rho_n)$ uniquely but we can find a unique and natural determination of the ϵ_P by comparing them for different sets of arguments. Let $\rho_1', \cdots, \rho_n'$ be any other set of n endomorphisms with ρ_i' not disjoint from ρ_i so that we have intertwining operators $\tilde{R}_i = (\rho_i ; R_i ; \rho_i')$. Then $\tilde{R}_1 \times \cdots \times \tilde{R}_n$ intertwines from $\rho_1 \cdots \rho_n$ to $\rho_1' \cdots \rho_n'$; $\epsilon_P(\rho_1', \cdots, \rho_n')$ from $\rho_1' \cdots \rho_n'$ to $\rho'_{P(1)} \cdots \rho'_{P(n)}$. Denoting for brevity by $R_1 \times \cdots \times R_n$ the operator which is the middle piece of the triple $\tilde{R}_1 \times \cdots \times \tilde{R}_n$ the operators $\epsilon_P(\rho_1, \cdots, \rho_n) R_{P(1)} \times \cdots \times R_{P(n)}$ and $R_1 \times \cdots \times R_{n-1} \times$ $\times R_n \, \epsilon_P(\rho_1', \cdots, \rho_n')$ both intertwine between $\rho_1 \cdots \rho_n$ and $\rho'_{P(1)} \cdots \rho'_{P(n)}$. One finds

<u>Lemma 2.5</u> It is possible to choose the operators $\epsilon_P(\rho_1, \cdots, \rho_n)$ in such a way that

(i) $\epsilon_P(\rho_1, \cdots, \rho_n) = 1$ whenever all the supports of the

ρ_i are mutually space-like.

(ii) For any other set ρ_i' and any choice of intertwining operators R_i from ρ_i to ρ_i' we have

$$R_1 \times \cdots \times R_n \, \epsilon_P(\rho_1', \cdots, \rho_n') = \epsilon_P(\rho_1, \cdots, \rho_n) R_{P(1)} \times \cdots \times R_{P(n)} . \tag{2.28}$$

The unitary operators $\epsilon_P(\rho_1, \cdots, \rho_n)$ are uniquely determined by these two requirements and they satisfy (2.27).

The proof of this lemma proceeds by a straightforward computation. An immediate consequence of the lemma (in particular of the uniqueness of ϵ_P) is the multiplication law

$$\epsilon_P(\rho_1, \cdots, \rho_n) \, \epsilon_Q(\rho_{P(1)}, \cdots, \rho_{P(n)}) = \epsilon_{PQ}(\rho_1, \cdots, \rho_n). \tag{2.29}$$

If we put all ρ_i equal, a more convenient notation is

$$\epsilon_P(\rho, \rho, \cdots, \rho) \equiv \epsilon_\rho^{(n)}(P) . \tag{2.30}$$

One then has by (2.29)

$$\epsilon_\rho^{(n)}(P) \, \epsilon_\rho^{(n)}(Q) = \epsilon_\rho^{(n)}(PQ) , \tag{2.31}$$

i.e. the $\epsilon_\rho^{(n)}$ form a unitary representation of the permutation group of n elements which is (up to unitary equivalence) characteristic of the sector ρ^n . Note that $\epsilon_\rho^{(n)}(P)$ intertwines ρ^n with itself. Hence one might be inclined to think that $\epsilon_\rho^{(n)}(P)$ should be trivial. This is however not so because the $\epsilon_P(\rho_1, \cdots, \rho_n)$ are defined to be $\mathbb{1}$ not for equal ρ_i but for space-like separation of the supports of the ρ_i and it turns out that the representation

$\epsilon_\rho^{(n)}$ is one-dimensional ($\epsilon_\rho^{(n)}(P) = \pm 1$) if and only if ρ is an automorphism. It is an instructive exercise to check that in this case, putting $n = 2$ and choosing P as the transposition $1 \leftrightarrow 2$ $\epsilon_\gamma(P)$ reduces to the sign factor in (2.7)

The representation $\epsilon_\rho^{(n)}$ may be characterized in terms of the Young tableaux it contains and we may compare the set of Young tableaux occurring for fixed ρ and varying n. This analysis is most satisfactory if the sector ρ has an adjoint sector ρ^* (related to the existence of anti-particles). Then one finds that associated with ρ there is a number λ, depending only on the sector generated by ρ, not on ρ itself. $\lambda^{-1} = p_\rho$ is a positive or negative integer. For positive p_ρ all possible Young tableaux appear in the sectors ρ^n for which the number of rows does not exceed p_ρ. For negative p_ρ one has all Young tableaux whose number of columns is limited by $|p_\rho|$. The first case is familiar as the para-Bose case of order p_ρ, the second as the para-Fermi case of order $|p_\rho|$.

Thus we conclude that a particle which has an anti-particle and belongs to a non-simple sector is a paraboson or parafermion of some fixed order.

III. Parastatistics.

Field theoretic models for the description of strange statistics (i.e. statistics which are neither of the Bose- or Fermi-type) have been given by H. S. Green [26] and further discussed by numerous authors. See e.g. the clear survey by Greenberg [27]. Here we shall analyse in terms of the concepts developed in section I the physical content of the simplest example of such a model, the case of a parafermi field of order 2. I shall sketch only the essential line of argument referring for details of proofs to [28]. The discussion can be carried through using only the field quantities at one time, say t = 0. Lorentz invariance plays no role. The model itself may be described as follows. Take two Fermi fields $\psi^{(1)}$ and $\psi^{(2)}$ commuting with each other. The commutation relations are

$$[\psi^{(i)*}(\underset{\sim}{x}), \psi^{(i)}(y)]_+ = \delta^3(\underset{\sim}{x}-\underset{\sim}{y}); \quad [\psi^{(i)}(\underset{\sim}{x}), \psi^{(i)}(y)]_+ = 0; \ i=1,2 \tag{3.1}$$

$$[\psi^{(i)}(\underset{\sim}{x}), \psi^{(j)}(y)]_- = [\psi^{(i)*}(\underset{\sim}{x}), \psi^{(j)}(y)]_- = 0 \ \text{for } i \neq j. \tag{3.2}$$

Then define the "parafield" ψ as the sum of these two:

$$\psi(\underset{\sim}{x}) = \psi^{(1)}(\underset{\sim}{x}) + \psi^{(2)}(\underset{\sim}{x}) \tag{3.3}$$

and demand that only such quantities which are expressible in terms of this parafield ψ should occur in the theory.

Before we can analyse the physical content we have to know what the observable algebras corresponding to various space regions are. There are several possibilities but the choice is limited by the following two restrictions:

i) The observable algebra $\mathfrak{A}(V)$ of the space region V shall be a subalgebra of the parafield algebra $\mathcal{F}_p(V)$ of the same region. The latter is the *-algebra generated by the $\psi(f)$ for all test functions f with support in V.

ii) If V_1 and V_2 are disjoint, then $\mathfrak{A}(V_1)$ and $\mathfrak{A}(V_2)$ shall commute (locality).

Of course the net of algebras $\mathfrak{A}(V)$ shall be covariant with respect to translations and the total algebra $\mathfrak{A}$ is defined as

$$\mathfrak{A} = \overline{\cup \mathfrak{A}(V)} .$$

The largest algebra satisfying these requirements is the algebra $\mathfrak{A}_0$, which consists precisely of all <u>even</u> elements of $\mathcal{F}_p$ (elements invariant under the substitution $\psi \to -\psi$). So $\mathfrak{A}$ must be contained in (or possibly be equal to) $\mathfrak{A}_0$. Now one remarks that $\mathfrak{A}_0$ can also be characterized in terms of the algebra generated by two Fermi fields, $\phi^{(1)}$, $\phi^{(2)}$ with normal commutation relations:

$$[\phi^{(i)}(\underset{\sim}{x}), \phi^{(j)}(\underset{\sim}{y})]_+ = 0, \quad [\phi^{(i)}(\underset{\sim}{x}), \phi^{(j)*}(\underset{\sim}{y})]_+ = \delta_{ij}\, \delta^3(\underset{\sim}{x}-\underset{\sim}{y}). \quad (3.4)$$

To see this we first embed $\mathfrak{F}_P$ in a large algebra $\tilde{\mathfrak{F}}$ generated by the "Green components" $\psi^{(1)}$, $\psi^{(2)}$ and one additional element K_2 (standing for "Klein transformation"). The relations of the $\psi^{(i)}$ are given in (3.1), (3.2). The relations involving K_2 are

$$K_2^{-1} = K_2 = K_2^{*} ;$$
$$\left[K_2, \psi^{(1)}(\underset{\sim}{x})\right]_- = 0 ; \quad \left[K_2, \psi^{(2)}(\underset{\sim}{x})\right]_+ = 0 . \tag{3.5}$$

It is readily seen that with the definition

$$\psi^{(1)}(\underset{\sim}{x}) = \phi^{(1)}(\underset{\sim}{x}) K_2 ; \quad \psi^{(1)*}(\underset{\sim}{x}) = \phi^{(1)*}(\underset{\sim}{x}) K_2 ;$$
$$\psi^{(2)}(\underset{\sim}{x}) = i\,\phi^{(2)}(\underset{\sim}{x}) K_2 ; \quad \psi^{(2)*}(\underset{\sim}{x}) = i\,\phi^{(2)*}(\underset{\sim}{x}) K_2 \tag{3.6}$$

we obtain the normal commutation relations (3.4). The general element of $\tilde{\mathfrak{F}}$ is of the form $F + F' K_2$ where F and F' are expressible in terms of the $\psi^{(i)}$. For the elements of $\mathfrak{A}_0$ F' is zero and F even. Rewriting such an element in terms of the $\phi^{(i)}$ and K_2 one sees that K_2 drops out because in each monomial the factors K_2 can be shifted to the right using (3.5) and an even power of K_2 is the identity.

Let us denote the algebra generated by the two normal Fermi fields $\phi^{(i)}(\underset{\sim}{x})$ (with $\underset{\sim}{x}$ ranging in V) by $\mathfrak{F}(V)$ and consider substitutions of the form

$$\phi^{(i)}(\underset{\sim}{x}) \longrightarrow \sum_{k} g_{ki}\, \phi^{(k)}(\underset{\sim}{x}) \,. \tag{3.7}$$

Such a substitution generates an automorphism of $\mathfrak{F}$ denoted by α_g as long as the 2 x 2 matrix g is unitary. One finds

Lemma 3.1 $\mathfrak{A}_0(V)$ is the subalgebra of $\mathfrak{F}(V)$ consisting of precisely those elements which are invariant under the automorphisms α_g when g runs through SO(2) (real, orthogonal matrices with determinant +1).

In other words: $\mathfrak{A}_0$ is the "gauge invariant" part of $\mathfrak{F}$ when we take SO(2) as the gauge group (acting on $\mathfrak{F}$ according to (3.7)).

Since the observable algebra $\mathfrak{A}$ must be contained in $\mathfrak{A}_0$ other possibilities for $\mathfrak{A}$ result if we take a larger gauge group $\mathcal{G}$ and again define $\mathfrak{A}$ as the gauge invariant part with respect to this group. We shall just consider one such example, the case where the gauge group is the largest automorphism group of the form (3.7), namely the group U(2). The resulting observable algebra will be denoted by $\mathfrak{A}_2$.

Summing up: instead of expressing the observables in terms of the parafield we can express them also as functions of two normal Fermi fields. The characterization of the observable algebra within this Fermi field algebra $\mathfrak{F}$ is best done by specifying the "gauge group" $\mathfrak{G}$ under which the observables are invariant. In our case (starting from a para fermi field of order 2) the minimal gauge group is SO(2), leading to the maximal observable algebra $\mathfrak{A}_0$.

Given $\mathfrak{F}$, $\mathfrak{G}$ and $\mathfrak{A}$ we have the following structure [23]: The different superselection sectors (families of states which are of interest) are in one-to-one correspondence with the "spectrum" of the gauge group.* If $\mathfrak{G}$ is Abelian we have only Bose- or Fermi statistics. Let us consider from this point of view the two examples mentioned above and compare the conclusions with the parafield description.

<u>Example 1.</u> $\mathfrak{G} = SO(2)$; $\mathfrak{A} = \mathfrak{A}_0$.

The gauge group is Abelian. Its spectrum consists of the integers $n = 0, \pm 1, \pm 2, \cdots$. We thus have just one ordinary charge quantum number, distinguishing the superselection sectors. States of charge $n = +1$ are obtained from the vacuum state vector Ω by applying $\phi^{(1)}(f) + i\,\phi^{(2)}(f)$ Note that $\phi^{(1)*}(f) + i\,\phi^{(2)*}(f)$ leads to the same sector. The sector $n = -1$ is reached from Ω by $\phi^{(1)} - i\,\phi^{(2)}$ or, equally well, by $\phi^{(1)*} - i\,\phi^{(2)*}$.

---

* The "spectrum of $\mathfrak{G}$" consists of the equivalence classes of irreducible representations of $\mathfrak{G}$.

The particles with $n = \pm 1$ are ordinary Fermions. How does that fit with the parafield description? We have, according to (3.6)

$$\psi(f) = (\phi^{(1)}(f) + i\,\phi^{(2)}(f))\,K_2 . \qquad (3.8)$$

The comparison between the two descriptions is simplest if $K_2\Omega = \Omega$. Then, according to (3.8), $\psi(f)\Omega$ is a state with charge $n = +1$. Suppose g and f are test functions with far separated supports. Then

$$\psi(g)\psi(f)\Omega = (\phi^{(1)}(g) + i\,\phi^{(2)}(g))\,K_2\,(\phi^{(1)}(f) + i\,\phi^{(2)}(f))K_2\,\Omega =$$
$$= (\phi^{(1)}(g) + i\,\phi^{(2)}(g))\,(\phi^{(1)}(f) - i\,\phi^{(2)}(f))\,\Omega$$

is a state with $n = 0$. Any polynomial of ψ applied to Ω will produce only states with $n = +1$ and $n = 0$ because in a product ψ will alternatingly raise or lower the charge, depending on its position. The fact that there are both symmetric and antisymmetric wave functions allowed for the states generated from the vacuum by two parafield operators has (in this example) nothing to do with parastatistics but results from the fact that the two operators produce different particles (the first a negatively charged, the second a positively charged one). The effect of ψ depends on the position it has within a product.

Example 2. $\mathfrak{G} = U(2)$; $\mathfrak{A} = \mathfrak{A}_2$.

We may first note that $\mathfrak{A}_2$ is generated by the bilocal densities

$$\rho(\underset{\sim}{x}, y) = \phi^{(1)*}(\underset{\sim}{x})\,\phi^{(1)}(y) + \phi^{(2)*}(\underset{\sim}{x})\,\phi^{(2)}(y) . \tag{3.9}$$

The irreducible representations of U(2) may be labeled by two quantum numbers (B , I) with the relation

$$B + 2I = \text{even}. \tag{3.10}$$

$B = 0, \pm 1, \pm 2$ may be conveniently interpreted as baryon number, $I = 0, \frac{1}{2}, 1, \cdots$ as isospin,

thinking of the theory of nuclei with strict charge independence. To each allowed pair (B , I) we have a sector. In a sector with isospin I each state appears with a multiplicity ($2I + 1$). Any vector in the ($2I+1$)-dimensional subspace spanned by an isospin multiplet gives exactly the same expectation values over the observable algebra and corresponds therefore to the same physical state as any other vector in this subspace. With the conventional choice of the three components of isospin (the Pauli matrices σ_1 and σ_3 real, σ_2 purely imaginary) one finds that the subgroup SO(2) is placed within U(2) in such a way that the charge quantum number n is related to the second component of the isospin

$$I_2 = 2n . \tag{3.11}$$

Applying the parafield algebra on the vacuum we get therefore only states with $I_2 = 0$ or $I_2 = \frac{1}{2}$. The former appear if we have an even number of ψ-factors, which leads to an even baryon number; the latter appear for odd B. According to (3.10) even B implies integer I and then we can find in each multiplet a vector with $I_2 = 0$. For odd B, half integer I, we have in each multiplet a state with $I_2 = \frac{1}{2}$. Hence the restriction to $n = 0, 1$ does not limit the selection of states. Applying the parafield algebra to Ω we obtain all relevant states over $\mathfrak{A}$, only the multiplicity is changed as compared to the representation space of the field algebra $\mathcal{F}$. We obtain with $\mathcal{F}_p$ each state only once instead of the ($2I+1$)-dimensional multiplets. In this second example the parafield model gives a complete description. The parastatistics of particles corresponds then to the fact that there is one hidden parameter (the charge I_2) which is not observable. We see that in this case the parastatistics may be reduced to ordinary (Bose-Fermi) statistics if one introduces this hidden degree of freedom as an additional distinctive quantum number. It appears that all reasonable parafield models can be reduced in this way to the Bose-Fermi case [28] although a general theorem to this effect based on the structure analysis described in section II has not yet been obtained.

Mathematical Appendix

The set of all bounded operators acting in a Hilbert space $\mathcal{H}$ is denoted by $\mathcal{B}(\mathcal{H})$. There are several topologies in this set $\mathcal{B}(\mathcal{H})$ which are important in our context. A topology means that we define what is a "neighborhood" of an element in the set. Actually it is sufficient here to define the neighborhoods of the origin. The three most important topologies in $\mathcal{B}(\mathcal{H})$ arising in the context of physics are:

a) Uniform topology. Since an operator $A \in \mathcal{B}(\mathcal{H})$ has a finite norm, $\|A\|$, we may define a neighborhood of the origin as the set of all $A \in \mathcal{B}(\mathcal{H})$ which have a norm less than ϵ. Any $\epsilon > 0$ gives us one such neighborhood. Convergence of a sequence $A_n \in \mathcal{B}(\mathcal{H})$ in this topology means that the $\|A_n - A_m\| < \epsilon_N$ for all n, m > N and $\epsilon_N \rightarrow 0$. Such a sequence is called "uniformly convergent" or a Cauchy sequence in the norm topology.

b) Strong topology. We pick an arbitrary vector $\Psi \in \mathcal{H}$ and an arbitrary number $\epsilon > 0$. The corresponding neighborhood of the origin, denoted by $\mathcal{N}(\Psi, \epsilon)$ consists of all $A \in \mathcal{B}(\mathcal{H})$ satisfying

$$\|A\Psi\| < \epsilon$$

where now the left hand side is the length of the vector $A\Psi$ (not the norm of the operator A). Convergence of a sequence A_n in this topology ("strong convergence") means that for every vector Ψ the sequence of image vectors $A_n\Psi$ satisfies

$$\| A_n\Psi - A_m\Psi \| \leq \epsilon_N \text{ for } n,m > N;\ \epsilon_N \to 0.$$

c) Weak topology. We pick an arbitrary pair of vectors $\Phi, \Psi \in \mathcal{H}$ and $\epsilon > 0$ and define a corresponding (weak) neighborhood $\mathcal{N}(\Phi, \Psi; \epsilon)$ as the set of all $A \in \mathcal{B}(\mathcal{H})$ satisfying $|(\Phi, A\Psi)| < \epsilon$. Weak convergence of a sequence A_n means the convergence of the sequences of matrix elements $(\Phi, A_n\Psi)$ for all pairs Φ, Ψ.

Example of a strongly convergent sequence which does not converge uniformly. Take a complete orthonormal basis Ψ_n in $\mathcal{H}$ and let E_n by the projector on the subspace spanned by the first n basis vectors. One easily sees that the strong limit of the sequence E_n as $n \to \infty$ is the unit operator (completeness relation). Yet for arbitrarily large n we still have $\| E_n - E_{n+1} \| = 1$, i.e. there is no uniform convergence.

Example of a weakly convergent but not strongly convergent sequence. Take Ψ_n as above and define A_n by

$A_n \Psi_m = \Psi_{n+m}$. This sequence A_n converges weakly to zero as $n \to \infty$ but obviously has no strong limit.

If A and B belong to $\mathcal{B}(\mathcal{H})$ and α, β are complex numbers then

$$\begin{aligned} &\alpha A + \beta B \\ &A \cdot B \\ &A^* \text{ (adjoint of } A) \end{aligned} \tag{A.1}$$

also belong to $\mathcal{B}(\mathcal{H})$. A subset of $\mathcal{B}(\mathcal{H})$ which is closed under the three operations (A.1) is called a *-algebra of bounded operators. If in addition it contains the unit operator and is closed in the strong topology it is called a "von Neumann ring." If it is closed in the uniform topology it is called a (concrete) C^*-algebra. Since the three topologies listed under a), b) and c) are decreasing in strength we have in general that the weak closure of a set is larger than the strong closure and this again larger than the uniform closure. A weakly closed set is always strongly closed, a strongly closed one is always uniformly closed. Hence any von Neumann ring is also a C^*-algebra but the converse is not true. It turns out, on the other hand, that for a *-algebra of bounded operators the weak and the strong closures coincide. Thus a von Neumann ring is also always weakly closed.

If S is any subset of $\mathcal{B}(\mathcal{H})$ one defines the "commutant" S' as the set of all operators from $\mathcal{B}(\mathcal{H})$ which commute with every element of S. Let S^* denote the set of all the adjoint operators of the members of S. One has

a) The commutant of any subset of $\mathcal{B}(\mathcal{H})$ which is closed under the *-operation is a von Neumann ring, i.e. $\{S \cup S^*\}'$ is a von Neumann ring.

b) If R is a von Neumann ring, then

$R = R''$ (R'' denotes the commutant of the commutant).

c) $\{S \cup S^*\}''$ is the smallest von Neumann ring containing S.

Let us now consider abstract *-algebras. Such an algebra $\mathfrak{A}$ consists of elements A, B, ... for which the three operations (A.1) are defined, satisfying the usual laws. As in the case of an abstract group the elements now are not regarded as operators in some space but just as objects which can be connected by the operations (A.1). It turns out that under rather general conditions the algebraic structure determines a natural norm $\|A\|$ for the elements of the algebra satisfying

$$\|A\| = \|A^*\| = \|AA^*\|^{1/2}; \quad \|\alpha A\| = |\alpha| \, \|A\| \qquad \text{(A.2)}$$

and the inequalities

$$\|A+B\| \leq \|A\| + \|B\| ; \quad \|AB\| \leq \|A\| \, \|B\| . \qquad \text{(A.3)}$$

If these conditions prevail, then the *-algebra may be equipped with this natural norm topology and completed in it. It is then called an abstract C^*-algebra.

Given an abstract C^*-algebra $\mathfrak{A}$ one considers the "positive linear forms" over it. Denote this set by $\mathfrak{A}^{*(+)}$. An element ω from $\mathfrak{A}^{*(+)}$ assigns to each $A \in \mathfrak{A}$ a complex number $\omega(A)$ subject to the two conditions

(i) linearity $\omega(\alpha A + \beta B) = \alpha\omega(A) + \beta\omega(B)$, (A.4)

(ii) positivity $\omega(A^* A) \geq 0$. (A.5)

The norm of such a form is defined as

$$\|\omega\| = \sup_{\|A\|=1} |\omega(A)| . \tag{A.6}$$

One has

$$\|\omega\| = \omega(1) . \tag{A.7}$$

The set of positive linear forms is a convex cone i.e.

$$\omega = \lambda_1 \omega_1 + \lambda_2 \omega_2 \tag{A.8}$$

belongs to $\mathfrak{A}^{*(+)}$ if ω_1 and ω_2 belong and λ_1, λ_2 are positive numbers. A positive linear form ω is called "extremal" if no non-trivial decomposition of the form (A.8) is possible.

A representation of $\mathfrak{A}$ (by operators in a Hilbert space $\mathcal{H}$) assigns to each $A \in \mathfrak{A}$ an operator $\pi(A)$ from

$\mathcal{B}(\mathcal{H})$ in such a way that the algebraic relations (A. 1) are conserved. In other words it is an isomorphic mapping from the abstract C^*-algebra to a concrete C^*- algebra of operators $\pi(\mathfrak{A})$. The representation is called faithful if $\pi(A)=0$ implies $A = 0$. In this case the norm of the operator $\pi(A)$ is the same as the C^*-norm of the abstract element A. We shall only consider faithful representations in these lectures. Given a representation one has immediately a family of positive linear forms over $\mathfrak{A}$, which are associated with this representation. Pick any vector Ψ in the representation space, then

$$\omega_{\Psi}(A) = (\Psi, \pi(A)\Psi) \tag{A.9}$$

is such a form. Also any positive trace class operator ρ in $\mathcal{B}(\mathcal{H})$ defines a positive linear form over the algebra by

$$\omega_{\rho}(A) = \mathrm{Tr}\,(\rho\,\pi(A)) . \tag{A.10}$$

Let us denote by $\mathcal{J}_{\pi}$ the family of forms associated with the representation π according to (A. 10) and by $\mathcal{J}_{\pi}^{V}$ the subset of "vectorial" forms (A. 9). The former are the convex combinations of the latter.

The connection between positive linear forms and representations can also be followed in the opposite direction. Given an $\omega \in \mathfrak{A}^{*(+)}$ one can construct a representation π_{ω} so that we have a vector Ψ in the representation space giving back the form ω by (A. 9). This vector Ψ is moreover a "cyclic vector", i. e. the vectors $\pi_\omega(\mathfrak{A})\Psi$ are dense in $\mathcal{H}$. This is achieved by the GNS-construction (the letters standing for Gelfand, Naimark, Segal). It proceeds as follows: First note that the algebra is itself a linear space and that an $\omega \in \mathfrak{A}^{*(+)}$ defines a semidefinite scalar product between the elements of $\mathfrak{A}$ by

$$(B, A) = \omega(B^* A) .$$

To obtain a positive definite scalar product one has to divide the algebra into equivalence classes modulo the set J which latter consists of all elements $Z \in \mathfrak{A}$ for which $\omega(Z^* Z) = 0$. Let us denote the class of $A \in \mathfrak{A}$ by $\hat{A}$. This contains all the elements of $\mathfrak{A}$ which differ from A by an element of the set J. The set of these classes is a linear space with the positive definite scalar product

$$(\hat{B}, \hat{A}) = \omega(B^* A) , \tag{A. 11}$$

where one checks that the right hand side is independent of the choice of the elements B, A in the respective classes $\hat{B}$, $\hat{A}$. Thus $\mathfrak{A}/J$ is a "pre Hilbert space," i. e., it may be considered as a dense set of vectors in a Hilbert space $\mathcal{H}$. We obtain a representation of $\mathfrak{A}$ by operators in $\mathcal{H}$ defining the operator $\pi(A)$ representing A by

$$\pi(A)\hat{B} = \widehat{AB} . \tag{A. 12}$$

The right hand side involves a choice of B from the class $\hat{B}$ but one checks again that the class of AB does not change when B varies within one class (the set J is a left ideal of $\mathfrak{A}$). The cyclic vector Ψ giving back the expectation value ω corresponds to the class of the unit element of $\mathfrak{A}$. We have, by (A.11) and (A.12)

$$(\Psi, \pi(A)\Psi) = (\hat{1}, \pi(A)\hat{1}) = \omega(1^* A 1) = \omega(A).$$

A representation π is called irreducible if there is no invariant subspace in the representation space $\mathcal{H}$. A criterion for irreducibility is Schur's lemma: π is irreducible if and only if $(\pi(\mathfrak{A}))'$ consists only of multiples of the identity. An equivalent criterion is $(\pi(\mathfrak{A}))'' = \mathcal{B}(\mathcal{H})$. The GNS-construction leads to an irreducible representation if and only if the form ω from which the construction starts is extremal.

If π is reducible we may consider the restriction of π to one of the invariant subspaces. This is called a subrepresentation of π. One calls two representations disjoint if they contain no subrepresentations which are unitarily equivalent. A representation is called "primary" or a "factor" if

$$(\pi(\mathfrak{A}))'' \cap (\pi(\mathfrak{A}))' = \{\lambda 1\},$$

i.e. if the von Neumann algebra generated by the representers $\pi(\mathfrak{A})$ has no nontrivial center. A form ω is called primary if the GNS-construction starting from it leads to a primary representation.

We frequently use the notion of an automorphism or an endomorphism of a C^*-algebra. In each case we mean a one-to-one mapping carrying $A \in \mathfrak{A}$ to $\rho(A) \in \mathfrak{A}$ and such that the algebraic structure and the norm are conserved. If the image set $\rho(\mathfrak{A})$ is equal to $\mathfrak{A}$, ρ is called an automorphism. If $\rho(\mathfrak{A})$ is smaller than $\mathfrak{A}$ then we call ρ an endomorphism.

References

[1] N. Bohr and L. Rosenfeld, Dan. Mat. Fys. Medd. XII, No. 8 (1933).

[2] E. P. Wigner, Group Theory and its application to the Quantum Mechanics of Atomic Spectra, Academic Press, New York, 1959.

[3] R. F. Streater and A. S. Wightman, "PCT, Spin Statistics and all that," Benjamin, New York, 1964.

[4] B. Schroer, private communication.

[5] H. Epstein, Nuovo Cim. 27, 886 (1963).

[6] R. A. Brandt, Am. Phys. 44, 221 (1967).

[7] W. Zimmermann, Brandeis Lectures 1970 and earlier work quoted therein.

[8] R. T. Powers, Commun. Math. Phys. 4, 145 (1967).

[9] K. B. Sinha, Univ. of Rochester Thesis, 1969.

[10] A. S. Wightman, Phys. Rev. 101, 860 (1956).

[11] R. Haag, Phys. Rev. 112, 669 (1959).

[12] D. Ruelle, Helv. Phys. Acta 35, 147 (1962).

[13] R. Jost, The General Theory of Quantized Fields, American Math. Soc., Providence, 1965.

[14] H. J. Borchers, Nuovo Cim. 15, 784 (1960).

[15] G. C. Wick, E. P. Wigner and A. S. Wightman, Phys. Rev. 88, 101 (1952).

[16] R. Haag and D. Kastler, Journal Math. Phys. 5, 848 (1964).

[17] H. Araki, J. Math. Phys. 5, 1 (1964).

[18] I. E. Segal in "Les Problemes Mathematiques de la Theorie Quantique des Champs," C.N.R.S. 1959.

[19] I. E. Segal, Ann. Math. 48, 930 (1947).

[20] J. M. G. Fell, Trans. Am. Math. Soc. 94, 365 (1960).

[21] H. Roos, Commun. Math. Phys. 16, 238 (1970).

[22] H. J. Borchers, Commun. Math. Phys. 1, 291 (1965).

[23] S. Doplicher, R. Haag, J. E. Roberts, Commun. Math. Phys. 13, 1, (1969).

[24] S. Doplicher, R. Haag, J. E. Roberts, Commun. Math. Phys. 15, 173 (1969).

[25] S. Doplicher, R. Haag, J. E. Roberts, "Locality and Statistics," forthcoming manuscript.

[26] H. S. Green, Phys. Rev. 90, 270 (1953).

[27] O. W. Greenberg, "Parafield Theory" in "Proceedings of the Conference on the Mathematical Theory of Elementary Particles," edited by R. Goodman and I. E. Segal. MIT Press, 1966.

[28] N. Drühl, R. Haag and J. E. Roberts, Comm. Math Phys. 18, 204 (1970).

Regge Models and Duality

Maurice Jacob
CERN, European Organization for Nuclear Research
Geneva, Switzerland

CONTENTS

1. INTRODUCTION

The purpose of these notes is to cover the recent developments in hadron physics referred to under the general name of duality. The duality between a Regge pole description and a resonance description of strong interaction reactions has proved to be an extremely useful phenomenological tool. During the last three years it has sharpened our understanding of reaction mechanisms which is still at a semi-quantitative level. As a result, it has increased the trust which we put in the Regge pole expansion, at least as a sound starting point, even though more precise experiments have contradicted many of the predictions boldly put forward as consequences of the leading role granted to Regge pole exchange. With the many self-consistency checks provided by duality, we may perhaps relax our insistence on good agreement with experiment during the process of collecting enough confidence in our theoretical Regge models in order to elaborate them to the stage where good agreement will be reached. Despite their predictions of dips and peaks at locations where dips and peaks are not observed, Regge models have thus gained through duality much attention and interest. This has even led to a bootstrap scheme in which the association of sets of observed resonances with linear rising Regge trajectories provides extremely interesting results. So far, however, duality is merely a new approach to hadron dynamics and, if a theory, it is still in the making. Nevertheless, the prominent results already obtained make developing such a theory a challenging and promising new theme of research. This would lead us beyond the too stringent dual Regge models which we have at present and which apply only in a model world of zero width resonances where we can itemize levels but have no realistic scattering. In his series of lectures, Professor Mandelstam describes the new and promising dynamical schemes which can be developed from the Veneziano model, the prototype of most dual models which have been devised so far. [1]

These notes will then have basically an introductory character, describing how duality, first brought into shape as a new idea from a phenomenological approach, has developed into a sufficiently precise formulation as to become a required property of theoretical models and a guide-line for a possible theory of strong interactions. Indeed, a list of tentative definitions of duality provides the basic framework of this series of lectures.

We shall first discuss global duality, a property which combines the assumed asymptotic Regge behaviour of a non-elastic two-body amplitude (say, charge exchange) with the empirical observation that its imaginary part is well approximated, at least in some average sense, by the sole contribution of resonances. This leads to a clear and most interesting connection between the leading and well-known low energy resonances and the leading Regge poles, the contribution of which dominates the observed high energy behaviour. [2]

This is readily expressed through finite energy sum rules, which will be discussed in some detail later:

$$\int_0^N \mathrm{Im}\, A(\nu,t)\, d\nu = \sum_i \beta_i(t)\, \frac{N^{\alpha_i(t)+1}}{\alpha_i(t)+1} \tag{1.1}$$

Where the scattering amplitude A is assumed to be such that its imaginary part can be replaced by the contribution from the resonances when integrated from 0 to a finite energy N.

The variable $\nu = \frac{s-u}{4m}$ is, at $t=0$, the laboratory energy of the incident particle, with m the mass of the target particle. The variables s, t and u are the Mandelstam invariants associated with the reaction considered. The high ν behaviour of the amplitude is here assumed to be given by a few Regge pole contributions. This reads:

$$\mathrm{Im}\, A(\nu,t) \simeq \sum_i \beta_i(t)\, \nu^{\alpha_i(t)} \qquad (1.2)$$

for $\nu > N$.

Replacing the integrand by a sum of resonance contributions, we obtain the specific connection between low energy resonances and high energy Regge poles which defines global duality.

Next we shall see that one may complement global duality in order to deal also with elastic amplitudes where the shadow of inelastic processes provides an important non-resonance "background". This may be done using the linear character of the finite energy sum rules by associating a non-resonant background with the leading Pomeranchon contribution at high energy and the resonances to specific Regge pole exchanges. Such a separation is compatible with present information and, used together with the empirical observation that resonances do not occur with certain possible internal quantum numbers, called "exotic", it leads to duality with its most generally accepted meaning.[3] It was the direct consequences of the assumptions so far brought into the concept of duality which led to enthusiasm for the scheme. As we shall see in detail later, the most striking consequence is the degeneracy between Regge trajectories of opposite signatures and even of different internal quantum numbers, a degeneracy which seems to be borne out remarkably well. Similarly striking is the equality in slope between baryon and meson trajectories, which holds to a remarkably good approximation. This is made evident in Figs. 1-1 and 1-2, where straight line trajectories have been drawn between several resonance points displayed on a Chew-Frautschi plot. This shows two clear examples of signature degenerate baryon trajectories and indicates that there appear to be four degenerate meson trajectories with both signatures and both allowed non-exotic quantum numbers.

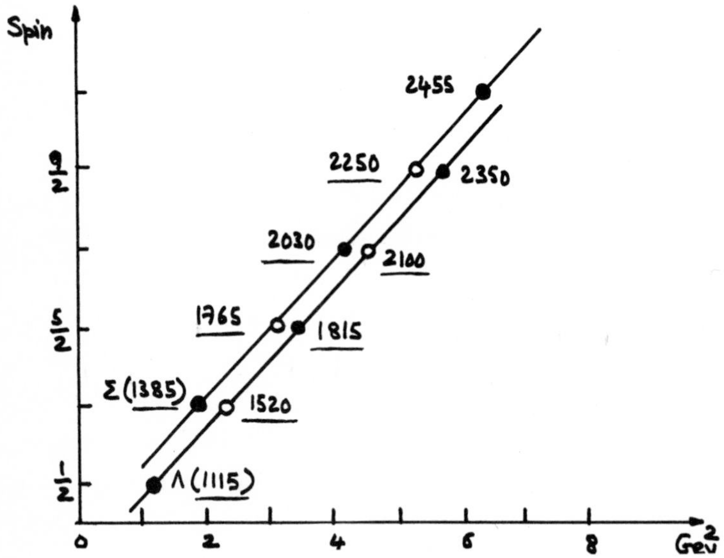

Figure 1-1. Regge trajectories associated with the $\Lambda(1115)$ and $\Sigma(1385)$ particles. The black dots indicate particles of positive parity, the white dots particles of negative parity. Underlined states have spin and parity experimentally determined.

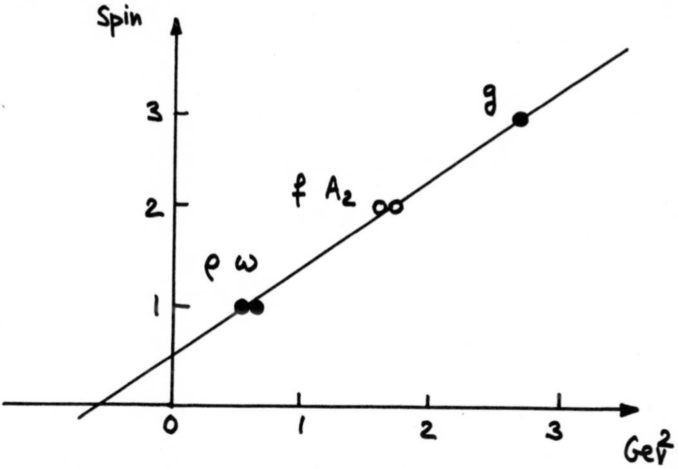

Figure 1-2. Regge trajectories drawn through vector and tensor meson points.

The degeneracy between trajectories with different signatures and different quantum numbers, as well as the observed equality of slope, may be considered coincidental. Nevertheless, one may suppose that so many coincidences must be the result of some specific properties of the underlying dynamics. As will be shown, duality indeed requires all these relations to occur, one of the main reasons for its present popularity. Duality also provides a rationale for the striking difference between the energy behaviour of scattering amplitudes which, as far as exchange mechanisms are concerned, would appear to be on the same footing. As shown in Fig. 1-3, the K^-p and K^+p total cross-sections behave in a different way in the 6-20 GeV range although both amplitudes have the same crossed channel $K\bar{K}\rightarrow p\bar{p}$, which is known to be the relevant one for a simple dynamical description at these energies.[4,5]

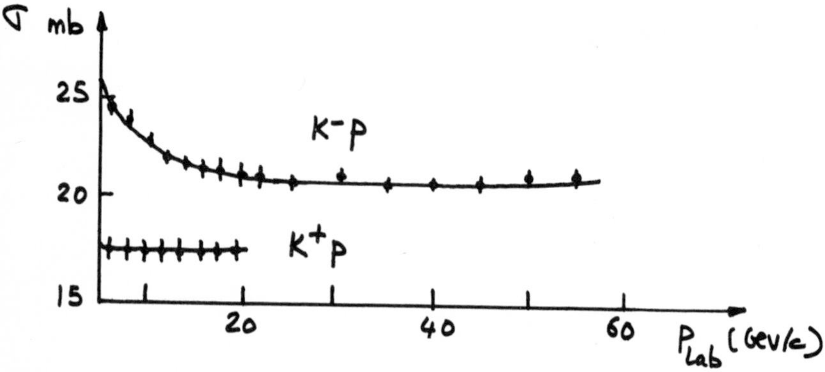

Figure 1-3. K^-p and K^+p cross-sections between 6 and 20 GeV.

This is one of the striking examples which show that, even though a simple exchange mechanism picture is found to hold, the direct channel quantum numbers remain highly relevant. In this particular case we have an exotic K^+p channel and a non-exotic K^-p channel; duality then accounts for the very different high energy behaviour which is observed. The general interplay found between the leading exchange mechanism and specific direct channel properties is another reason for present interest in duality. This leads to a dynamical framework in which the quark model and the bootstrap scheme both contribute in a constructive way. A few years ago such a collaboration might have been considered as highly improbable.

It is tempting to postulate duality in a more local sense and take the attitude that many specific features of reaction amplitudes could be easily explained in terms of Regge exchange or in terms of direct channel resonances. The two complementary descriptions would then be valid in relatively small energy intervals, with one of them simpler than the other, depending on whether we deal with the low or high energy domain. The connection would not need the resonance averaging imposed in (1.1). As will be discussed later in detail, this attitude meets with some success. Nevertheless two such descriptions limited to the few leading resonances or the few leading Regge trajectories which we may know to be present cannot coincide. Both may describe the amplitude in an asymptotic series sense in a certain domain, but for mathematical reasons they cannot be identified. Only an infinite Regge pole expansion and an infinite resonance expansion are mathematically flexible enough to provide equivalent descriptions for a scattering amplitude. This equivalence is possible in a model amplitude which incorporates infinite sets of resonances in two crossed channels which can be partially resummed into infinite sets of Regge trajectory contributions. The duality property is then expressed by the fact that the infinite resonance contribution in one channel already contains the infinite resonance contribution in the crossed channel and vice versa. The two descriptions (s channel resonance and t channel resonance) are equivalent. Duality is then true in a local sense and can be expressed graphically as in Fig. 1-4.

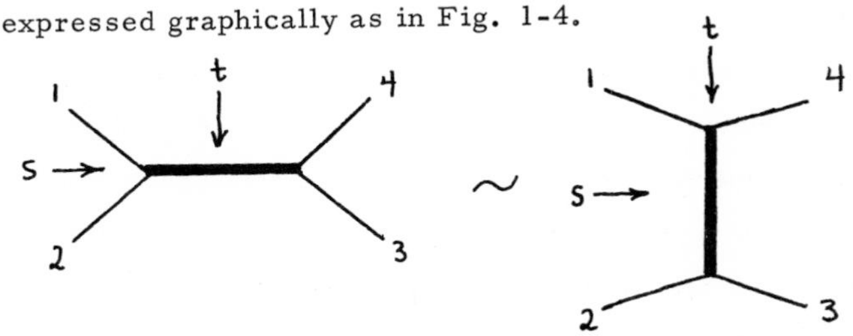

Figure 1-4.

The model amplitudes so far constructed apply only to a model world of zero width resonances lying on linear infinitely rising Regge trajectories. While not suitable for the description of two-body scattering, they do show a level pattern and a general behaviour which have so many points in common with the actual amplitudes as to make them worth considering in great detail. Joining simplicity to many other interesting properties, they are - to a large extent - at the origin of the present interest in duality.

A further and important reason for this interest is provided by the fact that these dual models can be generalized to production processes. Even though the only case yet amenable to a detailed and successful study corresponds to zero spin identical particles, the importance of having at long last a common and practical framework for two-body scattering and production reactions among hadrons cannot be overstressed. The Regge amplitudes so far written for two-body scattering and the multi-Regge model considered at the same time for production processes are built up in such a way as to be a priori compatible with unitarity. They are tentative expressions for high energy amplitudes which should mutually constrain themselves.[6] The unitarity relations connecting them all have been poorly exploited so far due to technical difficulties. On the other hand, dual models built out of zero width resonances cannot use these unitarity relations. Nevertheless, they directly connect the level pattern and coupling strengths of the elastic and inelastic amplitudes in a very stringent way and show a welcome asymptotic Regge behaviour. They further give crossing symmetric expressions for production processes.

All resonances in the many dual channels are already included in the infinite sum of resonances explicitly associated with the maximum tree graph drawn on Fig 1-5 for a $2 \to 4$-body reaction. This will appear as the straightforward generalization of the equality between the two terms represented on Fig. 1-4, where an infinite set of resonances in one channel already includes the resonances in the dual crossed channel.

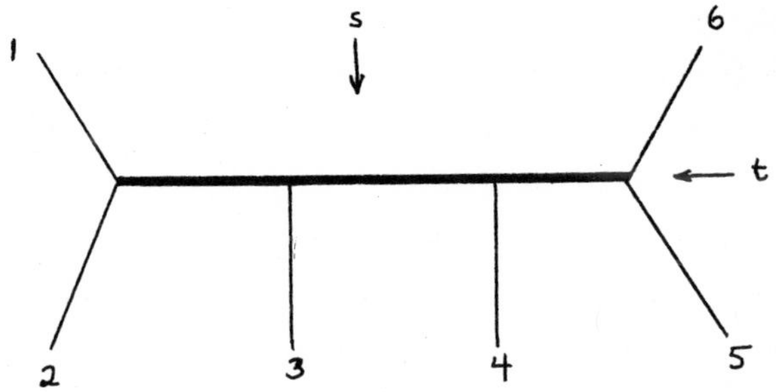

Figure 1-5.

At present these dual Regge models can be considered in several different ways. One may say that they are amplitudes for a model world of zero width resonances. Then we do not have actual scattering amplitudes but merely infinite collections of levels mutually sustaining themselves as the explicit solution of a bootstrap scheme. The general structure of the level patterns and coupling strengths thus obtained shows great similarities with those actually observed even though the strict linear trajectory pattern, out of which the models are built, should be lost in the actual hadronic world as soon as we analyze it with any scrutiny. In these models the only significant forces are those associated with resonance exchange. As a result, particles in different channels sustain themselves in a model world of zero width resonances as they may do in the actual one. Imposing unitarity would of course lead to different scattering amplitudes but would neither eliminate nor introduce new particles in any dramatic way.

One may try to go further and consider a dual model as the "Born term" of a theory of hadronic interactions.[7] This means that the actual reaction amplitudes should be obtained as the sum of "higher order" terms built out of the primitive terms of Figs. 1-4 and 1-5, a dual relation between resonances in the different channels being kept at each stage in the calculation. We shall indicate the progress recently made along this line and the problems which we face at present. We may speculate that the dual Regge models we aim at could eventually appear as the sum of an infinite set of field theory graphs in a similar way as a Regge amplitude may be associated with a ladder diagram.[8]

The purpose of these notes is to go from the primary phenomenological motivations for duality to these mathematical constructions.[9] In so doing I shall borrow from past reviews of duality [10,11] and in particular from my Schladming lecture notes [12] and the main part of my Lund talk [13] ; these notes can be looked upon as an updated and more detailed version. I shall assume some familiarity with Regge pole theory in particle physics, as recently reviewed by Collins and Squires [14] and will refer to it in any detail only in one instance of present technical interest, namely when discussing the still challenging question of dip mechanisms for which duality proposes a definite answer.

Fig. 1-6 summarizes the historical and logical developments of duality as general properties and empirical observations were synthesized and idealized into a general theoretical construction; it serves as the outline for these lectures.

Cumulative input	Formalism (code name)	Some results and applications
Analyticity . Regge asymtpotics .	Finite energy sum rules .	Better Regge phenomenology .
Resonance dominance .	Global duality .	Reciprocal relation between high energy Regge poles and low energy resonances. FESR bootstrap .
Special role for the Pomeranchon . Absence of exotic resonances .	"Duality" as generally quoted .	Exchange degeneracy, Nonet mixing, Nonsense choosing residues .
Closer relation between two asymptotic series expansions .	Semi-local duality .	Complementary description for local features. Argand loops from Regge poles. Special degeneracy.
Infinite sets of zero width resonances used in a closed dual form .	Dual models .	Simple scattering amplitudes; simple generalization to production amplitudes .
Dual calculation procedure for higher order terms .	A dual theory(?)	Many hopes and problems .

Figure 1-6.

2. DUALITY IN A GLOBAL WAY

A. Resonance and Regge Pole Descriptions

Two approximations appear to be useful in the analysis of two body hadronic processes. Such a process is represented on Fig. 2-1 The three Mandelstam invariants are as usual denoted by $s, t, \text{ and } u$. We again set $\nu = \frac{s-u}{4m}$.

$$s = (p_1 + p_2)^2$$
$$t = (p_1 + p_3)^2$$
$$u = (p_1 + p_4)^2$$

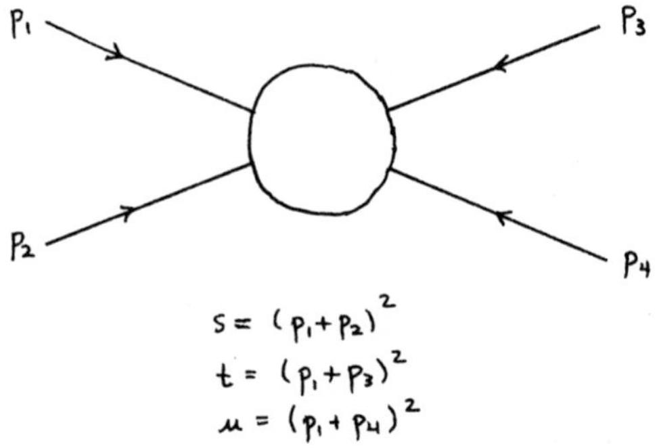

Figure 2-1.

One approximation is satisfactory at low energy $(\nu < 2 \text{ Gev, say})$ and consists in keeping only the contributions from resonances. It follows from the empirical observation that prominent resonance peaks almost exhaust the significant partial wave amplitudes. As a result the observed behaviour of each amplitude depends on the corresponding s-channel quantum numbers. In a narrow width approximation, which neglects the constraint of unitarity, the amplitude is approximated as

$$A^{RES}(s,t) = \sum_i \frac{G_i}{s - s_i} P_{J_i}(\cos \theta) \qquad (2\text{-}1)$$

where $\sqrt{s_i}$ and J_i are respectively the energy and the angular momentum of resonance i.

The second approximation is useful at high energy (ν > 5 GeV, say) and only retains the contributions of a few Regge pole exchanges in the t -channel. We write it as

$$A^{REGGE}(s,t) = -\sum_i \beta_i(t) \frac{\xi_i + e^{-i\pi\alpha_i(t)}}{\sin \pi\alpha_i(t)} \nu^{\alpha_i(t)} . \qquad (2\text{-}2)$$

Such an expression displays many important properties of the actual amplitude. The particular energy behaviour of each term is now associated with the t -channel quantum numbers of each Regge pole α_i . The differential cross section shows a strong forward peak unless β is forced to vanish at $t = 0$, as is the case in some helicity configurations. Relation 2-2 implies fixed t structure, as experimentally observed. Although not explicitly unitary, the model amplitude is not a priori incompatible with unitarity.

As an analytic function of the angular momentum J the scattering amplitude can be written as a contour integral. The contour may be drawn in such a way as to isolate some singularities and leave a "background integral " with a vanishing contribution at asymptotic energies.[14] As is well known the high energy behaviour is then approximated for each value of t by the contribution of the J- plane singularities which have the largest real parts; the so-called leading singularities. One frequently assumes that the J- plane singularities thus isolated are simple poles (as is the case in potential scattering) which, as a result of unitarity, should have positions α_i which move as t varies. In general the amplitude will get contributions from poles, with only a few significant ones at high energy, and also from angular momentum cuts. Nevertheless Regge pole theory, that is, the theory of high energy hadron physics as it would be if the leading J singularities were actually isolated poles, is appealing by its simplicity and some of its successes. [14,15,16]. Among them is the satisfactory interpretation which is reached for the observed π-nucleon charge exchange differential cross section. A single Regge trajectory, associated with the ρ meson, fits remarkable well the experimental results. [17]

The variable ξ_i as it appears in (2-2) is the signature of the Regge trajectory $\alpha_i(t)$. The signature factor $\left(\xi_i + e^{-i\pi\alpha_i(t)} \right)$ formally comes from the necessary separate continuation to complex J values of the even and odd parts of the amplitude. [14] With the simple leading power approximation retained in (2-2), it readily results from analyticity and crossing symmetry.

To some extent the contribution of a Regge cut can be approximated, in a certain domain, by one or a few effective poles. The clearest phenomenological distinction comes from the behaviour of the residue function $\beta(t)$. In the case of an isolated pole it should factorize in terms of the t-channel initial and final state properties (this includes polarization), whereas this behaviour would be lost if a pole approximation such as (2-2) is used to simulate the contribution from a Regge cut. Indeed it is the failure to observe constraints imposed by factorization which provides the best present experimental evidence for cuts.

The approximations (2-1) and (2-2) have proved to be most useful in two different energy domains. In either case one finds good reasons to neglect secondary contributions which respectively are a non-resonance background at low energy and the contributions from the non-leading Regge singularities (poles or cuts) together with the "background integral" contribution at high energy. If everything were kept in both cases one would have two complementary descriptions of the scattering amplitudes, whereas (2-1) and (2-2) limited to a few terms show prominent differences. If used at high energy (2-1) would not give a Regge behaviour but a fixed pole s^{-1} contribution. Only an infinite set of resonances could eventually give Regge behaviour. However the existence of such an infinite set cannot be inferred from an analysis of low energy scattering and requires a specific model. On the other hand only an infinite set of Regge contributions such as (2-2) could yield energy poles as present in (2-1), which, when unitarity is imposed, are displaced away from the physical sheet. When only a few terms are kept in order to analyze high energy scattering, such pole contributions are left in the "background integral". Nevertheless it has been realized that the two descriptions, even though they are far from equivalent when approximated by a few terms, show an important overlap. Many features which can be readily interpreted as due to Regge pole exchange, such as fixed structure in the differential cross section, can be fruitfully interpreted as co-operative effects of direct channel resonances and conversely the presence or absence of prominent resonance contributions in specific channels can be associated with relations among several Regge trajectories exchanged in the crossed channels. There is duality between the two descriptions.

B. Finite Energy Sum Rules

Such a duality is most clearly expressed in terms of finite energy sum rules [18,19]. The FESR translate the analyticity properties of the scattering amplitude and the validity of an asymptotic Regge behaviour such as (2-2). They are written as:

$$\int_0^N \nu^n \, \mathrm{Im}\, A(\nu, t)\, d\nu = \sum_i \beta_i(t) \frac{N^{\alpha_i(t)+n}}{\alpha_i(t)+n+1} \qquad (2\text{-}3)$$

with, as already defined: $\nu = \frac{s-u}{4m}$.

In (2-3) n is an integer and N is chosen such that the Regge pole expansion accurately describes the amplitude for $\nu > N$. There is however no specific way for choosing N. The derivation goes as follows [19,20]. Neglecting bound state poles which are readily introduced if needed, the contour integrals of Fig. 2-2 for the scattering amplitude A, written to avoid the two unitarity cuts drawn from ν_0 to $+\infty$ and $-\nu_0$ to $-\infty$, vanishes. The parameter ν_0 is the energy threshold for a fixed t value.

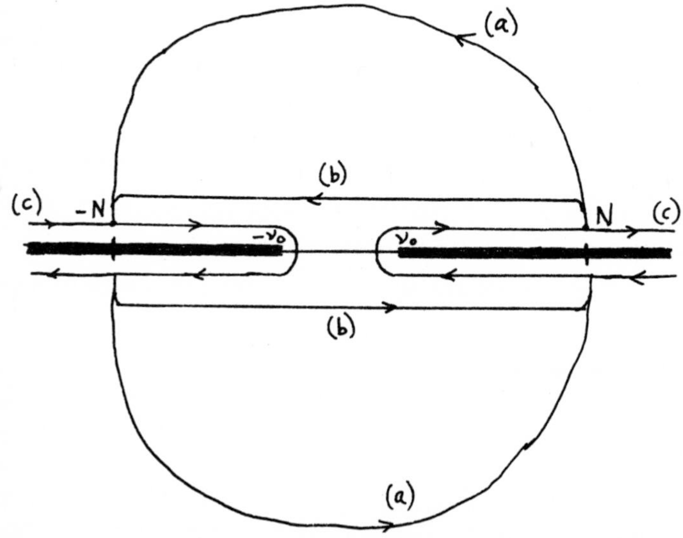

Figure 2-2. Contours in the ν-plane, used to write FESR.

The contour integral is evaluated in two steps (we consider the $n=0$ case). For $\nu < N$ one uses the fact that the discontinuity across the cut is equal to twice the imaginary part of the amplitude. One should consider only the case where the amplitude is odd since when it is even the contour integral vanishes in a trivial way. For the odd part the two integrals along the real axis give equal contributions which sum up to

$$4i \int_{\nu_0}^{N} d\nu \; \mathrm{Im}\, A(\nu,t) \tag{2-4}$$

At high energy, $|\nu| > N$, the amplitude is replaced by its Regge expansion (2-2), odd under s-u crossing:

$$A(\nu,t) = \sum_i \beta_i(t) \frac{1 - e^{-i\pi\alpha_i(t)}}{\sin \pi\alpha_i(t)} \nu^{\alpha_i(t)} .$$

In practice, one retains only leading poles with $\alpha_i > -1$. The contour of integration can then be deformed so that integration along the two semi-circles, (a), is replaced by an integration along the real axis from $+N$ to $-N$, (b), of the discontinuity of the Regge expansion. This gives

$$-4i \sum_{\alpha_i > -1} \beta_i(t) \frac{N^{\alpha_i(t)+1}}{\alpha_i(t)+1} . \tag{2-5}$$

Combining it with the expression already obtained one gets (2-3). The relation has of course no reason to be limited to those poles with $\alpha_i > -1$. For $\alpha_i < -1$ the contour along the two semi-circles, (a), has however to be deformed in a different way. The integration is now performed along the real axis from $+N$ to $+\infty$ and $-\infty$ to $-N$, (c), and two semi-circles at infinity which, since $\alpha_i < -1$, give a vanishing contribution. As readily checked one gets an expression identical to (2-5) but now restricted to poles with $\alpha_i < -1$. The limiting case of $\alpha_i = -1$ yields at least formally the same result. In this case there is no Regge discontinuity across the real axis and the integration is carried around the whole circle at $|\nu| = N$ to give $2\pi i$ times the residue at $\nu = 0$ which is given by $-\frac{2}{\pi}\tilde{\beta}$ with $\beta = \tilde{\beta}(\alpha(t)+1)$. The residue function must indeed vanish when $\alpha = -1$ in order to avoid a spurious singularity in the amplitude. Keeping all Regge poles we then obtain the finite energy sum rule as given by (2-3) . Multiplying the amplitude and its Regge expansion by ν^n does not modify its analyticity properties and a similar analysis may be carried out. The contour integral now does not vanish in a trivial way for an even (odd) amplitude depending on the choice of an odd (even) value for n. We thus obtain a whole family of sum rules .[21,22]

We stress that what went into the derivation is very little. Indeed the only physical content is the writing of a Regge expansion for $|\nu| > N$ where N is some large energy. The sum rules simply translate this property. In effect, as asymptotic relations for $N \to \infty$, they merely correspond to the l'Hôpital rule which equates the ratio of two vanishing functions to the ratio of their derivatives which in this case are the amplitude itself minus the Regge contributions with $\alpha \geq -1$ and the Regge contributions with $\alpha < -1$. The FESR are then trivial consequences of analyticity and asymptotic Regge behaviour. What they actually test is the validity of the Regge expansion for values of ν as low as N. If the amplitude at low energy $(\nu < N)$ can be directly determined, or approximated in some cases through the contribution of known resonances, one thus obtains a further check on the validity of a high energy Regge parametrization which can be independently obtained from the analysis of high energy scattering. If we wish N to be small enough so that the amplitude is known with some precision through a phase shift analysis we have to postulate the approximate validity of a Regge expansion, known to work at higher energies, down to N as small as 2 GeV. We have no precise way to know how small N can be. Nevertheless once N is chosen (2-3) has predictive value. The predictions hold in the best explored case at present, which is π-nucleon charge exchange. At high energy the experimental results are well described in terms of a single Regge pole exchange[15, 23] while at low energy a phase shift analysis gives a precise determination of the amplitude for all t values in the physical region [24]. The finite energy sum rule then gives [25, 26]

$$\int_0^N d\nu \, \mathrm{Im}\,(A' - A'_\rho) = 0 \qquad (2\text{-}6)$$

where only the ρ contribution to the A' amplitude has been retained. A relation such as (2-6) is obviously wrong when extended to all t values, but we may try it at $t=0$ where the validity of the Regge pole approximation is a priori the best. Furthermore at $t=0$ the imaginary part of the amplitude is simply related to the difference of the $\pi^- p$ and $\pi^+ p$ cross sections. As pointed out by Dolen, Horn and Schmid [19] the ρ contribution when extrapolated to low energy interpolates well between the prominent wiggles associated with resonances, which, according to their isospin, favor either the $\pi^+ p$ or the $\pi^- p$ cross section. This is illustrated by Fig. 2-3. As a result (2-6) is well satisfied even though Regge ρ exchange would give a very poor description of the reaction below 2 GeV.

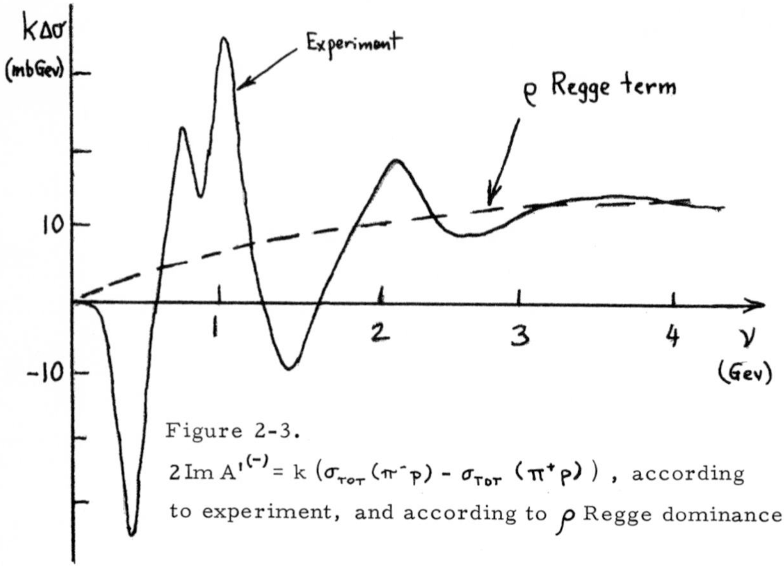

Figure 2-3.
$2\,\mathrm{Im}\,A'^{(-)} = k\,(\sigma_{TOT}(\pi^- p) - \sigma_{TOT}(\pi^+ p))$, according to experiment, and according to ρ Regge dominance.

A simple point is worth stressing. Even though one extrapolates a Regge contribution providing a good approximation at high energy down to low energy it should not be modified by the addition of further Regge terms of no importance at high energy but probably of greater effect at low energy. We have used analyticity to relate two different descriptions of the scattering amplitude and the only Regge expansion to consider is the one which is known to be good above N. The FESR weigh the different Regge contributions as high energy analysis does. One should not try to get something like a Regge expansion below N since it is certainly cumbersome and of little use.

At the same time the analytical form chosen for the Regge expansion is of no relevance. If a parametrization is used for high energy it should be kept in (2-3). The choice has to be made above N irrespective of any continuation which one might prefer below N.

Finite energy sum rules have proven to be very interesting tools for refining Regge phenomenology and many useful applications have been presented.[16,22] They are too approximate to fix precisely the type of Regge singularity present in the amplitude but, once a choice has been made, they are of great help in determining the best values for the parameters in an expansion.

C. Global Duality

The finite energy sum rules provide a simple framework in which to express the duality between the resonance and the Regge exchange description of a scattering amplitude. As the preceding figure shows, the resonance contributions play a prominent role at low energy and we may assume that to a good approximation they exhaust the imaginary part. This should of course be incorrect for an elastic amplitude where the shadow of inelastic processes provides a strong non-resonance contribution, but does seem legitimate when dealing with two-body amplitudes where only non-zero quantum numbers are exchanged as is the case for charge exchange. This is the content of global duality as introduced by Dolen, Horn and Schmid.[19,20] It assumes resonance saturation for the calculation of the finite energy sum rule up to an energy where Regge behaviour holds. One should of course stress that the resonance saturation is taken only for the imaginary part of the amplitude[27] and only in the average sense involved in integration up to N. Nevertheless global duality connects in a specific way the leading low energy resonances, that is, those known to give very large contributions to the imaginary part at lower energies, and the leading Regge poles at higher energy. One may say that the direct channel resonances thus "build up" the exchanged Regge pole. It again relates two approximations of the scattering amplitudes valid in two different domains.

Confidence in the value of such an approximate scheme can be gained from its success in some particular cases which are well analyzed. The πN channel resonances are known with great accuracy [24,28] and their dominant role can be checked at least up to 2 GeV. In particular, resonance saturation when introduced in the A' and B amplitudes of π-nucleon charge exchange [19] gives contributions with the same sign in B thus building up a large flip amplitude, while they cancel each other in A' [29] as shown at $t=0$ on Fig. 2-3, and give altogether a small non-flip amplitude. At the same time the Regge pole analysis at high energy gives a large spin flip amplitude and a small non-flip amplitude. [15] Duality relates these two observations. Starting from the resonance picture at low energy, and assuming the validity of a simple Regge parametrization at high energy, we are led to conclude that it should show a strong flip and a weak non-flip amplitude. The prominent N^* resonances have spin increasing with energy. Important contributions come from the 1240 $(3/2^+)$, 1520 $(3/2^-)$, 1688 $(5/2^+)$, and 1950 $(7/2^+)$ resonances and, as a result, the angular distribution shows a structure which has a remarkable stability in t. The contributions from the last three resonances to the B amplitude all vanish in the neighborhood of $t = -0.5\ \mathrm{GeV}^2$. So does, in particular, the finite energy sum rule written with $n=3$ which favors these higher resonances.[19,20] Duality then requires that the ρ contribution to B must vanish at the same t value. The residue of the flip amplitude has a zero at $t \simeq -0.5\ \mathrm{GeV}^2$, and the differential cross section, dominated except for $t \simeq 0$

by the flip amplitude, shows indeed a dip there. The vanishing of the residue function is on the other hand simply interpreted in terms of Regge pole exchange since $t = -0.5\ \text{GeV}^2$ corresponds to $\alpha = 0$. The flip residue function must then vanish. [30] A similar effect happens for the A' amplitude. However, the contribution of a resonance of spin J now involves $P_J(\cos\theta)$ whereas its contribution to the flip amplitude B is proportional to $P_J'(\cos\theta)$. It follows that all prominent resonance contributions are zero at almost the same t value but this value is now closer to 0. One finds that it corresponds to $t \simeq -0.2\ \text{GeV}^2$ and this is in particular the case for the sum rule with $n = 2$ which is written for A' [19] in order to favor the higher mass resonances.

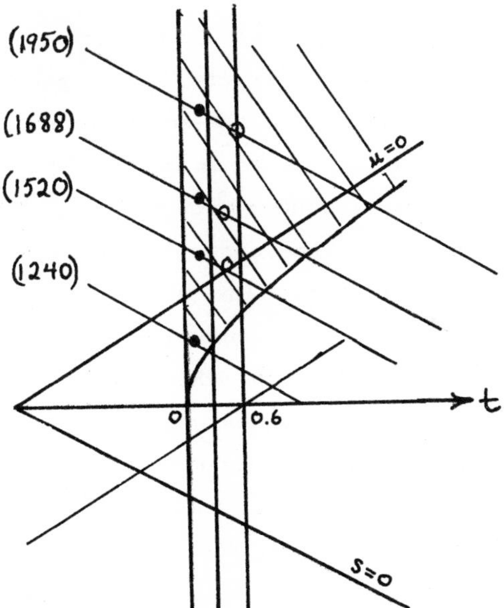

Figure 2-4. Zeros in A' (black) and B (open) from first prominent resonances. Dashed region is physical for πN scattering.

This zero could also be "explained" in terms of Regge pole exchange with a non-flip residue function vanishing at $t = -0.2\ \text{GeV}^2$. Such a zero used together with factorization gives a contradiction with experiment. The zero is considered at present to be a Regge pole Regge cut destructive interference which is not a factorizable effect and which can be successfully used in explaining the observed polarization in π -nucleon charge exchange. [31,32] This illustrates an important point. Duality, or more generally a finite energy sum rule (when a phase shift analysis is used instead of resonance saturation), does not determine the type of Regge singularity which dominates the high energy behaviour. It merely helps in determining the best parameters once a specific type of approximation has been chosen. Duality connects the low-energy direct channel resonances to the high energy Regge expansion whether it involves one pole, several poles or poles and cuts. Analyticity relates the low-energy amplitude to whatever dominates at high energy; in the non-flip amplitude the low-energy resonances appear to be dual to a Regge pole Regge cut combination whereas for the flip amplitude a single Regge pole seems satisfactory. The great predictive power of duality in the case of the A' amplitude is worth stressing. In the neighborhood of $t = -0.2\ \text{GeV}^2$ the flip amplitude dominates and the zero of A' is hard to ascertain from the analysis of high energy scattering. The polarization, which selects the interference between the pole contribution to B and the cut contribution to A', also does not show any structure there, since the zero is produced by the cut contribution cancelling the pole contribution. Duality requires that the zero be present.

Looked at the other way the success of a Regge pole parametrization of high energy scattering, with a fixed t

structure imposed in the right hand side of (2-3), implies through duality strong relations among the direct channel resonances. A fixed t structure corresponds to a fixed impact parameter and, as a result, an "effective" angular momentum increasing as:

$$\ell \sim \sqrt{s} \tag{2-7}$$

This means that the prominent resonances in elastic scattering must be peripheral. This is of course true for high energy resonances if the imaginary part of the amplitude at high energy is still dominated by resonance contributions. If this is the case we have however no way of itemizing them all, except in the framework of dual models as we shall see in detail later, and it is then easier to consider only their global effect represented to a good approximation by the simpler Regge exchange contribution. But peripherality is also true (and this is where duality has a predictive value) for the low-energy well-separated resonances, which are thus strongly correlated.

D. <u>The Interference Model</u>

Though it may not seem to be the case from the preceding section, duality did not gain immediate general acceptance. The fact is that the regions where resonance dominance and Regge pole dominance are generally recognized to be reliable approximations are somewhat dissociated. In between stands an "intermediate region" in which Regge exchange and direct channel resonances could well be thought of as simultaneously contributing. [33] As we saw this would be a double counting if one accepts duality, according to which <u>either</u> description can be used, at least in some average sense, but not both simultaneously. It is not at present possible to prove the interference model to be wrong or to prove that duality is a misleading approach. They are both flexible and approximate enough to escape critical tests. In the interference model, one assumes that there is a domain where the sum of the resonance and the Regge pole contribution can be used, even in a local sense, to approximate the amplitude. If assumed to provide a good approximation at intermediate energies it then leads to sum rules of the type

$$\int_0^N d\nu \, \mathrm{Im}\left(A - A^{RES} - A^{REGGE}\right) = 0 \,,$$

which, combined with a finite energy sum rule, gives a stringent relation for the resonance contribution :

$$\int_0^N d\nu \, \mathrm{Im}\, A^{RES} = 0 \,. \qquad (2\text{-}8)$$

This can be satisfied even for elastic amplitudes but only if an important background is present together with the resonances. [34] It is difficult to reconcile this view with present information in πN scattering where resonance saturation holds very well for the charge exchange amplitude . [35] On the other hand, as shown by Jengo [36], it is possible to build a specific model in which resonance and Regge behaviour are fully dissociated. If we consider two crossed channels s and t (the generalization to three brings inessential complications) we wish to decompose the amplitude into two terms: one, A_t, has resonances only in the t channel and Regge behaviour at large s and fixed t. It further goes to zero faster than any power of t when t goes to infinity at fixed s. The other, A_s, has the same property with the role of the channels interchanged. The full amplitude

$$A = A_t + A_s \qquad (2\text{-}9)$$

has resonances in both the t and s channels and asymptotic Regge behaviour. However, the high s (t) Regge behaviour has nothing to do with the s (t) channel resonances. Nevertheless it may still be associated with the t (s) channel resonances as explicitly done in some models. [37]

The construction of such an amplitude merely involves a function with decreasing exponential behaviour at large s (t) which respectively multiplies the terms containing the singularities associated with the s(t) channel resonances. The exponential function will increase in other directions in the s plane, but if one takes a fractional power in the exponent, $e^{-s^{1/4}}$ say, for the term multiplying the s resonance contribution, thus defining the amplitude in a cut plane, the pertinent contribution will decrease asymptotically in all directions at least on the physical sheet and "blow up" only on other sheets. This gives "super convergent resonance contributions" which are completely dissociated from the asymptotic Regge behaviour. Here again each resonance pole term is associated with an important background. Such models are possible. Nevertheless they would lead us to list as hitherto uninterpreted coincidences all the facts itemized in the introduction which find a simple explanation in the framework of duality and would not provide us with any new and specific insight of their own.

E. Finite Energy Sum Rule Bootstrap

With a resonance approximation the t dependence of the imaginary part of the amplitude is explicitly given in terms of a few Legendre polynomials. One can therefore use it in (2-3), according to duality, in order to continue both sides to positive values of t. The N^* resonances do not "build up" the ρ trajectory only in the scattering region but also in the crossed t channel where the ρ exists as a particle. The ratio of sum rules of different moments determine the trajectory function. One has for instance for the flip amplitude

$$\frac{N^2 \int_0^N d\nu \, \nu \, \mathrm{Im}\, B(\nu, t)}{\int_0^N d\nu \, \nu^3 \, \mathrm{Im}\, B(\nu, t)} = \frac{\alpha(t) + 4}{\alpha(t) + 2} , \qquad (2\text{-}10)$$

which can be used up to $\alpha = 1$ to determine the ρ meson mass. Continuing the amplitude beyond the limits of the Lehman ellipse relies fully on resonance saturation limited to a few prominent N^* states. Indeed it is only the convergence of values obtained from different sum rules which may attach significance to the result obtained. In any case, duality thus provides a way to construct meson resonances once baryon resonances are known. When continuing the resonance contribution to $t > 0$ the higher spin peripheral resonances have a much stronger contribution than those obtained from the lower spin resonances since all resonances are relatively scaled by the big factors given by the Legendre polynomials. This is quite different from what prevails in the scattering region where, due to the oscillations of the Legendre polynomials, the low partial wave resonances also play an important role in building up the amplitude. Conversely, if we know a Regge trajectory in the time-like region, drawing it through particle points on a Chew-Frautschi plot say, this imposes through duality strong constraints among the peripheral low energy resonances but teaches us almost nothing about the lower spin resonances.

One may further remark that resonance saturation, which may be directly tested in the scattering region, should also be excellent at very large positive t. [20] This corresponds to large values of the cosine of the scattering angle and the amplitude should then be well approximated by its leading direct channel Regge poles which are the peripheral resonances considered. FESR bootstrap assumes that resonance saturation remains valid in between, where the t channel resonances are to be found.

If one finally considers a reaction where the direct and crossed channels are identical such as $\pi\pi$ scattering, resonance saturation can be used in a self-consistent way through FESR in order to get a full bootstrap solution. [38] This however implies a special treatment for elastic scattering to which we now turn.

3. LINEAR TRAJECTORIES; DIFFRACTION SCATTERING, AND EXOTIC STATES

In this section we discuss three prominent facts which when used together with global duality as previously introduced will yield a predictive scheme for hadrons. This will lead us to duality in its most generally accepted meaning.

A. Rising Linear Trajectories

Many hadrons show striking regularities when set on a Chew-Frautschi diagram. Several resonances can be grouped together on rising linear trajectories. This is in particular the case for the Y^* resonances shown on Fig. 1-1. At the same time 7 resonance points can be located on a straight line with the same slope originating from the $\Delta(1240)$. [12]

The degeneracy in slcpe of all these baryon trajectories is a significant fact, but the linearity of the trajectories is already a remarkable property. Most model calculations giving Regge trajectory behaviour used unitarity and in particular elastic unitarity(which is easily imposed) as a principal input. This is in particular the case in potential scattering calculations. As a result Regge trajectories were expected to rise and fall, the centrifugal barrier preventing two low spin particles to get "bound" into a high spin particle. With linear trajectories, we have to conclude that inelastic channels play an increasing role as the energy rises. Whether trajectories keep rising is an open question. It is likely that with increasing energy new structures will become of importance and that eventually the trajectory will depart from a straight line. Nevertheless, if we attempt a bootstrap solution of hadron physics in which the lower mass resonances in the various channels should sustain themselves through crossing symmetry and analyticity, we should consider a rising linear trajectory as the proper input. A linear behaviour is strictly speaking not compatible with unitarity. On quite general grounds we can write a dispersion relation for the trajectory function $\alpha(t)$ which reads:

$$\alpha(t) = \alpha(0) + \alpha' t + \frac{t^2}{\pi} \int_{t_0}^{\infty} dt' \frac{\mathrm{Im}\,\alpha(t')}{t'^2 (t' - t - i\epsilon)} , \qquad (3\text{-}1)$$

with

$$\mathrm{Im}\,\alpha(t) \neq 0 \quad \text{for} \quad t > t_0 ,$$

t_0 being the lowest threshold.

The empirical fact that trajectories are to an extremely good approximation linear is a hint that unitarity is perhaps not so important in understanding the resonance pattern although obviously important when writing down a reaction amplitude.

A further prominent fact obvious on figure (1-1) is the degeneracy of trajectories of opposite signature. As already mentioned the signature results from separate continuation of the even and odd wave contributions to complex values of the angular momentum. Singularities and in particular Regge poles are found through each continuation but they should generally not coincide. They may however coincide if there are only direct or exchange forces at work, or in other words forces associated with exchange process in either the t or the u channel, but not both. In such a case the even and odd contributions are equal up to a sign and, as a result, the same Regge trajectory should be found with both signatures. This is not a natural attitude again if one associates a prominent role to elastic unitarity as done in the first bootstrap approaches.[39] Unitarity then creates significant forces in all channels. It should be stressed though that one uses with elastic unitarity only an approximate form of unitarity, so that the result thereby obtained is perhaps not very significant. A different attitude is to assume that elastic unitarity is misleading and that significant forces are obtained only in channels where prominent resonances can be exchanged. The absence of resonances in one particular channel would then lead to a degeneracy in signature for the Regge trajectories generated in the crossed channels. This is the attitude which is implied by duality where the imaginary part of the amplitude is assumed to be exhausted by the resonance contributions.

Using exchange degeneracy one might draw straight line

trajectories through the known meson points. Only in the case of the ρ (Fig. 1-2) are two resonances already associated with the same trajectory. Nevertheless the A_2 point is on the trajectory [40] as if one would have two signature degenerate trajectories with isospin 1. Indeed two further trajectories which are respectively associated with the ω and the f_0, also coincide with the first two. Missing mass spectrometer data [41] indicate that this trajectory may indeed continue to rise through masses of three to four GeV. Nothing is yet known however about the spin and parity of such states (Fig. 3-1). Only their narrow widths may lead one to suspect that their spins are increasing with mass.

Similar trajectories can be drawn through the K^* and φ points with again degeneracy in signature, and a slope equal to the ρ trajectory slope. It is a further striking property that, as already mentioned, the same slope is found for the baryon and meson trajectories. [42]

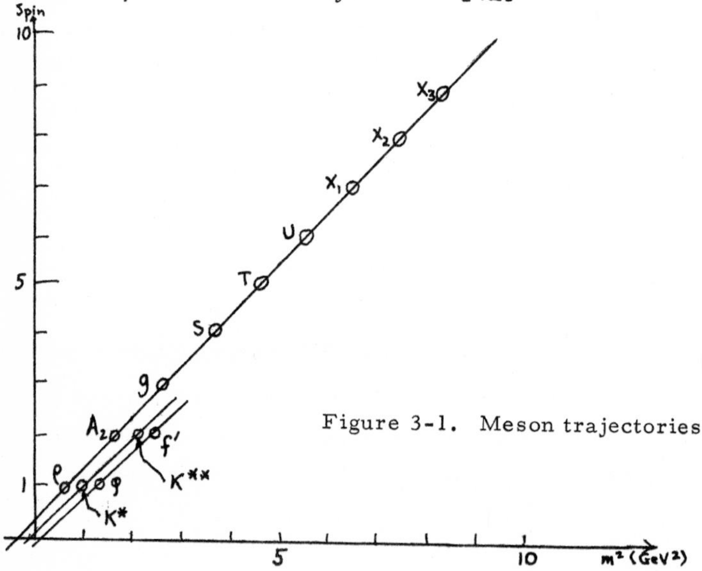

Figure 3-1. Meson trajectories.

$$\alpha(t) \simeq \alpha(0) + \alpha' t \quad \text{with } \alpha' = 0.9 \text{ GeV}^{-2} \tag{3-2}$$

This again suggests a bootstrap scheme in which resonances located on rising linear Regge trajectories in two channels mutually sustain themselves with the equality between the slopes as a self consistency condition. This assumes though that no significant forces should result from non resonance intermediate states which nevertheless generate through unitarity an important fraction of the imaginary part of the scattering amplitude.

B. Diffraction Scattering and Duality

Duality between s channel resonances and t channel Regge poles cannot be extended to the leading vacuum singularity often referred to as the Pomeranchuk trajectory and associated with diffraction scattering. If it were so duality would impose a similar resonance pattern in very different channels such as K^+p and K^-p where Pomeranchon exchange give identical contributions. The corresponding degeneracy in internal quantum numbers is obviously excluded. On the other hand it would be very hard to consider a resonance dominance scheme for a contribution to the elastic amplitude which basically results from the shadow of all the inelastic processes. In the framework of the finite energy sum rule there is however no compelling reason to consider the Pomeranchon and the standard Regge poles on the same footing. The relations being linear, one may separate the contributions associated with the Pomeranchon on the one hand and with the other poles on the other hand. The first one is associated with the non resonating background while the second is associated with the resonances.[43,44]

This distinction is illustrated on Fig 3-2 for two π-nucleon partial waves.[35] There is a marked difference between the isospin 0 and isospin 1 t channel amplitudes. For isospin 1 resonance saturation is very well satisfied whereas for isospin 0 the resonance contributions "ride" over an increasing background.

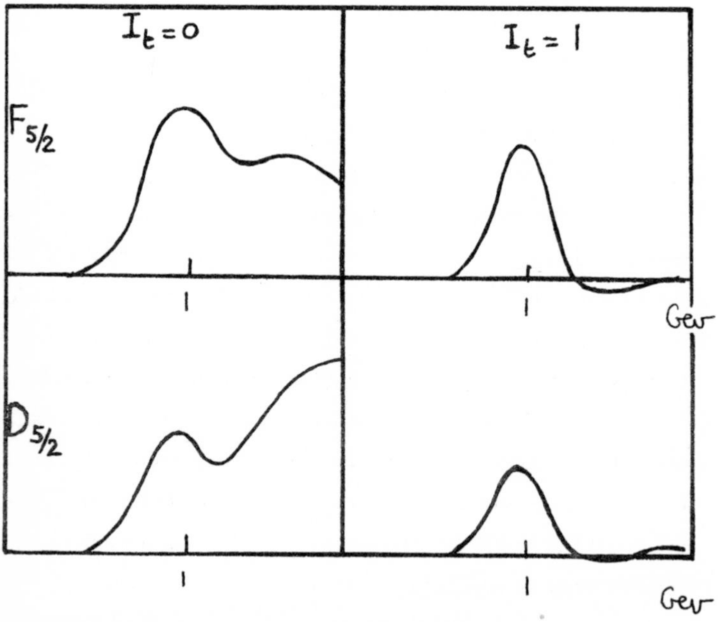

Figure 3-2. 2k Im f for π-N partial waves, crossed channel contributions separated.

It can however further be shown that the resonance approximation actually also holds for the isospin 0 amplitude if one considers only the s -channel helicity flip amplitude [45], as if the Pomeranchon were present only in the s -channel non-flip amplitude. [46]

The nature of the Pomeranchon is irrelevant. In effect, in order to clearly separate its contribution from those of the standard Regge poles one would welcome a complicated structure in terms of angular momentum singularities. As stressed later if it is represented by an effective pole it should have a slope definitely smaller than the common slope of all leading trajectories. This seems to be born out by the facts. [47]

Going slightly beyond global duality, we now assume that the relation between resonance and Regge contributions implied in (2-3) should be satisfied separately when integrating from ν_0 to N and from $-N$ to $-\nu_0$. These separate integrals are respectively related to the sum and the difference of the even and odd signature Regge contributions. If resonances are absent in one channel, the discontinuity across the cut is due to Pomeranchon exchange alone and should not be related to Regge pole exchange. As a result the standard Regge contributions have to cancel themselves out. This requires the degeneracy of trajectories with opposite signatures. This is in particular the case in K^-p scattering where prominent resonances exist in the direct channel but where no resonance is known in the crossed u channel. [48] As a result the Regge contribution to the integral should vanish when evaluated on the left hand cut but give a contribution on the right hand cut. This will be the case for the degenerate ρ and A_2 trajectories which have opposite signatures if their residue functions in this particular reaction are equal. This should also be the case for the ω and f^0 trajectories which can also be exchanged. The relation between the residues should separately hold for both isospins since prominent resonances are also present in K^-n while absent in K^+n scattering.

Such degeneracy between Regge trajectories has important implications. The leading Regge contributions cancelling their imaginary part in the K^+N channel while adding them up in the K^-N channel yield very different energy behaviours. In the first case the imaginary part of the amplitude comes from the Pomeranchon alone and gives through the optical theorem a constant total cross section. In the second case the Regge contribution gives a decreasing positive (since related to elastic resonances in the direct channel) contribution to the imaginary part and the cross section is predicted to fall towards its asymptotic limit.[43] This is illustrated on Fig 1-3 and 3-3 and more generally the result can be stated as follows: Whenever a reaction shows prominent low energy resonances it is expected to have a high energy total cross section decreasing toward its asymptotic value. If on the contrary there are not prominent low energy resonances it should be constant from a very few GeV onwards.[49]

A related remark can be made for the absence of t structure in K^+p and the presence of dips in K^-p and πN elastic differential cross sections.[50] At low energy such structure is directly associated with the resonances and is absent in K^+p scattering. It should be preserved at higher

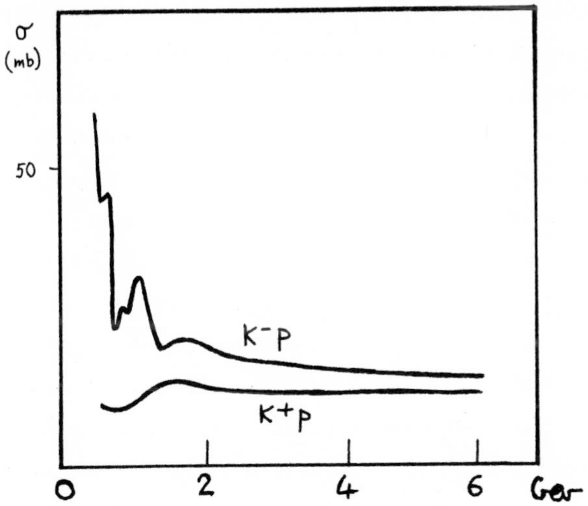

Figure 3-3. K^+p and K^- total cross sections.

energies where the Regge description holds. However and as already stressed, it is the pole and cut contributions which should reproduce it, if cuts are at all important, not a priori the Regge poles alone.

Our separation of the Pomeranchon contribution from the other Regge poles has been motivated on phenomenological grounds. It has however very important consequences. Resonances in one channel are now related through duality to resonances in other channels irrespective of forces resulting from complicated intermediate states. Using the linear character of finite energy sum rules we can therefore try a bootstrap scheme in which resonances sustain themselves irrespective of the detailed constraints imposed by unitarity. We are close to a model world of zero width resonances which we shall later consider and in which the solution of such a bootstrap scheme can be guessed. Once a tentative expression for the scattering amplitude is obtained the contribution due to the shadow of all inelastic channels should be added. This should however become important only after many channels become open.

C. Exotic States and Exchange Degeneracy

It is a prominent fact that all resonances so far known can be considered to be built up of three quarks (for baryons) or a quark and an anti-quark (for mesons). This corresponds to SU_3 multiplets $\underline{1}$ and $\underline{8}$ for mesons and $\underline{1}$, $\underline{8}$ and $\underline{10}$ for for baryons, and provides the best justification for the quark model[51,52]. All other sets of quantum numbers are called "exotic". This is in particular the case for the K^+N and $\pi^+\pi^+$ channels. If the only important forces which build up resonances in crossed channels are associated with observed resonances we have to conclude that no significant forces should be expected to originate from exotic channels. Of course our as yet unsuccessful search for exotics refers to relatively low mass states. Nevertheless following a bootstrap scheme based on nearby singularities we may exclude them from the start. The prominent differences between K^+p and K^-p scattering which we mentioned and discussed is a further hint at doing so. This leads us to expect exchange degeneracy to be a common phenomenon since there are many exotic channels where no resonance should be found. In particular it can then be used as a theoretical tool in meson-meson scattering.

$\pi^+\pi^-$ scattering has an exotic $\pi^+\pi^+$ channel, which can be taken as the u channel. The s and t

channel then both involve states connected to $\pi^+\pi^-$ and have therefore non-exotic contributions. Indeed, excluding the Pomeranchon, the ρ and the f^0 trajectories are exchanged in both channels. The absence of resonances and therefore of an imaginary part in the u channel imposes degeneracy between the ρ and the f^0. The residue functions are equal and the two trajectories exchanged in the t channel give a non-vanishing imaginary part in the s channel but, having opposite signatures, cancel themselves out in the u channel. The equality between the residue functions will of course not hold in a different reaction such as $\pi^+\pi^- \rightarrow \pi^0\pi^0$ where there are no exotic channels. Nevertheless the identity between the ρ and f^0 trajectory functions remains as a general property. A similar reasoning applied to $K\pi$ scattering with an exotic isospin $3/2$ channel and to $K\bar{K}$ scattering with exotic KK channels leads to degeneracy between the $K^*(890)$ and $K^*(1420)$ trajectories and to a separate degeneracy between the ρ and A_2 and the ω and f^0 trajectories. This gives a pattern of degenerate trajectories which is remarkably well met by the facts as shown on Fig 3-1.

The equality in mass between the ρ and the ω on the one hand and the A_2 and the f^0 on the other hand results from nonet structure with the standard mixing angle $\tan^2\theta = 1/2$. [52] Duality, since it requires trajectories of opposite signatures (which implies different isospins in $\pi\pi$ scattering) to coincide, leads directly to this degeneracy in mass. It now results from the absence of exotic resonances. We thus obtain a link between two prominent properties which is independent of the former link supplied by the quark model. One may show directly [53] that the same standard mixing angle is required for the vector and tensor mesons.

We consider to this end the coupled channels $\pi\pi$ and $K\bar{K}$. The ρ and the f^0 trajectories are exchanged in both these channels and their residue functions should be related so as to eliminate possible "exotic" contributions in the isospin 2 $\pi\pi$ channels, the isospin $3/2$ $K\pi$ channel and both the KK channels.

The absence of $\pi^+\pi^+$ resonances imposes opposite residues for the ρ and f^0 couplings in the crossed $\pi^+\pi^-$ channel whereas the absence of $K^+\pi^+$ resonances requires that the product of the ρ and f^0 couplings to the $\pi^+\pi^-$ and K^+K^- channels are also the same. Using the factorization property of residue functions [14] we are led to identical couplings to $\pi^+\pi^-$ and K^+K^- for the ρ and for the f^0.

$$\beta_{\rho\,\pi^+\pi^-} = \gamma^2_{\rho\,\pi^+\pi^-} = \gamma^2_{\rho\,K^+K^-} = \gamma^2_{f^0\pi^+\pi^-} = \gamma^2_{f^0\,K^+K^-} . \qquad (3\text{-}4)$$

This is however incompatible with what we would find with two octet members since the ρ coupling has then to be pure F with the f^0 coupling pure D !

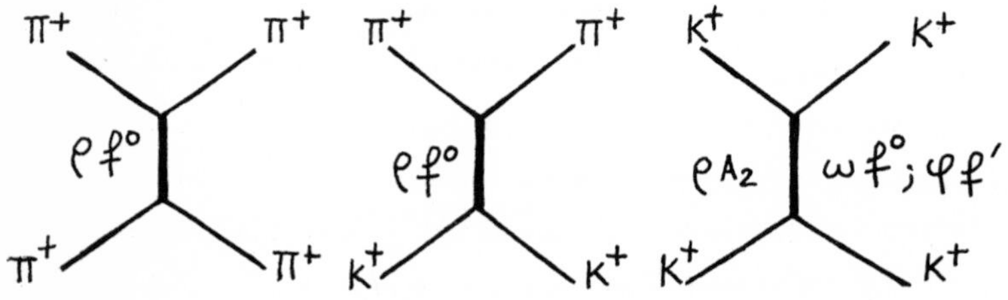

Figure 3-4. Reactions related by factorization.

Relation (3-4) is incompatible with an octet structure for both the vector and tensor mesons. Mixing with at least a singlet tensor is required. Such a mixing is indeed natural since there is an extra tensor meson, the f', which can mix with the f^0 in order to satisfy (3-4). Since we want to keep a single ρ, the f' should decouple from $\pi\pi$. The f^0 and the ρ are then exchange degenerate and cancel their contributions to the imaginary part of the $\pi^+\pi^+$ amplitude and the f', with no isospin 1 partner, does not contribute to it. This condition, used together with (3-4), fixes the two mixing parameters, namely the tensor meson mixing angle θ_T and the ratio of the coupling of the singlet and octet members to the meson meson channel. One readily finds

$$\tan^2\theta_T = \frac{1}{2} \qquad (3\text{-}5)$$

Canceling the imaginary part in the KK channel for both isospins imposes exchange degeneracy between the ρ and A_2 trajectories on the one hand and the ω and f^0 trajectories on the other hand. This further requires exchange degeneracy between the φ and f' trajectories. The ω and f^0 residues are then related but, using factorization, the ρ and f^0 residues should also be related in the same way. It then follows that

$$\gamma^2_{\rho K^+K^-} = \gamma^2_{\omega K^+K^-} \qquad (3\text{-}6)$$

The singlet state cannot couple to K^+K^- since it could do it symmetrically only when we have two mesons in a P wave. The ω can then couple only through its octet component and this imposes

$$\tan^2\theta_V = 1/2 , \qquad (3\text{-}7)$$

that is, the same nonet structure for the vector and tensor mesons.

This nonet structure is indeed well known for the tensor and vector mesons and was considered as a success of the quark model.[52] The standard mixing angle combines the octet and singlet members into two states which respectively have non strange and only strange quarks. The first one is then degenerate in mass with the isospin one member of the octet. It should be stressed though that this results from some symmetry breaking still forbidding exotic states, which is also what we require with duality. In both cases we want to bring an isovector and an isoscalar to couple in a similar way with a symmetry breaking which involves only mixing between multiplets. This is indeed very restrictive and the solutions should be the same. Duality further requires however that the nonet structure should always prevail. In an exotic channel with non strange mesons we need degeneracy between trajectories of opposite charge conjugation so that they may cancel their effects in one channel and collaborate in another channel (non trivial solution). Since they should have the same G parity this imposes different isospins. At the same time in an exotic channel with strange mesons we need separate degeneracy for both isospins. As a result 4 trajectories with both signatures and both isospins should always coincide. This is indeed what one observes for the leading meson trajectories. This should however also hold in principle for the pseudoscalar and pseudovector trajectories exchanged in pseudoscalar-vector scattering, $\pi\rho$ and KK^*. Here the pion trajectory should be exchange degenerate with an isospin 0 trajectory starting with a 1^+ meson with negative charge conjugation, the H, and, at the same time, with the B trajectory which should, on the other hand, be exchange degenerate with the one associated with the η when we use vector vector scattering. [54] This is in clear contradiction with the almost pure octet structure observed for the pseudoscalar and pseudovector mesons [55], when on the other hand a trajectory drawn through the η and B shows the standard slope (Fig. 3-5). It seems that the small mass of the pion, which is related to the approximate validity of chiral symmetry, is at odds with the weaker constraints which duality would impose.

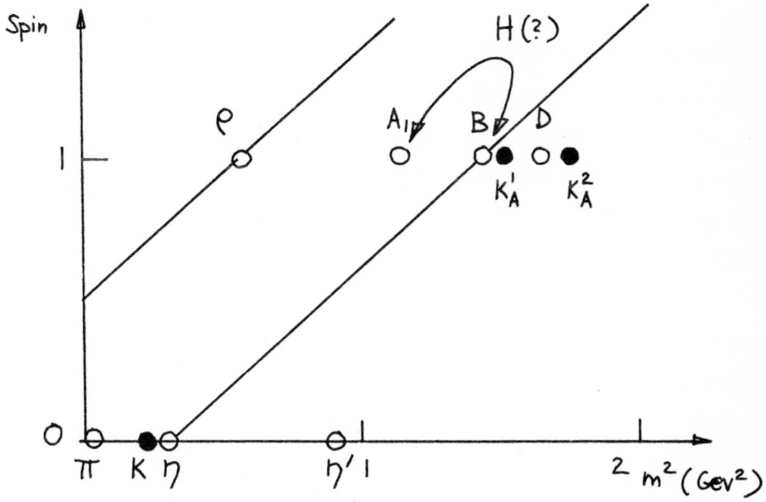

Figure 3-5. Pseudo vector and axial vector mesons.

As remarked by Schwimmers,[56] duality already leads to the mass degeneracy between the isovector and an isoscalar tensor meson from the absence of "second class exotics". This is a further implication of the quark model when the spin of a meson is obtained by adding the total quark spin to the orbital angular momentum, denying any spin property to whatever constituent with the quantum numbers of the vacuum helps at "building up" the meson. As a result, positive C mesons $\left(C=(-1)^{\ell+s}\right)$, with natural parity $\left(P=-(-1)^{\ell}\right)$ must have positive parity. The $\eta\pi$ channel, with positive C, must then show an s-u symmetry which requires a degeneracy of the A_2 and f^0 trajectories respectively exchanged in the crossed u and t channels. It should be stressed though that the absence of second class exotics is a much stronger assumption than the absence of straight exotics. They are furthermore required in the Regge expansion of the Toller pole which can well refer to only non-exotic channels.[57]

As a further example of exchange degeneracy we shall study the $\bar{K}N$ problem.[58] Since the KN channel is exotic we have only direct forces (meson exchange) and no exchange forces (baryon exchange). The Y^* resonances should then be exchange degenerate. This is a very strong duality statement again related to the linear character of the finite energy sum rules. If we were considering at the same time all coupled channels such as $\Sigma\pi$ and $\Lambda\pi$ they would show direct and exchange forces which, through unitarity, should be relevant in the $\bar{K}N$ channel. We have however a linear set of equations which allows our considering a particular channel at a time.

We find indeed two sequences of exchange degenerate resonances which are shown in Fig. 1-1. We have on the one hand the $\Lambda_0(1115, \frac{1}{2}^+)$ together with the $Y_0^*(1520, \frac{3}{2}^-)$, the $Y_0^*(1815, \frac{5}{2}^+)$, the $Y_0^*(2100, \frac{7}{2}^-)$ and the $Y_0^*(2350, \frac{9}{2}^+?)$ and, on the other hand, the $Y_1^*(1385, \frac{3}{2}^+)$ together with the $Y_1^*(1765, \frac{5}{2}^-)$, the $Y_1^*(2030, \frac{7}{2}^+)$, the $Y_1^*(2250, \frac{9}{2}^-?)$ and the $Y_1^*(2455, \frac{11}{2}^+?)$.

This however does not exhaust the Y^* resonances known at present.[28] This leaves aside several hyperon resonances which we cannot associate with exchange degenerate trajectories as one a priori should. They are the $Y_0^*(1405, \frac{1}{2}^-)$, the $Y_0^*(1690, \frac{3}{2}^-)$ the $Y_0^*(1830, \frac{5}{2}^-)$, and the $Y_0^*(1660, \frac{1}{2}^-)$ with, on the other hand the $\Sigma(1190, \frac{1}{2}^+)$ and the $Y_1^*(1910, \frac{5}{2}^+)$. This leads us to two important remarks. Since we selected a particular channel we should rightfully consider only those resonances which are strongly coupled to this channel. The other ones have no reason to be constrainted by the absence of force in the KN channel. We should leave aside from the analysis all states relatively weakly coupled to the $\bar{K}N$ channel and more strongly coupled to the $\Sigma\pi$ or $\Lambda\pi$ channels. This is indeed the case for all the resonances left apart except for the $Y_0^*(1405, \frac{1}{2}^-)$ and the $Y_0^*(1660, \frac{1}{2}^-)$ states.

These two resonances have however a relatively low spin for their mass. As already stressed they therefore contribute weakly to building up trajectories in the crossed channels and conversely a particular relation among such trajectories (or the absence of resonances in a particular channel) does not affect them appreciably.

Exchange degeneracy does not apply in general for the low partial waves; instead it relates the peripheral resonances to the leading Regge trajectories. It is therefore dangerous to use duality arguments in channels about which we know nothing (such as the scalar-pseudoscalar channels). We cannot obtain from duality very detailed information, such as knowledge of polarization. In particular we should not conclude that KN charge exchange should not show any polarization even though our dual amplitude is purely real. At most we can say that the interference effects between almost real higher waves and complex lower waves will result in a polarization with no prominent structure.

The separation between the isospin 0 and 1 Y^* trajectories requires degeneracy between the isospin 0 and 1 meson resonances Indeed degeneracy between the ρ, ω, A_2 and f^0 has already been required. We further conclude that the φ and f' should decouple from $N\bar{N}$, as they actually do to a very good approximation.

The 4 fold degenerate meson trajectory can be computed [20] from the resonance contributions, extrapolating to large enough positive values of t [59]. The agreement between various moment sum rules is very good. The degeneracy between the coupling of the baryon resonances can also be verified. [20,58]

To conclude: such a FESR bootstrap is far from complete nevertheless selecting some particular channels it allows an easy derivation of interesting relations which should be preserved to a very good approximation when a more complete dynamical calculation can be made.

D. Line Reversal Invariance

Two exchange degenerate trajectories contribute to scattering amplitude terms

$$-\beta_1 \frac{1+e^{-i\pi\alpha(t)}}{\sin \pi\alpha(t)} \nu^{\alpha(t)} \quad \text{and} \quad -\beta_2 \frac{-1+e^{-i\pi\alpha(t)}}{\sin \pi\alpha(t)} \nu^{\alpha(t)}$$

with eventually $\beta_1 = -\beta_2$ in some specific channels. If the incoming and outgoing lines at one vertex only are reversed the even signature trajectory will give the same contribution while the contribution from the odd signature trajectory will change sign. A particular example, shown in Fig. 3-6 is:

$$K^- p \longrightarrow \pi^- \Sigma^+ \quad \text{and} \quad \pi^+ p \longrightarrow K^+ \Sigma^+ . \tag{3-8}$$

The pertinent trajectories are associated with the $K^*(890)$ and the $K^*(1420)$ and are indeed degenerate (Fig. 3-1)

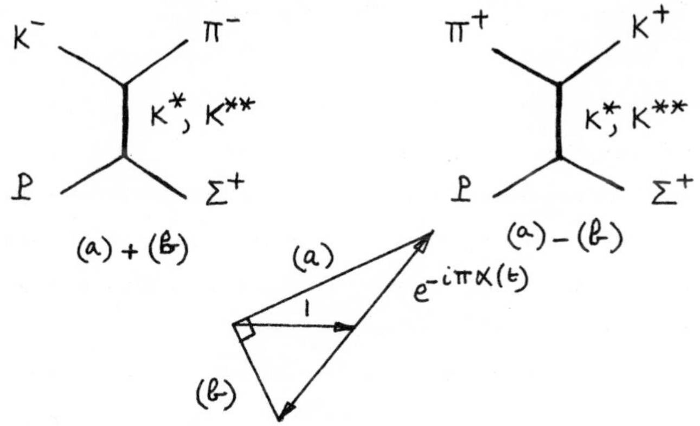

Figure 3-6.

Their contributions are however pure imaginary with respect to one another (Fig. 3-6) and their contribution to the differential cross section add incoherently. As a result the two reactions (3-8) should show the same differential cross section <u>if</u> the Regge pole approximation with exchange degeneracy is reliable. At the same time the polarization should change sign from one reaction to the other. As shown later however, they should both vanish (the first amplitude is purely real while the second one is purely imaginary). This however should hold only for the peripheral and spin flip waves and the actual polarization may well be large.

The differential cross sections have been compared at several energies [60] and are found to differ by a factor 2 although showing a similar behaviour :

$$\frac{d\sigma}{dt}(K^- p \to \pi^- \Sigma^+) \simeq 2 \frac{d\sigma}{dt}(\pi^+ p \to K^+ \Sigma^+) .$$

It is however too early to conclude that the predictions obtained are contradicted by experiment. Smaller contributions, which could be associated with cuts or scalar trajectories, [61] may not appear very important when analyzing separately the two angular distribution, yet may still produce, through their interference effects with the leading terms, a large difference between the two differential cross sections. I would consider a more reliable interpretation to be that exchange degeneracy works well for the flip amplitude, which should be correctly described by K^* exchange, and does not work as well for the non-flip amplitude where Regge cuts are important. Line reversal applied to the differential cross section should then fail but should work when used for the flip amplitude. It should then be very interesting to know separately the flip and the non-flip amplitude as obtained in an R experiment [62] where the target nucleon is polarized in the plane of scattering. One has

$$P^f_{x'} = A P^i_z + R P^i_x , \tag{3-9}$$

with

$$\frac{d\sigma}{d\Omega} R = \left(| F_{++} |^2 - | F_{+-} |^2 \right) ,$$
$$\frac{d\sigma}{d\Omega} = \left(| F_{++} |^2 + | F_{+-} |^2 \right) , \tag{3-10}$$

where P^i (P^f) is the initial (final) baryon polarization. The z (z') axis is chosen along the initial (final) meson momentum and the y axis is taken normal to the reaction plane.

This leads us to the spin question for which duality also has specific predictions

E. Nonsense Choosing Residues

When two trajectories of opposite signature are exchange degenerate so as to cancel their effects in an exotic channel, the residue functions are equal. If α goes through a negative integer value (this includes zero) the residue of the trajectory for which it is a right signature point (a point where the signature factor does not vanish) must vanish. We would otherwise have a singularity in the physical region. Duality tells us further that the residue of the trajectory for which it is a wrong signature point should also vanish even though this is not required in order to eliminate spurious singularities [63].

In particular the f^0 residue function in $\pi^+\pi^-$ scattering has to vanish when $\alpha(t) = 0$ and the ρ residue function should also vanish there. Using factorization we then expect that both the flip and the non-flip residue functions in π-nucleon charge exchange will vanish at $\alpha(t) = 0$ which corresponds to $t = -0.6\ \mathrm{GeV}^2$. As already stressed, cut contributions, in particular in the non-flip amplitude, will not leave this prediction as strong as it looks [17]. Nevertheless a marked dip should be found and it is indeed observed.

For the sake of completeness the origin of the sense-nonsense terminology is briefly recalled [14, 30]. The partial wave expansion which is continued to complex J values after proper separation of the two signature contributions is

$$A_{\{\lambda\}} = \sum_J (J+\tfrac{1}{2}) A^J_{\{\lambda\}} d^J_{\lambda\mu}(\theta) .$$

It corresponds to a partial wave expansion in the t -channel where $\{\lambda\}$ represents the helicities of all 4 particles. The total helicities in the initial and final states are represented by λ and μ . The helicity partial wave amplitudes properly continued to complex J values have Regge singularities. The d function which is present as a factor to these singular terms also has singularities of its own in J which should be matched by compensating terms in the residue function. One may write the d function as [14]

$$d^J_{\lambda\mu}(\theta) = \left(\sin\frac{\theta}{2}\right)^{|\lambda-\mu|} \left(\cos\frac{\theta}{2}\right)^{|\lambda+\mu|} \sqrt{\frac{\Gamma(J+M+1)\,\Gamma(J-N+1)}{\Gamma(J+N+1)\,\Gamma(J-M+1)}}\; R^J_{\lambda\mu}(\cos\theta)$$

where $R^J_{\lambda\mu}$ is a regular function of J , and M and N are respectively defined as

$$M = \mathrm{Max}\,(|\lambda|, |\mu|) ,$$
$$N = \mathrm{Min}\,(|\lambda|, |\mu|) .$$

If $J \geq M$ the amplitude is sense-sense $(s\text{-}s)$. For a physical J value it makes sense as an exchanged process. If $N \leq J < M$ the amplitude is nonsense-sense $(n\text{-}s)$ and if $J < N$ the amplitude is nonsense-nonsense $(n\text{-}n)$.

The d functions have square root singularities which should be properly compensated for through the residue function so as to disappear from the full amplitude and leave only "dynamical" singularities. In order to study the behaviour of the residue in the neighborhood of a sense-nonsense point we take the asymptotic expansion of the d function for large $\cos\theta$, valid for $J \geq -\frac{1}{2}$. It reads [14]

$$d^{J}_{\lambda\mu} \sim \left(\frac{\cos\theta}{2}\right)^{J} \frac{\Gamma(2J+2)\,(-1)^{N-\mu}\,(i)^{M-N}}{\left(\Gamma(J+M+1)\,\Gamma(J+N+1)\,\Gamma(J-M+1)\,\Gamma(J-N+1)\right)^{1/2}} .$$

In the neighborhood of an integer (or half integer) J value this expression is regular for a sense-sense point. The last term in the denominator becomes however singular for sense-nonsense points and both the last two terms become singular for a nonsense-nonsense point.

If we consider the particular case of $J = \alpha(t) = 0$ (meson trajectory function crossing the t-axis), we find a $\sqrt{\alpha}$ behaviour for an $(n\text{-}s)$ amplitude and an α behaviour for an $(n\text{-}n)$ amplitude. The corresponding factor in the $(s\text{-}s)$ amplitude is 1. The leading terms show factorization properties, notwithstanding a singular residue at a nonsense wrong signature point [64] which cannot be ruled out. The $(n\text{-}s)$ residue function should then vanish as $\sqrt{\alpha}$ in the neighborhood of $\alpha(t)=0$ and, according to factorization, we have the choice between 1 for the $(s\text{-}s)$ and α for $(n\text{-}n)$ or α for $(s\text{-}s)$ and 1 for $(n\text{-}n)$. The first possibility is referred to as "sense choosing". At a right signature point, the $(s\text{-}s)$ amplitude is singular, as expected from the presence of a particle. The second possibility is referred to as "nonsense choosing" according to Gell-Mann. At a right signature point only the $(s\text{-}s)$ amplitude does not vanish but it is not singular.

At a wrong signature point no amplitude is singular and only the $(s\text{-}s)$ amplitude is not zero in the choosing sense mechanism. Considering again the neighborhood of $\alpha = 0$ the $(n\text{-}s)$ amplitude behaves as α and the $(n\text{-}n)$ amplitude as α^2. If we take however the Gell-Mann mechanism all amplitudes vanish as α. The Gell-Mann mechanism is what we are led to according to duality.

Instead of imposing a $\sqrt{\alpha}$ behaviour on the $(n\text{-}s)$ residue function we could have taken $\alpha\sqrt{\alpha}$ with again a sense choosing and a nonsense choosing possibility respectively referred to as the Chew mechanism and the no-compensating mechanism [30]. We shall see later how the dual model indeed selects the Gell-Mann mechanism with a single zero in all amplitudes. Again this should lead to a strong prediction but only to the extent that a single Regge pole description with factorization is an accurate approximation. As we already stressed this seems to be the case for the flip amplitude $(n\text{-}s)$ at $\alpha = 0$ in π-nucleon charge exchange but certainly not for the non-flip $(s\text{-}s)$ amplitude [17]. We therefore expect a dip at $\alpha(t) = 0$, not a zero; the dip is indeed observed.

4. DUALITY IN A SEMI-LOCAL WAY

A. Argand Loops

Duality supposes that there is an overlap region where the Regge expansion and the resonance expansion are both good approximations. Nevertheless their complementary aspect is required only in some average sense and one may then probe how far in detail it may actually hold. As we already stressed the cumulative effects of resonances can give many features readily associated with Regge exchange such as a strong forward peak and a fixed t structure. This is the case for the low energy resonances as imposed by duality but this should be even more accurate for intermediate energy resonances in a region where the Regge description is already the simplest one.

The dual description of the total amplitude can be extended

to the partial wave amplitudes calculating the Regge contribution in each partial wave

$$A_\ell^{REGGE}(\nu) = \frac{1}{2}\int_{-1}^{+1} A^{REGGE}(\nu, t)\, P_\ell(\cos\theta)\, d\cos\theta \qquad (4\text{-}1)$$

The partial waves should have phase varying with energy. The phase of the Regge amplitude is given by the signature factor and is fixed in t. The relative contribution of the partial waves at fixed t changing with energy, the phase of each partial wave should vary so that all partial waves can maintain a fixed phase for each t value. At the same time the average spin has to increase with s so as to give the fixed t structure (up to a logarithmic shrinking) of the Regge contribution.

In order to study this one has to take an accurate Regge expression and perform the integration (4-1). Nevertheless one may readily obtain several points which are then confirmed by a more careful analysis [65].

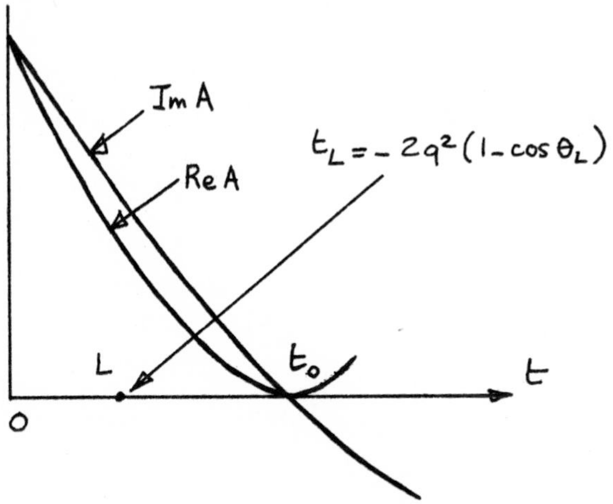

Figure 4-1. Behavior of the real and imaginary parts of the Regge amplitude. L stands for the first zero of the Legendre polynomial $P_L(\cos\theta)$.

As shown on Fig. 4-1 the first zero of a Legendre polynomial will move towards the first t zero of the Regge amplitude as the energy increases. As a result the partial wave will first be small, will then strongly increase when the two zeros match, and will eventually decrease. If we consider a negative signature Regge pole with intercept $1/2$, e.g. the ρ, the real and imaginary parts are equal at $t=0$ but the real part drops faster with t. Therefore the maximum will occur earlier for the real part than for the imaginary part and the partial wave will start an Argand loop. If we consider a higher partial wave the same features will appear but at a higher energy.

In order to be more quantitative we can make a partial wave expansion of the complex factor $e^{-i\pi\alpha(t)}$ in the signature factor. As shown by Chiu and Kotanski [65] it contains most of the pertinent t dependence at low t values provided that $\log(\alpha' s) \ll \pi$. We further assume that the exchanged trajectory is linear

$$\alpha(t) = \alpha(0) + \alpha' t \quad .$$

$$\int_{-1}^{1} e^{-i\pi\alpha(t)} P_\ell(\cos\theta)\, d\cos\theta =$$

$$= e^{-i\pi(\alpha(0) - 2\alpha' q^2)} \int_{-1}^{+1} e^{-2i\pi\alpha' q^2 \cos\theta} P_\ell(\cos\theta)\, d\cos\theta \quad ,$$

with $t = -2q^2(1-\cos\theta)$.

The integral gives a spherical Bessel function

$$(-i)^{\ell} j_{\ell}(2\pi\alpha' q^2)$$

and the phase of the partial wave amplitude is given by

$$\varphi = -i\pi\alpha(0) + 2i\pi\alpha' q^2 - \frac{i\pi}{2}\ell . \qquad (4-2)$$

The phase goes then through $\frac{\pi}{2}$ as if one would have a resonance and the partial wave describes an Argand loop in the correct direction.

If we associate the phase of $\frac{\pi}{2}$ with the presence of a resonance [66] with spin ℓ then another resonance with spin $\ell+1$ should be obtained when q^2 has increased by $4\alpha'$ or s by α'. Therefore if a linear trajectory with slope α' is exchanged in the t-channel, exchange degenerate "resonances" located on a trajectory with the same slope appear in the s- channel. The equality between the slopes appear as a consequence of the duality picture.

In our heuristic argument we label as a resonance spin only the lowest value of ℓ for which (4-2) gives a phase of $\frac{\pi}{2}$. It is obvious that resonance phase $\frac{\pi}{2} + n\pi$ will also be obtained for all the values of ℓ equal to $\ell+2, \ell+4, \cdots$. Such states have been called "ancestors" [67] in clear opposition to the daughter state expected for values of ℓ equal to $\ell-2, \ell-4, \cdots$ The presence of these ancestors is however but a spurious effect. They correspond to Legendre polynomials with an odd number of zeros inside the forward peak and they are therefore extremely sensitive to the details of the residue function whereas the first term is not. Indeed the Bessel function decreases very fast with ℓ and they correspond to very small diameter loops at even low energy. This would be incompatible with unitarity.

It has been bitterly argued whether or not these Argand loops should be associated with resonances[67,68]. Resonance amplitudes have indeed a very similar behaviour but the Regge pole term one starts with does not have the second sheet poles specific to direct channel resonances. The dual point of view is that the Regge description, which is good on the real axis and reproduces many of the features more readily associated with the direct channel resonances, becomes very bad as one goes into the second sheet, as is the case for many mathematical expansions which are defined originally outside of a wedge. As a result one should then put a one to one correspondance between the large Argand loop thus obtained (lowest ℓ value) and direct channel resonances. As already stressed, the flip amplitude is the Regge amplitude to project on partial waves for this purpose. Schmid could obtain this way a sequence of N^* resonances from the ρ approximation to B^- and match it with the 1920-2190-2920 series.[69]

The resonance widths thus obtained are too large since they have to correspond to a smooth Regge energy behaviour. They are also globally displaced in energy. Nevertheless crude knowledge can be obtained from the analysis. We reach a likely set of resonances which could match one singled-out Regge pole, even though this is by no means an accurate description of the amplitude.

B. Resonance Poles from Infinite Daughter Series

Since a single Regge pole is not satisfactory to represent the amplitude on the second sheet, even if it gives a reliable description on the real axis, one may try an infinite series of daughters, which have regular s contributions but which sum up to a series which diverges on the s singularity associated with the resonance. The leading Regge pole would then determine where the resonances have to be through the phase variation of its partial wave amplitudes but the resonances themselves would result from the contributions of an infinite series of daughters of no significance for the high energy Regge behaviour. We can represent such a daughter series as

$$A(\xi) = \sum_n \beta_n e^{\alpha_n \xi} \qquad (4\text{-}3)$$

$$\text{with} \quad \xi = \log(\alpha' s) .$$

Writing the Regge expansion (without signature) as

$$R(s,t) = \sum_n \beta_n(t) (\alpha' s)^{\alpha_n(t)}$$

we have $0 > \alpha_1 > \alpha_2 \cdots > \alpha_n$ and ξ a complex variable

$$\xi = \eta + i\zeta .$$

The fact that the highest value of α is negative is irrelevant since only the asymptotic part of the series is important for the present analysis.

One introduces an asymptotic daughter density as

$$D = \lim_{n\to\infty} \frac{n}{\alpha_n(t)}$$

Then $D < 1$ (>1) if the trajectories are spaced by more (less) than one unit. One assumes that the series converges for a value of η or s large enough, $\eta > \eta_c$, where

$$\eta_c = \lim \frac{\log|\beta_n|}{|\alpha_n|}$$

This supposes that all the resonances are to be found at finite energy. The series will then have singularities in ξ (and s) for lower values of η.

There is then an interesting theorem which applies to series such as (4-3) which are called Dirichlet series. It states the following property : [70]

In any strip $\zeta_0 < \zeta < \zeta_0 + 2\pi D$, limited in η according to the limiting spacing of the daughter trajectories, there should be at least one singularity. This is translated in the s plane (Fig. 4-2) by the presence of at least one singularity in the section of a spiral which could extend over several sheets .

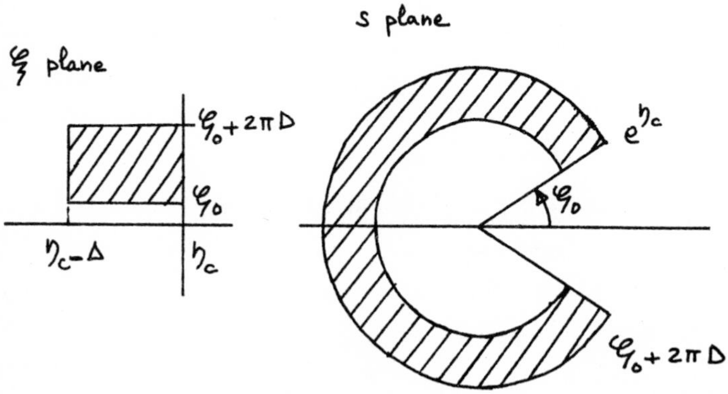

Figure 4-2.

We require however that on at least one sheet, defined as the physical sheet, no singularity should be found. This is possible only if the smallest section of the spiral goes across more than one sheet. This corresponds to a daughter density greater than 1.

We realize that such a construction is possible. Nevertheless it is not practical. We ought to understand the Regge expansion as an asymptotic series expansion, not as a convergent series expansion. If we are nevertheless going to take the second approach it should only be when dealing with a closed form expression where all daughters are specified once the parent trajectory is defined.

Dealing with the Regge expansion as an asymptotic series expansion we may expect it to give many features of an amplitude more easily described in terms of resonance contributions, but not all of them. Conversely a finite sum of resonances can give a good description of the amplitude, and show features more easily expressed in terms of simple Regge exchange; however such a sum is not reliable when extended to high energies.

C. Two Approximate and Complementary Descriptions

We may therefore try to analyze a reaction both ways and get valuable information from matching the two descriptions as much as possible. As an example we may consider $\rho^+\pi^-$ elastic scattering. The ρ and f^0 are exchanged in the t $(\pi\pi \to \rho\rho)$ channel and as a result we expect a sharp forward peak (Regge picture). If we now consider charge exchange scattering $\rho^-\pi^+ \to \rho^+\pi^-$ the t- channel is exotic and we should not have a forward peak. On the other hand the ρ and f^0 are now exchanged in the u- channel and we therefore expect a sharp backward peak (Regge picture again) .

In the s channel we find the ω and A_2 resonances and all the resonances on their respective trajectories. The A_2 and ω having respectively even and odd spin, they may interfere coherently as to give a strong forward peak. This will be the case in $\rho^+\pi^-$ elastic scattering (resonance description) .

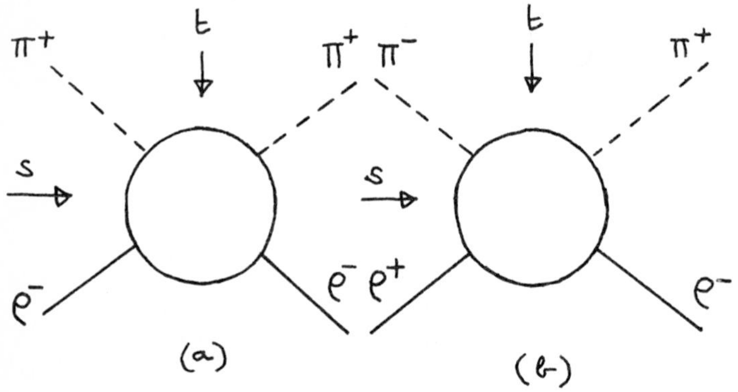

Figure 4-3. $\pi\rho$ scattering.

This will of course not be the case if we define the energy to be precisely the ω or the A_2 meson mass (or the mass of one of the other particles on their trajectory) . This will however be the case if we are taking an average over several resonance contributions.

We go from elastic scattering to charge exchange taking the charge conjugate of the final state, namely $\rho^-\pi^+ \to \rho^+\pi^-$. Under charge conjugation the ω contribution changes sign while the A_2 contribution remains the same. If the two contributors were initially interfering to give a forward peak they will now interfere to give a backward peak (Resonance picture) . Both pictures are equally adequate at describing what takes place. Nevertheless the Regge picture does not give the cross section variations associated with prominent resonances while the resonance picture gives forward and backward peaks only if we average over several resonance contributions. The two pictures converge but in a semi-local way.

One may ask then how local should this semi-local picture be. In order to have forward and backward peaks at each energy as obtained in the Regge picture one would need to have resonances of opposite parity and charge conjugation (and therefore of different isospin) respectively degenerate with the ω and the A_2. As secondary effects there should be daughter resonances one unit of spin lower than the parent ω and A_2. As the energy increases the "effective" spin should increase as $\sqrt{s}$ although the spin increases as s on each trajectory. This implies in the resonance picture (this could also follow from a nonresonance background) a family of parallel daughters. The first representative should be a 1^- state with isospin 0 (and odd under C) at the mass of the A_2 since scalar states are not coupled to $\rho\pi$. At the same time the leading ωA_2 Regge trajectory is expected to give a reliable description of the higher peripheral waves but not to be very accurate for the low waves. If the lower wave contributions are suppressed through the destructive interference of a secondary Regge pole or more likely of a cut contribution, the balance between partial waves, which has been keeping the total cross section smooth, will be destroyed and "resonance bumps" will appear. With a more complicated Regge picture and a more complicated resonance picture semi local duality could become more local, the equivalence between the two descriptions requiring of course an infinite set of Regge poles and an infinite set of direct channel resonances. It should be stressed though that we do not know how local duality should be expected to be and that at present only experiment can give us information.

Applying similar arguments to $\pi\pi$ scattering we are led to expect an S wave isospin 0 resonance at the ρ mass and a P wave isospin 1 resonance at the f^0 mass. They are not required by global duality which only demands ρf^0 exchange degeneracy. They would be required however by a more local form of duality, yet to be to be tested. [71]

Concluding these first four sections we may say that we have accumulated enough experimental evidences in favor of an approach to hadron dynamics in which the only significant forces pertinent at building up resonances in the framework of the bootstrap scheme are those associated with resonance exchange. This gives us confidence at building up dynamical models based on an idealization: a model world of zero width resonances mutually sustaining themselves in the different channels.

5. DUALITY AND UNITARY SYMMETRY

A. Exchange Degeneracy and Hadron Patterns

One of the prominent implications of duality is the degeneracy between trajectories of opposite signatures and often of different quantum numbers which results from the constraints imposed by the absence of resonances in exotic channels. We now turn to a more general survey of duality implications in the hadron patterns.

The absence of resonances in the multiplet i in channel x leads to a relation between the contribution of the Regge poles in the multiplets j exchanged in any of the crossed channels y as follows:

$$\sum_j M_{xy}^{ij} \ \mathrm{Im}\, R_y^j = 0 \ , \tag{5-1}$$

where R^j_y is the Regge contribution in the multiplet j in channel y and M stands for the SU(2) or SU(3) crossing matrix, depending on our choice of isospin symmetry or unitary symmetry.

We write each t- channel Regge contribution as :

$$R = -\beta(t)\,\frac{\zeta + e^{-i\pi\alpha(t)}}{\sin\pi\alpha(t)}\;\nu^{\alpha(t)}$$

so that the imaginary part at large $\nu\,(s)$ reads

$$\mathrm{Im}\,R = \beta(t)\,\nu^{\alpha(t)}\,, \qquad (\text{high } s)\,.$$

The imaginary part at large u is the same or the opposite depending on the signature

$$\mathrm{Im}\,R = \zeta\,\beta(t)\,(|\nu|)^{\alpha(t)}\,, \qquad (\text{high } u)\,.$$

We can then rewrite (5-1) as

$$\sum_j M^{ij}_{st} \sum_n \beta^j_n(t)\, s^{\alpha^j_n(t)} = 0 \qquad (5\text{-}2)$$

in order to impose the absence of an exotic multiplet of resonances i in the s- channel, and similarly as

$$\sum_j M^{ij}_{ut} \sum_n \zeta^j_n\, \beta^j_n\, u^{\alpha^j_n(t)} = 0 \qquad (5\text{-}3)$$

in order to exlude the same multiplet in the u - channel.

Duality imposes a relation for the average contribution of the resonances. Nevertheless, if it holds for a relatively large range of N values, the cancellations implied by (5-2) and (5-3) have to work for each power, leading to exchange degenerate trajectories with the same trajectory function α and with relations among the residue functions

$$\sum_j M_{st}^{ij} \sum_n \beta_n^j(t) = 0 \tag{5-4}$$

$$\sum_j M_{ut}^{ij} \sum_n \xi_n^j \beta_n^j(t) = 0 \tag{5-5}$$

for channels i containing no exotic resonances. As already stressed, exchange degeneracy is expected to be the better, the stronger the trajectories are coupled. It also supposes Regge pole dominance and except for some amplitudes (spin flip charge exchange), relations such as (5-4) should be less and less satisfied as $|t|$ increases in the physical region for scattering and Regge cut contributions win over the leading Regge pole contributions.

The implications of (5-4) in related channels have been extensively studied [54,72] and recently reviewed by Mandula, Weyers and Zweig [73].

We already saw in the case of meson-meson scattering how the absence of exotic states in the $\pi^+\pi^+$, π^+K^+, K^+K^0, and K^+K^+ channels could impose the observed degeneracy between the leading vector and tensor trajectories and at the same time imply relations among the couplings. These relations imply in turn, through factorization, the observed nonet structure. Proceeding in a more general way, one selects all the exotic channels offered by a two body process $AB \to CD$, imposes either (5-4) or (5-5) and looks for solutions of the system of linear equations thus written. The solution is further required to factorize, namely

$$\beta^j_{n\,A\bar{C}\to\bar{B}D} = \gamma^j_{n\,A\bar{C}}\,\gamma^j_{n\,\bar{B}D} \tag{5-6}$$

as a pole contribution should. This in general requires a further degeneracy between Regge trajectories $\alpha_n(t)$ of opposite signatures and of different internal quantum numbers belonging to the same or to different multiplets. As an example, the st SU(2) crossing matrix for $\pi\pi$ scattering is

$$M_{st} = \frac{1}{6}\begin{pmatrix} 2 & 6 & 10 \\ 2 & 3 & -5 \\ 2 & -3 & 1 \end{pmatrix} . \qquad (5\text{-}7)$$

The absence of exotic resonances in the isospin 2 channel, from (5-4), implies that

$$\beta^1_{\pi\pi} = \frac{2}{3}\beta^0_{\pi\pi} \qquad (5\text{-}8)$$

where the isospin superscripts 1 and 0 eventually refer to the ρ and f^0 trajectories. Projecting on the $\pi^+\pi^-$ channel one obtains the relations previously written (3-4).

This is readily generalized to SU(3) multiplets. The necessary SU(3) crossing matrices have been worked out by Rebbi and Slansky [74] and the full set of linear equations required by the absence of exotic resonances is easily written down.

As an example meson-meson scattering has 5 different SU(3) channels, namely

$$1 \qquad 8_{SS} \qquad 8_{AA} \qquad 10 \oplus \overline{10} \qquad 27 \qquad (5\text{-}9)$$

The 10 and the $\overline{10}$ representations must both occur in a self-conjugate combination. The symmetric (S) or antisymmetric (A) forms of the octet coupling being related to the spin of the exchanged meson, only the coupling structures SS and AA are relevant. This is however not the case in meson-baryon scattering or baryon-baryon scattering where a mixed SA coupling is also possible.

We have to require that no exotic resonance belonging to $10 \oplus \overline{10}$ or 27 multiplets exist in any channel and that no Regge term associated with them is exchanged in another channel. This gives 2 conditions among 3 unknowns and admits one non-trivial solution. This implies degeneracy between two octets with symmetric and antisymmetric couplings and a singlet. The singlet must here correspond to an even signature trajectory since a vector singlet cannot couple to a P-wave meson-meson system odd under charge conjugation. We have therefore 9 degenerate trajectories respectively associated with 8 vector mesons and 9 tensor mesons. As worked out in Section 3, the solution is factorizable and even holds when mass degeneracy is not required among all the nonet members. Factorization imposes standard mixing $\tan^2\theta = \frac{1}{2}$ and decoupling of one of the tensor trajectories from the $\pi\pi$ channel.

The same analysis of course holds for meson vector scattering where the same leading trajectories are also exchanged. However it is now the tensor singlet which cannot couple in an antisymmetric way to a meson-vector system, odd under charge conjugation. As a result we now have a nonet of vector mesons and an octet of tensor mesons and again standard mixing with a decoupling of one of the vector mesons (the φ) from the $\rho\pi$ channel. Altogether this gives two nonets with the same mixing angle.

As seen in the preceding section, all these results are in remarkable agreement with experiment and give strong support to duality. Conversely, observation of important deviations from the predicted scheme would indicate that at least one of our hypotheses is not legitimate. This already occurs with the pseudoscalar and pseudovector mesons which are also exchanged in meson-vector scattering. These abnormal parity mesons should also show a nonet pattern, but the π, η and K on the one hand, and the A, D and K_A^* on the other hand, show an almost exact octet structure. One may argue that these trajectories have lower intercepts and that isolating them in our analysis as factorizable Regge contributions is perhaps too strong an approximation. Their contribution could be difficult to dissociate from those from Regge cuts which are not factorizable. The pion contribution is furthermore almost real for low t values where this separation might be best attempted and therefore is of no relevance to the analysis. It can also be argued that resonance saturation should fail as the threshold increases in energy[54]. It should be imposed only for low threshold channels such as meson-meson $(\pi\pi)$ or meson-baryon (πB), but not considered as reliable for high threshold channels such as vector-vector $(\rho\rho)$ or annihilation channels$(\pi\pi \to N\bar{N})$. [75]

Such a distinction is important in meson-baryon scattering [72]. The s-u baryon resonances can be connected through duality, as already discussed in the $\bar{K}N$ case. This is first done notwithstanding the absence of exotic states in the t- channel. There are in this case many more amplitudes than conditions and several solutions are possible [72,74] even after imposing factorization for all meson-octet and meson-decuplet amplitudes. These solutions correspond however to different degeneracy patterns among the baryon octets and decuplets which should be compared to the actual resonances.

There is in particular an interesting solution [72] which requires an octet with negative signature degenerate with the

Δ decuplet $(3/2^+)$ of positive signature. It corresponds to the $\frac{5}{2}^-$ octet with an N member of 1688 MeV. At the same time the nucleon trajectory with positive signature could be degenerate with a singlet, an octet and a decuplet with negative signature. This may be associated with the $\frac{3}{2}^-$ octet and singlet including the $N(1518)$, the $\Lambda(1520)$ and $\Lambda(1700)$ and the $\Sigma(1660)$ with on the other hand the $\frac{3}{2}^-$ decuplet with a Δ member at 1690 MeV. [28]

The obtained pattern:

$$\begin{array}{l} 10\left(\tfrac{3}{2}^+\right) - 8\left(\tfrac{5}{2}^-\right) \\ 8\left(\tfrac{1}{2}^+\right) - \left(1\left(\tfrac{3}{2}^-\right) \quad 8\left(\tfrac{3}{2}^-\right) \quad 10\left(\tfrac{3}{2}^-\right)\right) \end{array} \qquad (5\text{-}10)$$

is very interesting since it corresponds to an SU(6) 56 exchange degenerate with an SU(6) 70 to which one unit of angular momentum would have been added (SU(6) × 0(3) symmetry pattern). The SU(3) × SU(2) decomposition of the 56 and 70 SU(6) representations are respectively

$$8(2) \oplus 10(4) \quad \text{for the } 56,$$

$$1 \oplus 8 \oplus 10(2) \oplus 8(4) \quad \text{for the } 70.$$

They correspond to the required degeneracy. As a result of duality such giant multiplets should alternate with increasing mass. It is impressive to see again the close connection between duality and quark model predictions. The $56\ (L=0)$ has to be associated with a $70\ (L=1)$.

The degeneracy (5-10) imposes an F/D ratio of $-1/3$ for the $\frac{5}{2}^-$ octet which is indeed compatible with the observed couplings.

New constraints have to be imposed if exotics are also ruled out in the t- channel. This can be satisfied with non-exotic baryon resonances but demands a degeneracy pattern which cannot be now reconciled with the known resonance scheme. In particular an octet with spin parity $\frac{3}{2}^+$ is now needed besides the decuplet, but no evidence for it has been yet gathered. However, as already stressed resonance saturation should not be strictly enforced there because of its high threshold channel. This leads to the broken duality scheme of Mandula, Weyers and Zweig [54] in which duality is granted an important but restricted role; the duality constraints (5-4,5-5) of the low threshold channels are the only ones kept.

B. The $B\bar{B}$ Puzzle

The question of resonance saturation in the annihilation channel also applies to the baryon-antibaryon $B\bar{B}$ elastic channels. In this case there is no solution which could exclude exotics [76]. For instance resonance dominance in $\Delta\bar{\Delta}$ scattering requires exotic states. There are 4 amplitudes corresponding to the SU(3) decomposition

$$10 \otimes \overline{10} = 1 \oplus 8 \oplus 27 \oplus 64 . \tag{5-11}$$

Excluding exotics in both the 27 and the 64 representation is not possible with only non-exotic singlet and octet contributions. A similar result is already true for SU(2) alone, in which case exotic mesons with isospin 3 and 2 have to be excluded.

As stressed by Lipkin [77] global duality and the absence of exotic states already requires the decoupling of the leading isovector trajectories from $\Delta\bar{\Delta}$ scattering, which would be a strong case against vector dominance. The $\Delta\bar{\Delta}$ charge exchange reaction

$$\Delta^{+} + \bar{\Delta}^{+} \longrightarrow \Delta^{++} + \bar{\Delta}^{0}$$

is exotic in both the s and u- channels. The s- channel has isospin 2 and 3 and the u- channel corresponds to a BB exotic system, and there is furthermore no Pomeranchon exchange. The ρ and A_2 contributions in the t- channel must therefore separately vanish in the left hand side of the corresponding FESR.

We are therefore led to doubt the utility of resonance saturation in the $B\bar{B}$ case, or to expect exotic mesons strongly coupled to $B\bar{B}$ but relatively weakly coupled to the meson-meson channels so as not to spoil exchange degeneracy between the leading meson trajectories. Such exotic mesons were first proposed by Rosner [76].

C. Duality Diagrams

As shown by Harari [78] and Rosner [79] it is possible to translate graphically the required absence of exotics and the specific decoupling which we already found as a result of factorization. The SU(3) properties of each particle are represented by its quark content. The exchange of a Regge trajectory which corresponds to a particular matching of the SU(3) indices of the initial and final particles is then represented by the corresponding matching of the quark lines. This is graphically translated by quark exchange.

As an example $\pi^{+}\pi^{-} \longrightarrow \pi^{0}\pi^{0}$ is represented by the graph of Fig. 5-1. Such a graph is called a duality diagram.

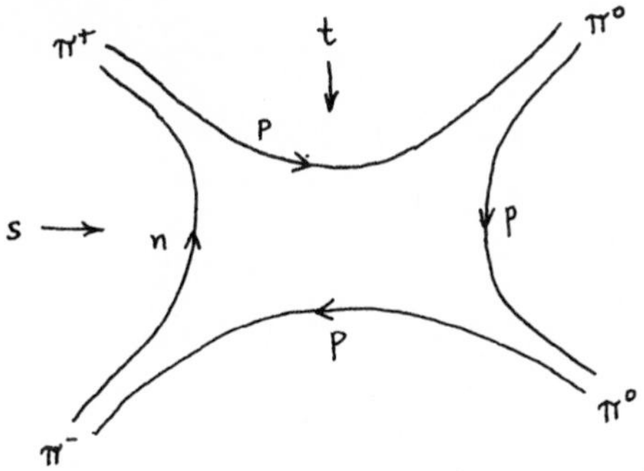

Figure 5-1. Duality diagram for $\pi^+ \pi^- \to \pi^0 \pi^0$.

The exchanged object consists of an $n\bar{p}$ system which has the internal quantum numbers of the ρ. The connection between the quark lines further shows that a $q\bar{q}$ system can indeed be exchanged in both the s and in the t channels. We can therefore have duality between the two channels, the s- channel resonances "building up" the t-channel Regge pole or vice versa. If we now draw a similar diagram for $\pi^+\pi^-$ elastic scattering (Fig. 5-2-a), we again explicitly show a possible dual connection between the s and t channels with exchanged $p\bar{p}$ and $n\bar{n}$ systems. They are associated with both the ρ and the f^0. However, if we cross two external particle lines so as to bring the exotic $\pi^+\pi^+$ system in the s- channel (Fig. 5-2-b) , we exhibit a $qq\bar{q}\bar{q}$ system which is exotic and not related through duality to any exchange in the t or u channel. The first graph is called a planar diagram, the second is called a non-planar diagram.

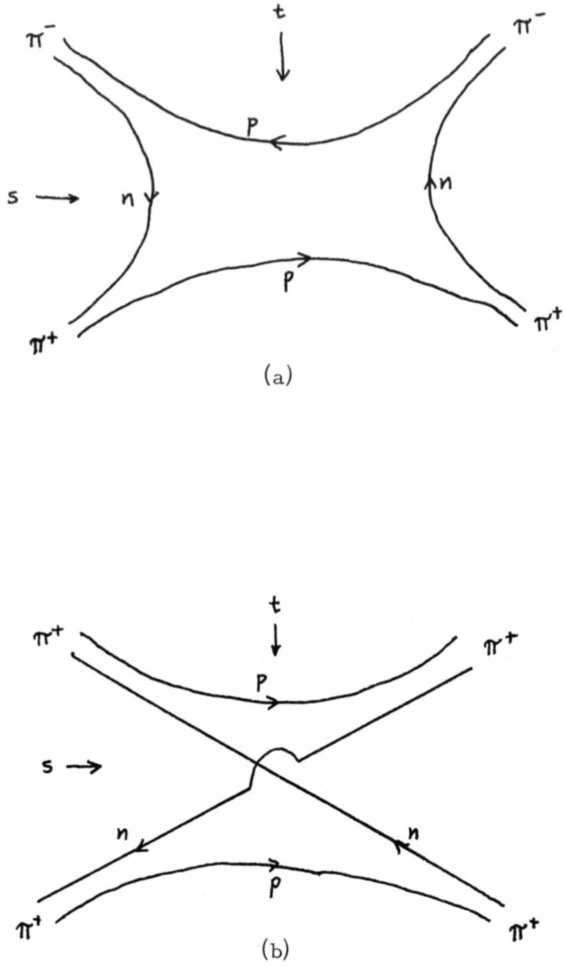

Figure 5-2.

Once we exhibit the quark content of the external particles it is easy to see through the possible matching of the quark lines whether a planar graph can be drawn or not. When it is possible we may have a duality relation between the two corresponding channels. When the only possible graph is non-planar we cannot have such a duality relation. We may however have it between two other channels, such as the t and u channel, for instance, in Fig. 5-2-b. We also meet unconnected graphs which need special consideration (Fig. 5-3).

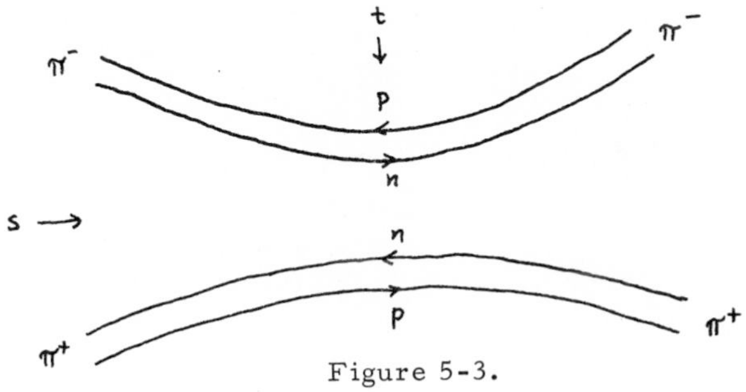

Figure 5-3.

They correspond to contracting separately the SU(3) indices of the initial and final t- channel particles, and stand for the exchange of the quantum numbers of the vacuum. Consequently such diagrams have to be associated with diffraction scattering. The reason is the following: the vertex couplings are of three types. We can write them as

$$\mathrm{Tr}\,(M\,[V, M]) \tag{5-12}$$

and

$$\mathrm{Tr}\,(M\,\{T, M\}) + \alpha\,\mathrm{Tr}\,(T)\,\mathrm{Tr}\,(MM) \tag{5-13}$$

M is a 3×3 matrix associated with the meson octet [80]

$$M = \begin{pmatrix} \sqrt{\tfrac{1}{2}}\,\pi^0 + \sqrt{\tfrac{1}{6}}\,\eta & \pi^+ & K^+ \\ \pi^- & -\sqrt{\tfrac{1}{2}}\,\pi^0 + \sqrt{\tfrac{1}{6}}\,\eta & K^0 \\ K^- & \bar{K}^0 & -2\sqrt{\tfrac{1}{6}}\,\eta \end{pmatrix}, \tag{5-14}$$

where V and T are 3×3 matrices associated with the vector and tensor nonets. They are readily taken from (5-14) through the substitutions $\pi^\pm \to \rho^\pm$, $K^\pm \to K^{*\pm}$, $K^0 \to K^{*0}$ except that the diagonal elements are given by:

$$V_1^{\,1} = \sqrt{\tfrac{1}{2}}\,\rho^0 + \sqrt{\tfrac{1}{2}}\,\omega \qquad V_2^{\,2} = -\sqrt{\tfrac{1}{2}}\,\rho^0 + \sqrt{\tfrac{1}{2}}\,\omega \qquad V_3^{\,3} = -\varphi$$

$$T_1^{\,1} = \sqrt{\tfrac{1}{2}}\,A_2^0 + \sqrt{\tfrac{1}{2}}\,f^0 \qquad T_2^{\,2} = -\sqrt{\tfrac{1}{2}}\,A_2^0 + \sqrt{\tfrac{1}{2}}\,f^0 \qquad T_3^{\,3} = -f' \tag{5-15}$$

This gives the general SU(3) coupling scheme with nonet mixing. (5-12) is an antisymmetric F coupling (commutator) while (5-13) is a symmetric D coupling (anticommutator) together with a (symmetric) scalar coupling. The unconnected graph is obviously associated with this last type of coupling.

Using (5-12 to 5-15) it is now simple to write down explicitly the individual meson couplings and it is easily found that the f' decouples from $\pi\pi$ only if $\alpha = 0$. This furthermore correctly relates the ρ and f^0 couplings. Keeping only connected graphs we therefore maintain all duality constraints including those required by factorization. We are therefore led to the following rule: [78,79] "Duality may hold between two channels only when they can be related through a planar connected duality diagram." The unconnected diagrams can be independently associated with Pomeranchon exchange. The presence of two uncrossed $q\bar{q}$ lines in any channel indicates that the resonance contribution to the imaginary part does not vanish. As an example the duality diagrams which can be written for $K\pi$ scattering lead to duality between the $K\pi$ and $\pi\pi \to K\bar{K}$ channels but not between the two $K\pi$ channels (Fig. 5-4).

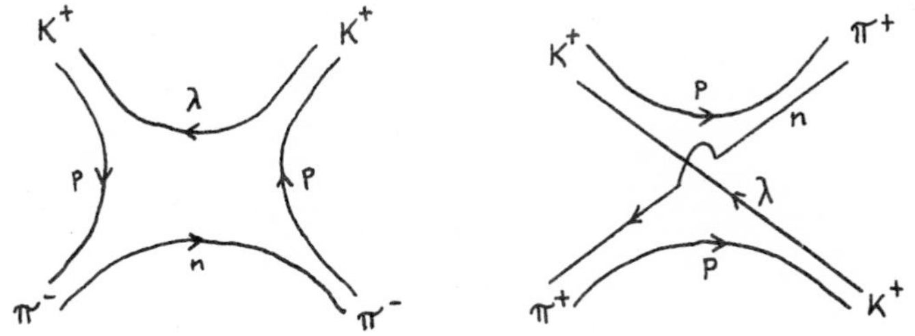

Figure 5-4.

This is indeed readily understood. The $K\pi$ channel, with resonances of isospin $\frac{1}{2}$ only, cannot be dual to a $K\pi$ channel without generating resonances of isospin $\frac{1}{2}$ and $\frac{3}{2}$. On the other hand it can be dual to the $\pi\pi \to K\bar{K}$ channel where it generates resonances of both possible isospins; conversely, the isospin 0 and 1 contributions may add in the isospin $\frac{1}{2}$ channel and cancel in the isospin $\frac{3}{2}$ channel.

The construction of duality diagrams can be readily generalized to meson baryon scattering. Eliminating unconnected diagrams corresponds to decoupling the φ and f' from $N\bar{N}$ (the φ and f' contain only strange quarks while the nucleon contains none). Fig. 5-5 shows duality diagrams associated with meson baryon scattering with s-t duality

($\pi^- p \rightarrow \pi^0 n$) and s-u duality ($\pi^+ p$ backward elastic scattering). It also shows a graph associated with K^+p elastic scattering with an exotic s- channel but still u-t duality as seen through twisting the upper part. A planar graph selects $q\bar{q}$ and qqq non-exotic systems and allows a possible duality relation between them.

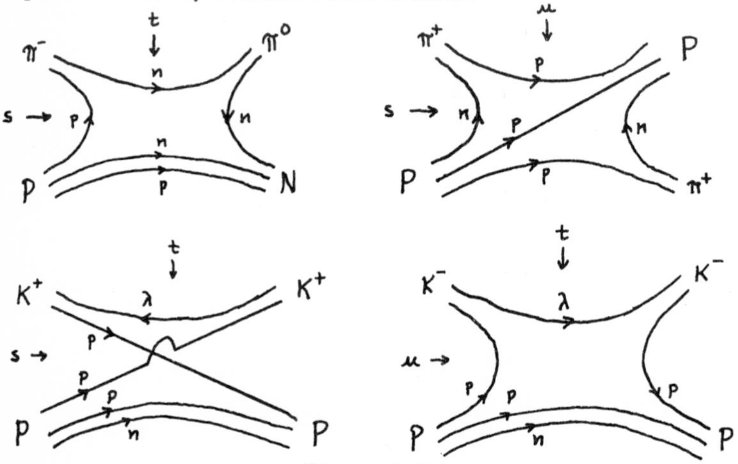

Figure 5-5.

One sees then immediately how troubles arise in the case of $B\bar{B}$ scattering. Matching of SU(3) indices now requires a $qq\bar{q}\bar{q}$ system in one channel (Fig. 5-6). The duality relation cannot be satisfied when unconnected graphs are excluded, so exotic mesons coupled to $B\bar{B}$ are required.

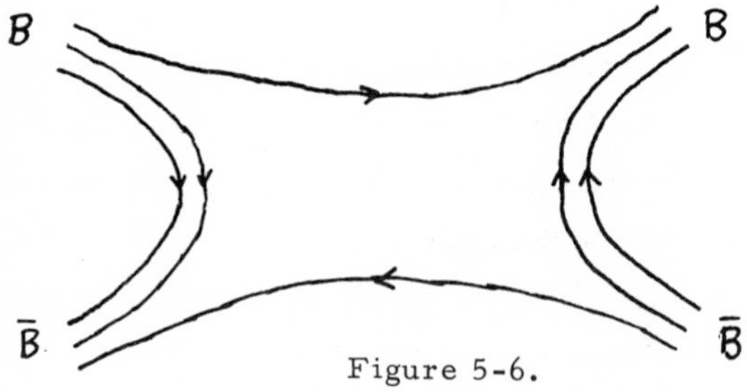

Figure 5-6.

In some cases no duality diagram can be drawn even though meson exchange is allowed. This is in particular the case for the amplitude $K^- p \to \pi^- \Sigma^+$. In such a case we conclude that the amplitude, or at least its peripheral part, is purely real.

D. Are There Exotic States ?

Two types of exotic states may be considered. The first type corresponds to resonances of low spin for their mass (say spin $1\,(\frac{1}{2})$ when a particle at a similar mass on the leading trajectory has spin $3\,(\frac{5}{2})$. Such exotics should a priori be wide highly inelastic resonances and so should hardly show up except for interference effects in production or exchange mechanisms with non-exotic prominent contributions. Such exotics are of almost no relevance to duality. The second type corresponds to those exotics strongly coupled to $B\bar{B}$ which are needed when extending resonance dominance to $B\bar{B}$ channels. They are a priori relatively narrow resonances since they are weakly coupled to the lowest mass meson channels and they should be on trajectories not too far down from the leading ones. They could be seen in missing mass experiments.

Producing an exotic meson in the near forward direction requires however either exotic exchange (supposed to be of no significant importance at high energies (Fig. 5-7-a)) or the introduction of an unconnected diagram as shown on Fig. 5-7-b. The unconnected diagram violates our duality rules in meson-meson scattering and so we don't expect it to play any important role in exotic production.

On the other hand backward production of an exotic meson state involves the same intermediate states as those found in the backward production of non-exotic mesons (Fig. 5-4) .

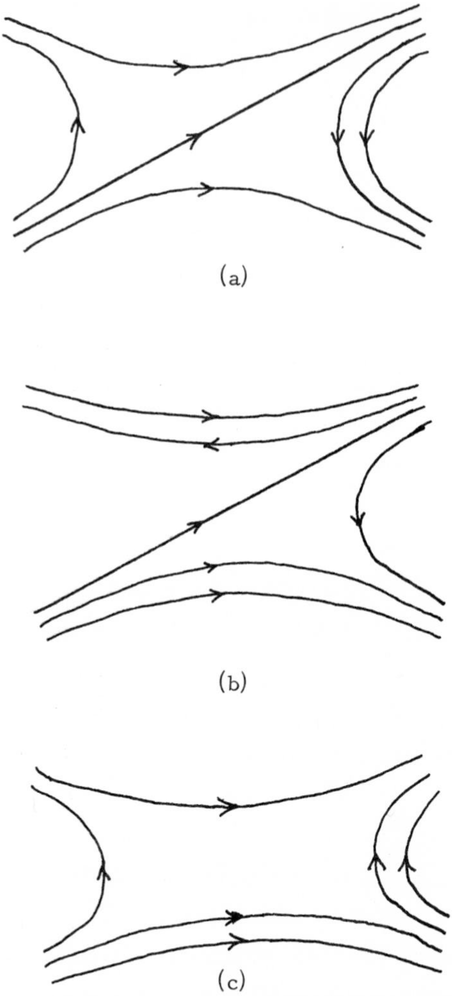

Figure 5-7.

This is shown on Fig. 5.7c. Such an amplitude is therefore expected to show the same energy behaviour as backward production, for example. Using duality diagrams as a guide we are therefore led to search for possible exotic states in backward meson-nucleon collisions. [81]

6. DUAL MODELS FOR MESON-MESON SCATTERING

A. The Veneziano Model

The preceding sections have provided a rationale for an attempt to build models in which resonance exchange would provide the only relevant forces in a bootstrap and where a simplifying pattern of narrow width resonances, located on infinitely rising exchange degenerate trajectories, could be used. We saw how such ideas are useful in correlating many prominent facts. Our next attempt is therefore to study the properties of a model world of zero width resonances and linear exchange degenerate Regge trajectories. Such a world is simple enough that bootstrap solutions can be readily guessed. It may also be sufficiently close to physics that basic properties such as resonance locations and widths can be compared with the actual values. It is simple enough that the infinite sets of resonances and Regge trajectories necessary for duality to hold in a local sense are easily dealt with.

We first take an amplitude with resonances in only two channels. This is the case for

$$\pi^+\pi^- \longrightarrow \pi^+\pi^-$$

where the u - channel $(\pi^+\pi^+)$ is exotic. In both the s and t- channels we have only the $\rho(f^0)$ exchange degenerate trajectory with equally spaced resonances. The scattering amplitude should then have infinite sets of poles in s and t for all positive integer values of $\alpha(s)$ or $\alpha(t)$ with

$$\begin{aligned} \alpha(s) &= \alpha(0) + \alpha' s \,, \\ \alpha(t) &= \alpha(0) + \alpha' t \,. \end{aligned} \tag{6-1}$$

This would be the case for the product of two gamma functions: $\Gamma(1-\alpha(s))\,\Gamma(1-\alpha(t))$. This is however not suitable since the residue of a pole in s (t) has to be interpreted in terms of particle exchange and should therefore be a polynomial in t (s). It cannot therefore be singular as would be the case when both $\alpha(s)$ and $\alpha(t)$ are positive integers. In order to exclude this possibility we should include a further term which vanishes then. This gives therefore the following simple expression [82]

$$V(s,t) = -\lambda \frac{\Gamma(1-\alpha(s))\,\Gamma(1-\alpha(t))}{\Gamma(1-\alpha(s)-\alpha(t))} \,, \tag{6-2}$$

where λ is a constant. The residue of the pole at $\alpha(s)\,(\alpha(t)) = J$ is now a polynomial of order J in $\alpha(t)\,(\alpha(s))$, since

$$\frac{\Gamma(1-\alpha(t))}{\Gamma(1-J-\alpha(t))} = \prod_{n=1}^{J} (1-n-\alpha(t)) \,,$$

and a polynomial in $t(s)$, as follows from the linear character of the trajectories.

In the neighborhood of $\alpha(s_J) = J$ we can write $\Gamma(1-\alpha(s))$ as

$$\frac{(-1)^J}{(J-1)!\,\alpha'(s-s_J)}$$

The leading term in t then reads

$$R_J = \lambda \frac{(\alpha')^{J-1}}{(J-1)!\,(s_J - s)} t^J , \qquad (6\text{-}3)$$

which gives a leading term in $\cos\theta$, readily associated with the highest spin contribution

$$\frac{\lambda}{s_J - s} \frac{(2qq')^J}{(J-1)!} (\alpha')^{J-1} \cos^J\theta , \qquad (6\text{-}4)$$

writing, with obvious notation

$$t = m_A^2 + m_C^2 + 2qq'\cos\theta - 2\omega\omega' .$$

Expanding the pole residue in terms of Legendre polynomials we get contributions which we may associate with spins $J, J-1, \cdots, 0$. This corresponds to at least $J+1$ particles [83]. Together with the leading spin one therefore obtains a family of equally spaced daughters.

The singularity structure of (6-2) is represented on Fig. 6-1 where the heavy lines stand for the poles in s and t. The dashed lines stand for the zeros associated with the vanishing of the argument of the denominator which eliminates forbidden coincident poles.

If we now consider the sequence of daughters obtained for each s pole we obtain a series of parallel trajectories as shown on Fig. 6-2.

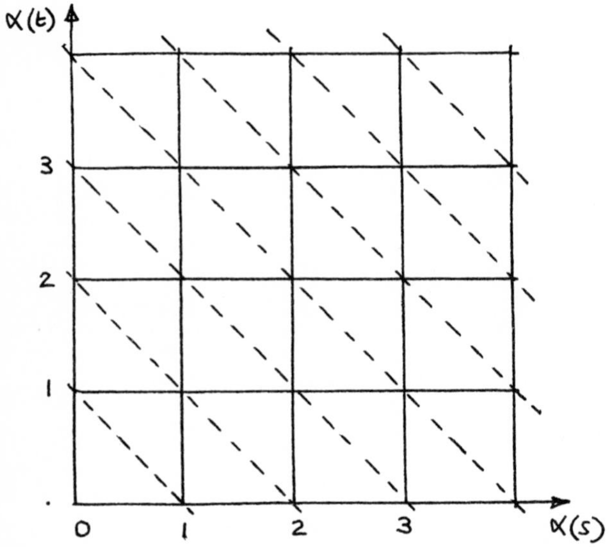

Figure 6-1.

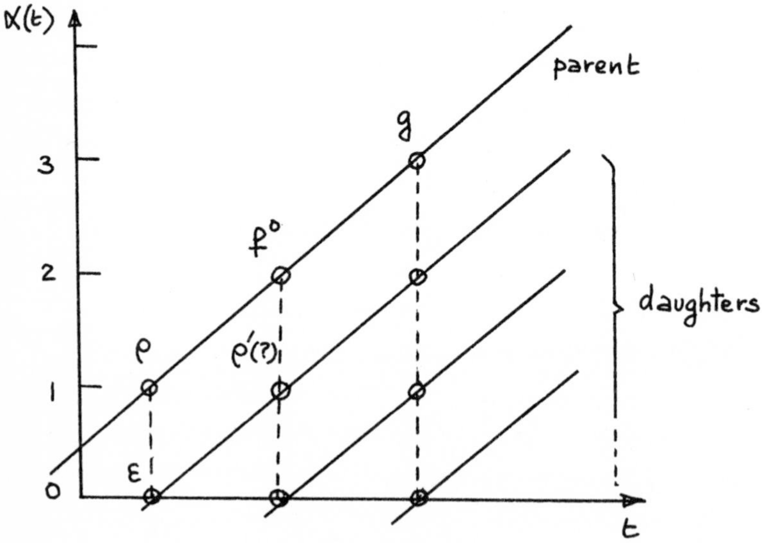

Figure 6-2.

$V(s,t)$ can be written in terms of the Euler beta function

$$V(s,t) = -\lambda\left(1-\alpha(s)-\alpha(t)\right) B\left(1-\alpha(s),\, 1-\alpha(t)\right), \qquad (6\text{-}5)$$

with

$$B(x,y) = \frac{\Gamma(x)\,\Gamma(y)}{\Gamma(x+y)} . \qquad (6\text{-}6)$$

The beta function would be the amplitude which we should write instead of (6-2) if the lowest spin on the exchanged trajectory were zero instead of one. Its arguments would then be $-\alpha(s)$ and $-\alpha(t)$. The beta function can also be written as

$$B(x,y) = \int_0^1 u^{x-1}(1-u)^{y-1}\,du , \qquad (6\text{-}7)$$

defined for $\mathrm{Re}\,x > 0$ and $\mathrm{Re}\,y > 0$, or, with the change of variables

$$v = \frac{u}{1-u} ,$$

as

$$B(x,y) = \int_0^\infty v^{x-1}(1+v)^{-x-y}\,dv .$$

The poles in s are associated with the end point singularity of the integral at $u=0$, when $\mathrm{Re}\,x \leq 0$. In order to see this and to obtain an expression valid for $\mathrm{Re}\,x \leq 0$ one may expand the second term in the integrand of (6-7) and integrate term by term. One gets

$$B(x,y) = \sum_{n=0}^{\infty} (-1)^n \binom{y-1}{n} (x+n)^{-1} =$$

$$= \sum_{n=0}^{\infty} \binom{-y+n}{-y} (x+n)^{-1} \qquad (6\text{-}8)$$

Writing this expression for $x = -\alpha(s)$ and $y = -\alpha(t)$ one finds

$$B(-\alpha(s), -\alpha(t)) = \sum_{J=0}^{\infty} \binom{\alpha(t)+J}{J} \frac{1}{J-\alpha(s)} . \qquad (6\text{-}9)$$

But we can also write it as:

$$B(-\alpha(s), -\alpha(t)) = \sum_{J=0}^{\infty} \binom{\alpha(s)+J}{J} \frac{1}{J-\alpha(t)} , \qquad (6\text{-}10)$$

now expanding in terms of the t - channel poles.

Both expressions are equivalent! The t- channel poles (s- channel poles) are already contained in the sum of the s- channel pole (t- channel pole) contributions even though each term in the expansion is separately regular in t (s). We have therefore an exact expression of duality; conversely duality can be defined as the salient property of the model amplitude. We have a model amplitude which is fully defined by its s- channel contributions or its t- channel contributions. The knowledge of both would be redundant. This can be graphically translated as shown on Fig. 6-3. The two graphs are equivalent when the internal line represents an infinite sum of resonances as explicitly determined by (6-9) or (6-10).

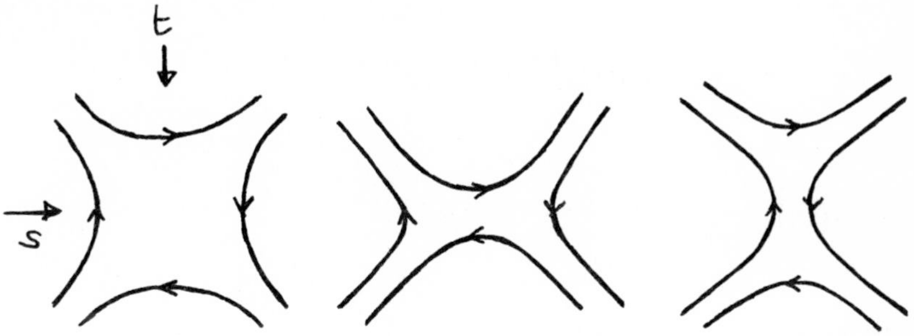

Figure 6-3.

A similar dual expansion can be written for the model amplitude written for $\pi\pi$. Combining (6-5) and (6-9) it reads:

$$V(s,t) = \lambda \left(\alpha(s) + \alpha(t) - 1\right) \sum_{n=0}^{\infty} \binom{\alpha(t)+n+1}{n} \frac{1}{n+1-\alpha(s)} . \qquad (6\text{-}11)$$

When $\alpha(s) = 1$, the pole residue is indeed linear in t. Expanding (6-9) or (6-11) one readily gets the different daughter couplings. The parent trajectory's residues (6-4) are all positive but we can a priori say nothing about the positivity conditions for the daughter residues; ghosts do occur.

The ratio of two gamma functions for large values of the argument is given by:

$$\frac{\Gamma(x+a)}{\Gamma(x+b)} = x^{a-b}\left(1 + \frac{(a-b)(a+b-1)}{x} + O(x^{-2})\right) . \qquad (6\text{-}12)$$

We therefore get:

$$V(s,t) \underset{s\to\infty}{\sim} -\lambda\, \Gamma(1-\alpha(t))\, (-\alpha' s)^{\alpha(t)} ,$$

$$V(s,t) \underset{t\to\infty}{\sim} -\lambda\, \Gamma(1-\alpha(s))\, (-\alpha' t)^{\alpha(s)} . \qquad (6\text{-}13)$$

The behaviour of B is:

$$B(x,y) \underset{x\to\infty}{\sim} \Gamma(-\alpha(t))\, (-\alpha' s)^{\alpha(t)} \qquad (6\text{-}14)$$

We therefore obtain Regge behaviour. However, it should be stressed that the Stirling formula does not hold within an infinitesimal wedge around the positive real axis where the poles are. We therefore do not get Regge behaviour in the physical high energy region but only off the real axis.

The Veneziano model amplitudes thus written have the following prominent properties [82, 84, 85]

1) Crossing symmetry
2) Desired singularity structure in s and t
3) Local duality
4) Asymptotic Regge behaviour

The Regge singularity properties of (6-2) can be studied quite generally. One indeed finds only moving poles except for fixed poles at wrong signature odd negative integer values of J. This analysis is carried through in the paper of Yellin and Sivers [86]. One may also study the Lorentz pole structure of the Veneziano term [87]. It has no specifically interesting features. The Veneziano daughter series, which leads to singularities in s, is not simply related to any Toller daughter series, which is regular in s at $t = 0$.

However, a series of real axis pole terms cannot be considered as a scattering amplitude since this obviously violates unitarity. We should consider a Veneziano model not as an amplitude but as a collection of levels which represents a bootstrap solution for a model world of zero width resonances and linear Regge trajectories fully connected through duality. We shall however often refer to it as a Veneziano amplitude.

B. The Two Meson Problem

Once we have such a bootstrap solution we can calculate couplings and relations as they hold in the model world it refers to and compare them eventually with the actual physical values. We assume that a model world of zero width resonances is actually not too different from physics, as far as bootstraps are concerned, and so is worth a detailed consideration. A second step will be to write tentative expressions for unitary scattering amplitudes which retain as many as possible of the properties obtained from the dual models (resonance location, resonance width. . .)

The resonance widths are easily computed from (6-4) (for the parent) and (6-11) for the daughters. The amplitude written as (6-2) gives a ρ pole with residue

$$\lambda (2qq') \cos\theta = G_\rho^2 \, 4qq' \cos\theta \, ,$$

writing with obvious notation the $\rho\pi\pi$ coupling as $G_\rho \epsilon (q+q')$. We therefore get:

$$G_\rho^2 = \frac{\lambda}{2} \quad \text{with} \quad \Gamma_\rho = \frac{1}{12 m_\rho^2} \frac{\lambda}{8\pi} (m_\rho^2 - 4m_\pi^2)^{3/2} \, .$$

A width of 120 MeV corresponds to $\lambda = 60$. The ratio of the f^0 and ρ widths, namely 1.2, is readily obtained from (6-4). The important point is that the only length in the model is $(\alpha')^{-1/2}$. This fixes the ratio of coupling constants with different dimensions, corresponding to a short "effective" range parameter. The ρ daughter, the ϵ, has a width 4.5 times larger than the ρ width. At the same time the P-wave ρ' daughter of the f^0 should have an elastic width similar to the ρ width. The model does not give any information about the total width of the ρ' so it could be large and have helped the ρ' to escape detection [88].

Using (6-2) for the $\pi^+\pi^-$ case (no u-channel resonance), we may combine similar terms interchanging the roles of s, t and u in order to write terms suitable for the three amplitudes. The s-channel amplitudes are then written as

$$S_0 = \beta \left(V(s,t) + V(s,u) + \gamma V(t,u) \right)$$

$$S_1 = \alpha \left(V(s,t) - V(s,u) \right)$$

$$S_2 = V(t,u) \qquad (6\text{-}16)$$

We have only used obvious symmetry relations and insisted on writing the full amplitude as a combination of bilinear terms [89] so that the absence of resonances in an exotic channel could be readily incorporated by the presence of only one term. Using the crossing matrix

$$M_{ts} = \frac{1}{6} \begin{pmatrix} 2 & 6 & 10 \\ 2 & 3 & -5 \\ 2 & -3 & 1 \end{pmatrix}$$

we may exclude poles in the isospin 2 t- channel. This gives

$$\gamma = -\tfrac{1}{2} \quad , \qquad \beta = \tfrac{3}{2}\alpha \quad .$$

There should furthermore be no pole with odd spin in the iso-spin 0 t-channel amplitude, as explicitly written down for the isospin 0 s-channel one. This finally gives $\alpha = 1$, thus fixing up all terms. The $\pi\pi$ amplitudes then read:

$$\begin{aligned} S_0 &= \tfrac{3}{2}\left(V(s,t) + V(s,u)\right) - \tfrac{1}{2} V(t,u) \\ S_1 &= V(s,t) - V(s,u) \\ S_2 &= V(t,u) \end{aligned} \tag{6-17}$$

which gives in particular

$$\begin{aligned} A(\pi^+\pi^- \to \pi^+\pi^-) &= V(s,t) \\ A(\pi^+\pi^0 \to \pi^+\pi^0) &= V(s,t) + V(u,t) - V(s,u) \\ A(\pi^0\pi^0 \to \pi^0\pi^0) &= V(s,t) + V(u,t) + V(u,s) \end{aligned} \tag{6-18}$$

All $\pi\pi$ amplitudes are expressed in terms of a single constant λ once the ρ trajectory parameters ($\alpha(0)$ and α') are known. We may then even use (6-17) to continue the amplitude slightly off the mass shell considering s, t and u as independent variables. This is justified on the ground that the Regge trajectories should not depend on the external masses; only λ should be allowed to vary.[84] At $s = t = u = m_\pi^2$, which corresponds to one of the pion momenta vanishing, the other three being kept on the mass shell, the amplitude goes to zero provided that:

$$\alpha_\rho (m_\pi^2) = 1/2 \quad . \tag{6-19}$$

This is compatible with the present determination of the $\rho\,(f^0)$ trajectory and we may impose it so that the $\pi\pi$ amplitudes will satisfy an important constraint imposed by PCAC and current algebra [90]. This was first pointed out by Lovelace [84] and produced a great deal of the initial popularity of the Veneziano model. Nevertheless its lack of generalization makes it still appear as a happy coincidence rather than as a basic property.

The amplitude vanishing at $s = t = u = m_\pi^2$, we may expand it linearly aroung this point and write:

$$V(s,t) \sim \lambda \pi \alpha' (s + t - 2m_\pi^2) \, , \tag{6-20}$$

with α' now given by:

$$\alpha' = \left[2(m_\rho^2 - m_\pi^2)\right]^{-1} = 0.88 \text{ GeV}^2 . \tag{6-21}$$

One may then compare (6-20) to the linear expressions obtained from crossing symmetry, current algebra, the Adler condition and the absence of isospin 2 σ terms. They are the Weinberg amplitudes written as [90]

$$M_{\alpha\beta;\gamma\delta} = \frac{1}{f_\pi^2} \left((s - m_\pi^2)\,\delta_{\alpha\gamma}\delta_{\beta\delta} + (t - m_\pi^2)\,\delta_{\alpha\beta}\delta_{\gamma\delta} + (u - m_\pi^2)\,\delta_{\alpha\delta}\delta_{\beta\gamma} \right)$$

where $\alpha\beta\,(\gamma\delta)$ are the isospin indices of the t-channel initial (final) pions, and f_π is the pion decay constant. The constant f_π has the dimension of a mass and is found to be of the order of 90 MeV. [91] (6-22) is a linear approximation to the $\pi\pi$ amplitude which results from the properties just listed. Its first derivatives at $s, t, u = m_\pi^2$ are given by current algebra and it is these derivatives which we may identify with (6-17) written in the approximation given by (6-20).

The Weinberg and the linear approximation to the Veneziano amplitudes respectively read

$$S_0^W = \frac{1}{f_\pi^2}(3s+t+u-5m_\pi^2) \ ; \ S_0^V = \lambda\pi\alpha'(3s+t+u-5m_\pi^2)$$

$$S_1^W = \frac{1}{f_\pi^2}(t-u) \ ; \ S_0^V = \lambda\pi\alpha'(t-u)$$

$$S_2^W = \frac{1}{f_\pi^2}(t+u-2m_\pi^2) \ ; \ S_2^V = \lambda\pi\alpha'(t+u-2m_\pi^2) \ .$$

Their identification requires

$$\lambda = (\pi f_\pi^2 \alpha')^{-1} \simeq 50 \ .$$

This is close to the value obtained from the experimental ρ width. The Veneziano amplitude may be used to relate the ρ coupling constant to f_π. This gives

$$G_\rho^2 = \frac{m_\rho^2 - m_\pi^2}{\pi f_\pi^2}$$

which improves significantly over the KSFR formula derived from the dominance of ρ exchange.

The possible identification between the two expressions is not surprising. In both cases we impose that isospin 2 exchange is irrelevant. In the one case we cancel isospin 2 resonances in a crossing symmetric way. In the other one we cancel isospin 2 σ terms in a crossing symmetric way.

As emphasized by Lovelace [92] (6-17) could be tried as a K-matrix $\pi\pi$ amplitude. It should then give the Weinberg small scattering lengths (positive a_0 and negative a_2 with the ratio $\frac{a_0}{a_2} = -\frac{7}{2}$). It should have a large effective range since the range parameter f_π^{-1} corresponds to 2 fermis. It should also have ρ and f^0 resonances with the correct width and furthermore a wide S-wave $I=0$ resonance at the mass of the ρ, which is required experimentally.

If (6-2) shows many interesting features, so also does

$$\sum_{N,M\geq 1} \sum_{R\geq M,N}^{R\leq M+N} \beta_{NMR} \frac{\Gamma(N-\alpha(s))\,\Gamma(M-\alpha(t))}{\Gamma(R-\alpha(s)-\alpha(t))} \tag{6-22}$$

The extra terms are called "satellites". They give poles for spin values greater than $M(N)$ and affect only daughter levels if $M(N) > R$. Their asymptotic Regge behavior is given by $(-\alpha(s))^{\alpha(t)+N-R}$ or $(-\alpha(t))^{\alpha(s)+M-R}$ Such a series of satellite terms may be resummed as:

$$\int_0^1 u^{x-1}(1-u)^{y-1} f(u)\, du \tag{6-23}$$

where f is a regular function in the interval $[0,1]$. From the structure of the two body amplitude we cannot therefore select (6-7) against (6-23) even on grounds of simplicity. The generalization to many particle terms will favor (6-7) by introducing a new objective criterion. However, the Adler condition is satisfied by the leading term in (6-2) ; it would involve a relation among satellite terms which do not comply with it identically unless they are zero, a hint that (6-2) is the $\pi\pi$ amplitude.

Similar amplitudes can be tried for $K\pi$ and $K\bar{K}$ scattering.[93] In this case we also have exchange degenerate trajectories but they are not identical. In the $K\pi$ case we have the $\rho\,(f^0)$ in one channel and the $K^*\,(K^{**})$ in the other channel. We know that in order to obtain a factorizable solution we have to write a dual amplitude only when we can draw a planar duality diagram. As is easily verified this leads us to write only $V^{K^*\rho}$ terms for $K\pi$ scattering, where the $\pi\pi$ Veneziano term would read $V^{\rho\rho}$, and only $V^{\varphi\rho}$ terms for $K\bar{K}$ scattering. As duality relations between the s and t - channel they read

$$V^{K^*\rho}(s,t) = -\lambda' \frac{\Gamma(1-\alpha_{K^*}(s))\,\Gamma(1-\alpha_\rho(t))}{\Gamma(1-\alpha_{K^*}(s)-\alpha_\rho(t))}$$

$$V^{\varphi\rho}(s,t) = -\lambda'' \frac{\Gamma(1-\alpha_\varphi(s))\,\Gamma(1-\alpha_\rho(t))}{\Gamma(1-\alpha_\varphi(s)-\alpha_\rho(t))} \tag{6-24}$$

with the duality diagrams of Figure 6-4.

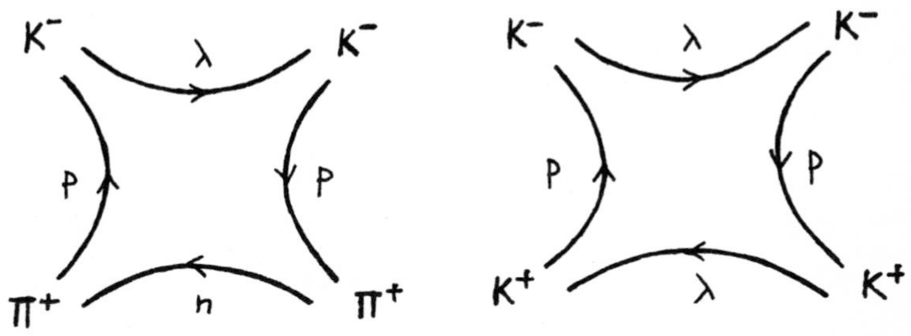

Figure 6-4.

The Adler condition is still satisfied in both cases with the conditions:

$$\alpha_{K^*}(m_K^2) = \tfrac{1}{2} \quad , \qquad \alpha_\varphi(m_K^2) + \alpha_\rho(m_K^2) = 1 \ .$$

We require the same slope,[95] if not identical trajectories. The $\rho(\omega f^0 A_2)$, $K^*(K^{**})$, and $\varphi(f')$ trajectories should then be related as follows:

$$\begin{aligned} \alpha_\rho(s) &= \tfrac{1}{2} + \alpha'(s - m_\pi^2) \\ \alpha_{K^*}(s) &= \tfrac{1}{2} + \alpha'(s - m_K^2) \\ \alpha_\varphi(s) &= \tfrac{1}{2} + \alpha'(s - 2m_K^2 + m_\pi^2) \end{aligned} \tag{6-25}$$

We have therefore three equally spaced trajectories as implied by the mass formula with nonet mixing. The external masses m_π and m_K have to obey the same relations as the internal masses m_ρ and m_{K^*}. This leads to SU(6) relations

$$\begin{aligned} m_{K^*}^2 - m_\rho^2 &= m_K^2 - m_\pi^2 \\ m_\varphi^2 - m_{K^*}^2 &= m_{K^*}^2 - m_\rho^2 \end{aligned} \tag{6-26}$$

which are known to be well satisfied.

Veneziano terms (6-24) written for different dual channels can be combined by imposing obvious symmetry relations and exotic states can then be eliminated using the pertinent crossing matrices [12,96]. Factorization and SU(3) symmetry are then used to relate λ, λ', and λ''. [93] As has already been seen, all amplitudes are expressed in terms of one single constant.

The solutions are

for $K\pi$

$$T_0 = \tfrac{1}{2}\sqrt{\tfrac{3}{2}}\left(V^{K^*\rho}(s,t) + V^{K^*\rho}(u,t)\right) \qquad (6\text{-}26')$$

$$T_1 = \tfrac{1}{2}\left(V^{K^*\rho}(s,t) - V^{K^*\rho}(u,t)\right)$$

with a crossing matrix

$$M_{st} = \frac{1}{2}\begin{pmatrix} \sqrt{\tfrac{2}{3}} & 2 \\ \sqrt{\tfrac{2}{3}} & -1 \end{pmatrix}$$

and for $K\bar{K}$ (KK in the t-channel)

$$T_0 = \tfrac{1}{2}\left(V^{\varphi\rho}(s,u) - V^{\varphi\rho}(u,s)\right) \qquad (6\text{-}27)$$

$$T_1 = -\tfrac{1}{2}\left(V^{\varphi\rho}(s,u) + V^{\varphi\rho}(u,s)\right)$$

with a crossing matrix

$$M_{st} = \frac{1}{2}\begin{pmatrix} -1 & 3 \\ 1 & 1 \end{pmatrix}$$

The success of the relations (6-26) comes from the fact that they do not differentiate between octet and nonet structure. If we try to also describe the $\eta\pi$ and ηK amplitudes, Veneziano terms involving respectively the f^0 and the A_2 and the f^0, the f', and the K^* can be written but the Adler condition is no longer satisfied [97]. It would obviously imply the same pattern for the external and exchanged particles and as a result a nonet structure for the pseudoscalar mesons with $m_\eta^2 - m_\pi^2 = m_{f^0}^2 - m_{A_2}^2 = 0$.

This is too far from experiment to be retained even as an approximation. A simple and general solution exists but only with octet-singlet mass degeneracy. It is in this case that we may consider the Veneziano amplitude as a simple bootstrap solution. When we break SU(3) symmetry we keep most interesting expressions for $\pi\pi$, πK and $K\bar{K}$ scattering which, even if their satisfying the Adler condition is a pleasant coincidence, appear as a compact synthesis of many required properties.

It has been noticed[95, 56] that some SU(3) results actually follow from the Veneziano formula. This is in particular the case of the relation between the $K^{*}K\pi$ and $\rho\pi\pi$ coupling constant which we can obtain from the single $V(s,t)$ Veneziano term written for $K^{-}\pi^{+}$ scattering, where the u-channel is exotic. One should stress however that we impose crossing symmetry with internal particles which are those of an octet ($K\pi$ does not differentiate between octet and nonet), namely a single isospin 1 and a single strange meson with isospin $\frac{1}{2}$. It is therefore not surprising that the coupling constants are found in the ratio of the structure constants of the simplest unitary group for which these particles can be members of the same multiplet.

Our writing of meson-meson amplitudes using only combinations of two channel terms followed from the requirement of having similar expressions whether exotic channels appear or not. Each term corresponds as we saw to a particular duality diagram which selects a particular ordering of the particles. Taking into account the obvious reflection symmetry of the graph we have 3 such orderings, and $\frac{(N-1)!}{2}$ orderings for N external particles. The two channels are then those which we can obtain through partitions grouping adjacent external lines. This is shown on Figure 6-5.

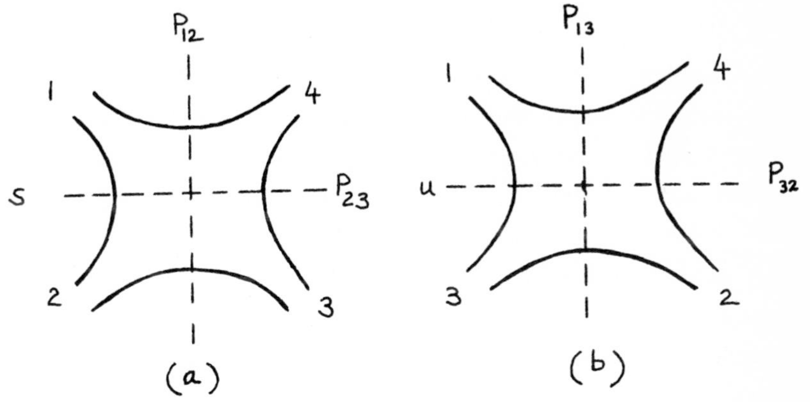

Figure 6-5.

This type of construction will be generalized to obtain production amplitudes.

It has been argued that the Veneziano amplitude is not dual on the ground that by integrating (6.7) from 0 to 1/2 (say) and 1/2 to 1 one can separate two terms, one with the s- channel resonances and one with the t- channel resonances, as is the case in an interference model.[98] This is however not the case since such a splitting produces two pieces which are separately not acceptable as scattering amplitudes. They separately increase exponentially in some directions and only their sum is well behaved and a possible amplitude. The beta function which can be written either as (6-9) or (6-10) is dual by definition. We have however not duality between s-channel resonances and t- channel Regge poles but between s- channel resonances and t- channel resonances. The model tells us how the Regge poles result from the exchange of an infinite set of resonances, as is done explicitly in the case of the Van Hove model [37]. It may also be misleading to obtain a Regge behaviour from a Veneziano amplitude to see whether or not it satisfies global duality or semi local duality. The Regge behaviour does not hold on the real axis, where we would need it in order to write finite energy sum rules, and we have therefore to use a limiting procedure which should no longer contain the real axis poles.[99]

The bootstrap solution which we then have for a model meson world of zero width resonances lacks a positivity property on the pole residue of the elastic scattering amplitudes. The question of ghosts will be discussed later when we deal with production amplitudes. In general ghosts cannot be avoided with the dual models which are presently considered and occur already at the second daughter level [100]. It is a happy coincidence of the $\pi\pi$ term (6-17) that ghosts occur only at a very deep daughter level [101]. As a result the Martin relations for $\pi^0\pi^0$ elastic scattering [102] which are derived from positivity are barely satisfied by the Veneziano amplitude. The $K\bar{K}$ amplitude has however nearby ghosts. [92]

At present this ghost question raises a very important problem. It may be solved only by a dual theory which would also cope with the constraints of unitarity.

C. Scattering Amplitudes from Dual Models

Once a dual model is written in order to implement a bootstrap, we may try to write from it a scattering amplitude, satisfying unitarity and retaining as many of the properties as we wish to stress (resonance position, width...). If the incompatibility of the Veneziano model with unitarity is looked at as a disease, all prescriptions so far proposed look like aspirin: they bring the fever down but do not remove the cause of the pain. The simplest way out is to introduce an imaginary part in the trajectory function, writing

$$\alpha(t) = \alpha(0) + \alpha' t + \frac{t^2}{\pi} \int_{t_0}^{\infty} \frac{\operatorname{Im} \alpha(t')}{t'^2 (t' - t - i\epsilon)} dt' . \qquad (6\text{-}28)$$

The amplitude has now cuts in s and t as required by unitarity and no poles on the real axis where it now has asymptotic Regge behaviour.

As an example a meson-meson amplitude even or odd under $s-u$ crossing would read:

$$A^{\pm} = -\lambda^{\pm} \left(V^{xy}(t,s) \pm V^{xy}(t,u) \right) =$$

$$= -\lambda^{\pm} \left(\frac{\Gamma(1-\alpha_x(t))\,\Gamma(1-\alpha_y(s))}{\Gamma(1-\alpha_x(t)-\alpha_y(s))} \pm \frac{\Gamma(1-\alpha_x(t))\,\Gamma(1-\alpha_y(u))}{\Gamma(1-\alpha_x(t)-\alpha_y(u))} \right) \qquad (6\text{-}29)$$

Using the relation

$$\Gamma(x)\,\Gamma(1-x) = \pi (\sin \pi x)^{-1}$$

we get

$$A^{\pm} = \frac{-\pi\lambda^{\pm}}{\Gamma(\alpha_x(t)) \sin \pi \alpha_x(t)} \left(\pm \frac{\Gamma(1-\alpha_y(u))}{\Gamma(1-\alpha_x(t)-\alpha_y(u))} + \frac{\sin \pi (\alpha_x(t)+\alpha_y(s))}{\sin \pi (\alpha_y(s))} \frac{\Gamma(\alpha_x(t)+\alpha_y(s))}{\Gamma(\alpha_y(s))} \right).$$

We can now take the asymptotic behaviour at large s and get

$$A^{\pm} \simeq -\frac{\pi\lambda^{\pm}}{\Gamma(\alpha_x(t))} \left(\frac{\cos \pi \alpha_x(t) \pm 1}{\sin \pi \alpha_x(t)} + \cot \pi \alpha_y(s) \right) (\alpha' s)^{\alpha_x(t)} . \qquad (6\text{-}30)$$

If $\text{Im}\, \alpha_y(s) \longrightarrow \infty$ with s, $\cot \pi\alpha_y(s) \rightarrow -i$ and we can write:

$$A^{\pm} \underset{s\to\infty}{\simeq} -\frac{\pi\lambda^{\pm}}{\Gamma(\alpha_x(t))}\left(\pm 1 + e^{-i\pi\alpha_x(t)}\right)(\alpha' s)^{\alpha_x(t)} . \qquad (6\text{-}31)$$

The amplitude has an asymptotic Regge behaviour. This further shows two important points. The scale of s is fixed to be α'^{-1}. The residue vanishes whenever $\alpha(t)$ vanishes or is equal to a negative integer. It must therefore choose nonsense if we extend this result through factorization to meson-baryon amplitudes. This we know is a general result of duality. We further see here that the zero is simple and corresponds therefore to the Gell-Mann mechanism.

Adding an imaginary part to α (in effect a fixed non-zero argument) we can obtain a physical amplitude which is compatible with unitarity and which has Regge behaviour. This is satisfactory as an interesting approximation at high energy. However if we now consider the residue of the $t(s)$ channel resonances they are still polynomials in $\alpha(s)$ $(\alpha(t))$ but no longer polynomials in $s(t)$ since (6-28) now replaces (6-1). Expanding the residues in terms of Legendre polynomials we get arbitrarily high powers and then arbitrarily high spruious spins. We have therefore traded a good feature for a bad one. Nevertheless we can say that (6-31), kept as a high energy approximation to the scattering amplitude written from the dual model, has important features (scale and vanishing residue) which readily result from the structure of the dual model.

At lower energies one may prefer to get a unitary amplitude with resonance widths closely related to those obtained from the residues of the dual model. It is then better to consider the Veneziano amplitude as the K- matrix [92] and to write the elastic partial wave amplitudes as

$$A_\ell(s) = \frac{V_\ell(s)}{1 - ik V_\ell(s)} \tag{6-32}$$

where $V_\ell(s)$ is the partial wave projection of $V(s,t)$ and k the center of mass momentum. Unitarity is now satisfied. The resonance widths are closely related with those obtained from the pole residue but crossing symmetry is lost [103]. Such amplitudes, or standard generalizations written for coupled channels [92] offer very interesting models to compare with experimental information on meson-meson scattering.

A prescription such as (6-32) still neglects diffraction scattering or Pomeranchon exchange which is excluded from the dual amplitude. If we need the amplitude at relatively high energy we have to add a diffractive contribution to V_ℓ and write an amplitude compatible with unitarity using for instance the Baker-Blankenbecler formalism [104]. The Pomeranchon is however the reflection of all inelastic channels and in meson-meson scattering we do not expect much of an effect below 1 GeV where most of the presently available information is confined. The Pomeranchon contribution will become significant and eventually the leading one at high energy but at relatively low energies it can be neglected. As a result, (6-32) is a good approximation to the meson-meson amplitude at low energy.

Information on $\pi\pi$ amplitudes comes from pion peripheral production in πN collisions, η and τ decay and K_{e4} decay. The type of amplitude obtained from duality has small scattering lengths and large effective ranges. It is much larger in the isospin 0 state (where there is a broad resonance at the mass of the ρ) than in the isospin 2 state (where the amplitude is real, notwithstanding unitarity constraints). These general features were already known from a combination of current algebra results and tentative experimental conclusions, even though not too widely admitted. An analysis based on the unitarized Veneziano model will not bring strikingly new results but reaches equally good fits without any parameter, since the scattering lengths are now properly related to the resonance widths. This has been reviewed by Lovelace [92] at Argonne and, as an example among many we give the mass distribution in K_{e4} decay obtained by Roberts and Wagner [105]. This is shown on Fig. 6-6.

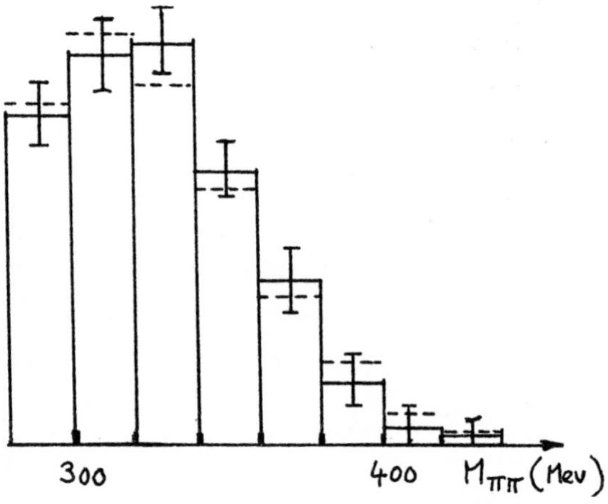

Figure 6-6. K_{e4} spectrum. Heavy line--Experimental; Dashed line--Veneziano.

This builds our confidence in looking at $\pi\pi$ scattering in terms of a small scattering length and a large effective range in the isospin 0 state instead of taking the opposite attitude prompted by an N/D bootstrap with smooth S-wave amplitudes. [106]

The Veneziano amplitude is amenable to an off mass shell

"extension" taking s, t and $u = 3m_\pi^2 + q^2 - s - t$ as independent variables. This is supported by the possible identification of the linear approximation of the Veneziano amplitudes with a linear model valid off mass shell and constrained by current algebra. This might be legitimized on the grounds that the exchange Regge trajectories which determine the basic structure of the model do not depend on the external masses. However, if an amplitude such as (6-2) is found suitable on the mass shell and slightly off the mass shell $(m_\pi^2 = 0)$ there is no reason to prevent satellites from being important far off mass shell. This is in particular the case for

$$\bar{p}n \to \pi^+\pi^-\pi^- \quad \text{, at rest,}$$

where the initial state is known to be the singlet S- state and can be considered as an "off mass shell pion" [84]. The decay amplitude now extends over the "double spectral function" region of $\pi\pi$ scattering with $\alpha(s)$ and $\alpha(t) > 0$. As obvious from Fig. 6-1 the Veneziano amplitude, once smoothed out, will have increasing maxima (polynomials in t of higher and higher degree) when moving to the large s (and t) region. They will however be separated by marked dips which give a very particular pattern on the Dalitz plot. This succession of dips and maxima [84] is indeed a specific prediction of the dual model and it agrees with experiment. The analysis requires however satellite terms [107]. One may hope that satellites should still be not too important when extrapolating from the pion mass to the K mass (or the η mass). The decay amplitudes associated with the graphs drawn on Fig. 6-7 have the desired zero imposed by current algebra (and not the undesired ones [108]) when the pertinent off mass shell $\pi\pi$

amplitudes are approximated by (6-18).

An important point when dealing with pion production in πN collisions is that the Veneziano amplitude gives S, P and D wave contributions which can be used all the way to the resonance mass and separately continued off mass shell. If nothing peculiar happens for the P (and D wave), the S waves have zeros at low t values which move with the $\pi\pi$ energy. The isospin 0 amplitude with one off-shell pion with mass q^2 vanishes when

$$q^2 = 2(s - m_\pi^2) .$$

This is an important consequence and is observed [109].

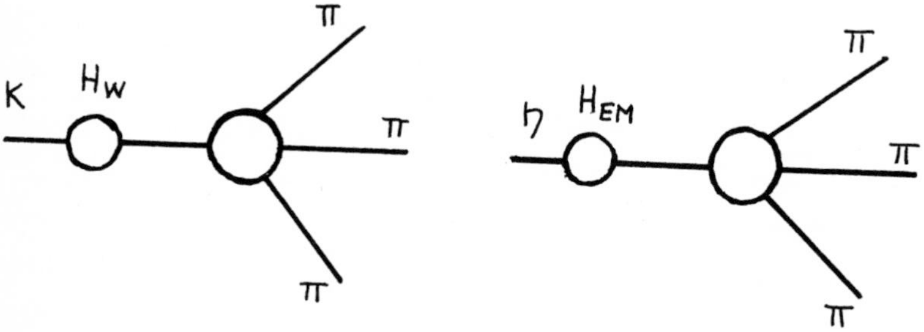

Figure 6-7.

This results however more generally from current algebra. We see that the dual model is useful in providing an amplitude up to the largest energy where experimental information can be used. Dual models for mesons are good enough to answer all phenomenological questions which can be put at present. This is however expected from the basic features they have (prominent resonances at the right place, reasonable width, Adler zeros. . .) ; the impressive aspect is that they can summarize all these properties in a very simple form.[110] We look at the dual model however as a tentative solution to the bootstrap problem and it is therefore of extreme importance to see how far the picture goes before being blurred by our neglect of unitarity as a prominent condition for a bootstrap. The existence of the daughter states, and in particular of the ρ', is of great importance. The ρ' should be a broad resonance and most easily observed in the 4π system. It could be displaced by a width or so from its predicted mass of 1250 Mev through unitarity. Experimental information which may a priori best come from colliding beam experiments [111], photoproduction [88] and $\bar{p}n$ annihilation in flight is still missing or controversial. It would also be very interesting to have information on the $\eta\pi$ channel where dual models do not comply so readily with the basic properties one wishes to have satisfied [97].

D. Dual Models for Other Reactions

The success of the dual model for meson-meson scattering has prompted many extensions to other types of reactions. This cannot be separated from the construction of dual models for production amplitudes where pole residues give scattering amplitudes for particle with spin. Nevertheless the dual models so far explored apply to very specific trajectories and useful insight can be gained trying dual terms for amplitudes free of kinematical singularities, defined for reactions with particles with spin. The situation is far from clear at present and we can merely mention a few points.

The next simplest case is pseudoscalar-vector scattering. There are four invariant amplitudes which have different asymptotic behaviours at large s (and u) but the same asymptotic behaviour at large t. [112] Dual terms with such properties can be obtained through multiplication of a common dual expression by suitable polynomials in s (u) or t. This procedure can be considered as an introduction of satellites, factoring out terms such as $N-\alpha(s)$, $M-\alpha(t)$, or $R-\alpha(s)-\alpha(t)$ in the polynomials. These satellites are also needed for the elimination of parity doublets which would otherwise generally occur [113]. The association of a definite parity to the exchanged mesons fixes the ratios of some invariant amplitudes and introduces relations which indeed require satellites in order to be obeyed. Tentative expressions can nevertheless be written [114] and are interesting attempts at relating asymptotic behaviours to resonance widths. It is also possible to obtain such an amplitude by taking the double pole residue of a dual amplitude with six external pions. Amplitudes with trajectories such as the ρ (f^0) or ω (A_2) with positive intercept have however still to be guessed and tested [114]. Interesting problems seem to come up in connection with the degeneracy of the leading trajectory [115] required by factorization. Parity doubling and ghosts at the daughter levels still present paramount problems.

The meson-baryon case is very interesting since many known results can be correlated. A dual model with linear trajectories cannot avoid the parity doubling of the baryon states imposed by the McDowell symmetry [116]. Such parity doublets are certainly absent at low energies. They may also be absent in any case [117] but a model with no Regge cuts is unable to dispose of them. In any case a dual model tried for the invariant amplitudes A' and B which respectively behave as $s^{\alpha(t)}$

and $s^{\alpha(t)-1}$ at large s must eliminate all spurious parity partners of the lower mass resonances before the parameters are fixed according to their decay widths. This is always possible for the first few resonances [118, 119] and the model has then predictive values at determining other resonance widths and asymptotic behaviours which are fully determined by the resonance parameters. This should of course be the case with duality, and the dual models thus constructed provide very interesting examples of "duality at work". Some smoothing procedure now common to low energy and high energy has to be applied. This can be done using the dual amplitude, together with a diffractive background, as the input of an eikonal approximation [119]. The resulting fits are not better than previous Regge fits but they involve fewer parameters and exhibit the duality connection between prominent low energy resonances and leading high energy Regge poles once the Pomeranchon contribution has been subtracted out. They are compatible only with the Gell-Mann nonsense choosing mechanism. Absorption corrections if used to describe Regge cut effects displace however the zero of the non-flip amplitudes toward the forward direction. Veneziano formula are therefore useful at parametrizing the amplitude but one cannot say that conversely such fits bespeak the power of the Veneziano formula when baryons are involved.

Dual amplitudes for currents have also been written [120]. The subject appears too controversial to be reported here. Nice features have been obtained but many problems are still open.

7. DUAL MODELS FOR PRODUCTION PROCESSES

A. Dual Channels

One of the important reasons for the very great interest raised by dual models is their relatively simple generalization to production processes. One obtains a framework in which two body collisions and many particle reactions can be analyzed together, excepting at present diffraction scattering. To do this has been a long standing challenge in hadron physics. Production processes are considered in a model world of zero width resonances located on linear infinitely rising Regge trajectories. The models obtained are tentative solutions of the bootstrap problem in which resonances in all the different channels sustain themselves in a self consistent way. We shall have to limit ourselves to the case of multi-meson reactions where the external particles are spinless and where exchange degenerate parent trajectories have negative intercepts

$$\alpha(s) = a + \alpha' s \quad \text{with} \quad a = \alpha(0) < 0.$$

Generalizations to more practical cases are still at an exploratory stage and are rich in challenging problems.

The construction of dual amplitudes for production processes follows a well defined procedure which may well be non-unique. It generalizes the basic steps itemized when discussing dual amplitudes for meson-meson scattering and may lead to too restricted a solution.

In particular dual expressions B_N are written for each planar graph, drawn as we saw according to a specific ordering of the external particles. There are $\frac{1}{2}(N-1)!$ possible orderings for N external particles and each of them can be associated with a planar and connected duality diagram. This is illustrated on Fig. 7-1. The graph represents one (out of 60) contributions to the production amplitude and, whatever way one chooses to partition it, one finds a $q\bar{q}$ pair which is readily associated with an infinite set of resonances with many different spins but the same internal quantum numbers which by construction exclude exotics.

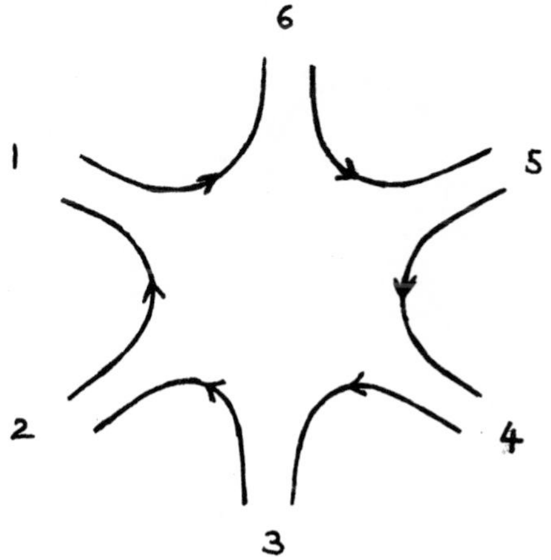

Figure 7-1. Duality diagram for a production amplitude.

The resonances which are thus selected are not all the possible ones which one wishes to include in the production amplitude but only those present in sets of adjacent channels. They are associated with partitions P_{ij} which select all external particles numbered from i to j, $i \leq k \leq j$. One isolates this way a particular Mandelstam variable

$$s_{ij} = \left(\sum_{k=i}^{j} p_k\right)^2 \qquad (7\text{-}1)$$

where p_k is the momentum of the external particle k and we of course require that the term chosen by our specific ordering should have poles whenever $\alpha_{ij}(s_{ij})$ is equal to zero or an integer. α_{ij} is the exchanged trajectory with the internal quantum numbers of the channel selected by the partition P_{ij}. This is illustrated on Fig. 7-2.

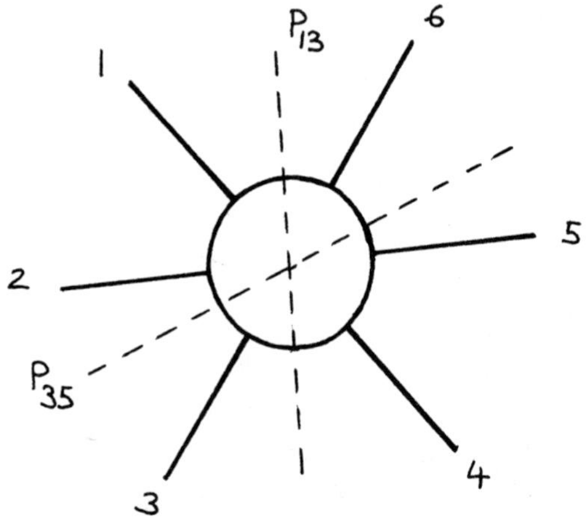

Figure 7-2. Partitions of a dual term with well-defined ordering.

It should also have all the poles associated with all the channels which can be selected by all such partitions. There are altogether $\frac{1}{2}N(N-3)$ possible partitions; that is, 2 for 4 external particles and already 5 for 5 external particles. These partitions are most easily displayed on the dual diagram of Fig. 7-3. The external momenta p_i are now drawn as an N-sided polygon (Fig. 7-3) and the partition P_{ij} is represented by the diagonal line joining the end points of the two extreme vectors p_i and p_j. A partition has to isolate at least two external particles and there are therefore $N-3$ partitions which start with each external particle p_i.

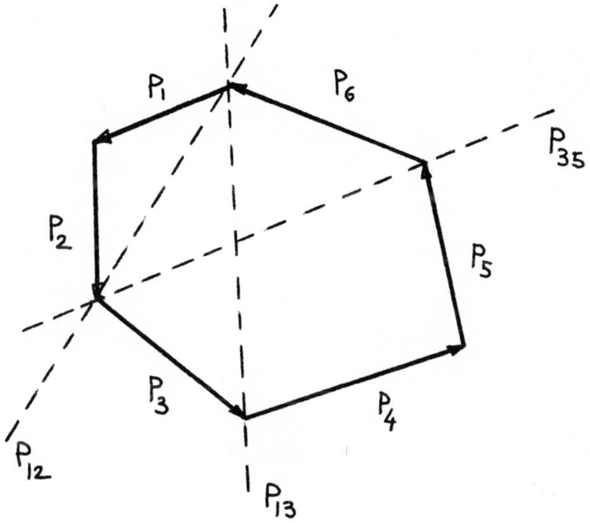

Figure 7-3. Momentum polygon used for partition specification.

The beta function used for the construction of the two body amplitude can be written as

$$B_4 = \int_0^1 du_{12}\, du_{23}\; u_{12}^{-\alpha(s_{12})-1}\; u_{23}^{-\alpha(s_{23})-1}\; \delta(u_{12}+u_{23}-1) \tag{7-2}$$

to show explicitly that the poles in s_{12} and s_{23} are associated with the end point singularity at $u_{ij}=0$ obtained when $\alpha(s_{ij}) \geqslant 0$.

The required singularities of B_N can be obtained by introducing as many variables as there are different partitions P and integrating down to zero all the u_P raised to the power $-\alpha_P(s_P)-1$, where α_P and s_P respectively are the Regge trajectory with the internal quantum numbers isolated by the partition P, and the Mandelstam variable which it defines. This reads:

$$B_N = \int_0^1 \cdots \int_0^1 \left(\prod_P u_P^{-\alpha_P(s_P)-1}\, du_P \right) R \tag{7-3}$$

R has however a complicated structure involving δ-functions (see 7-2). R should indeed be such as to prevent all u_P from vanishing at the same time. Even though we want to obtain all poles associated with all possible partitions we do not want them to be able to coincide as would be the case if all u_P could be zero simultaneously. Poles should be allowed to coincide only when they can all be associated with the same Feynman diagram. Poles associated with two partitions P_A and P_B can obviously be allowed to coincide when all the external particles isolated by the partition P_B are already among the groups chosen by the partition P_A, or vice versa. They cannot coincide however when the two partitions have particles in common but also particles which are not in common. This is illustrated on Fig. 7-4. Partitions such as P_{13} and P_{23} or P_{12} and P_{13} give allowed coincident poles whereas partitions such as P_{12} and P_{23} or P_{12} and P_{24} lead to unallowed coincident poles; they are instead dual to one another.

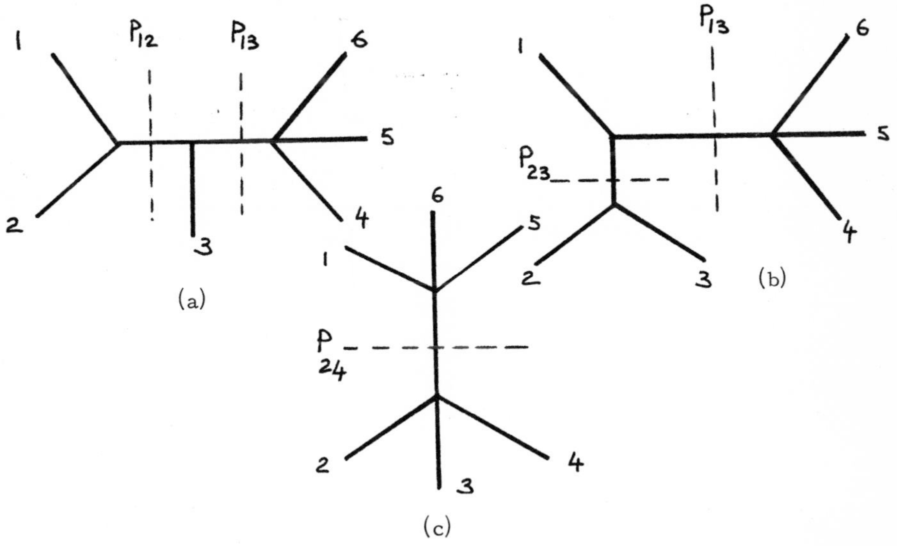

Figure 7-4.

On the diagram of Fig. 7-3, dual partitions correspond to intersecting diagonals. The maximum number of non-intersecting diagonals is $N-3$ and they specify the maximal tree diagram associated with the selected ordering.[121]

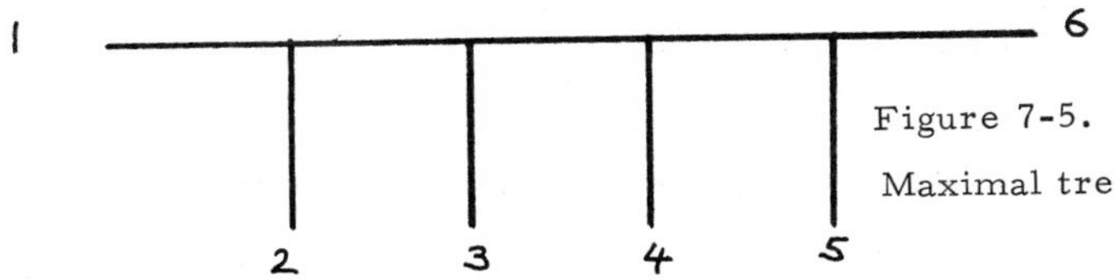

Figure 7-5.
Maximal tree diagram.

The diagram can be drawn as a duality diagram and all the specific pole terms represented on Fig. 7-4 can then be attained through deformation of the tree diagram keeping the same topological structure defined by the ordering (Fig. 7-6) .

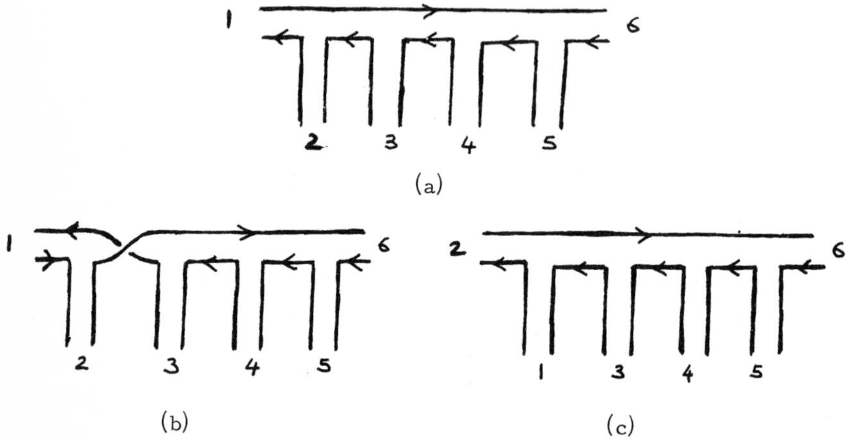

Figure 7-6.

If we twist two quark lines as on Fig. 7-6b we can deform the diagram into an untwisted one but the ordering is now changed (7-6c) . The twisted diagram is also present in the amplitude but among the $\frac{1}{2}(N-1)! - 1$ other B_N terms associated with different orderings. The ordering from the twist 12, 3456 is now inverted 12, 6543 . The new term (c) does not have poles in the channel s_{23} since there is no partition to isolate it. It has however poles in the channel s_{26} which were not present in the original term (a) . One may therefore regroup terms so as to have twist invariant dual terms. This selects leading trajectories with positive signature.

When calculating B_N one should therefore integrate only $N-3$ independent variables so as to generate a maximum of $N-3$ coincident poles

$$B_N = \int_0^1 \cdots \int_0^1 \prod_p u_p^{-\alpha(s_p)-1} \, dV^{(N-3)} \quad . \qquad (7\text{-}4)$$

We exclude through such a procedure coincident poles in all channels connected through duality transformations, where an infinite set of resonances exchanged in one channel can be also considered as an infinite set of resonances exchanged in another channel. This is readily seen by deforming duality diagrams as shown on Fig. 7-7.

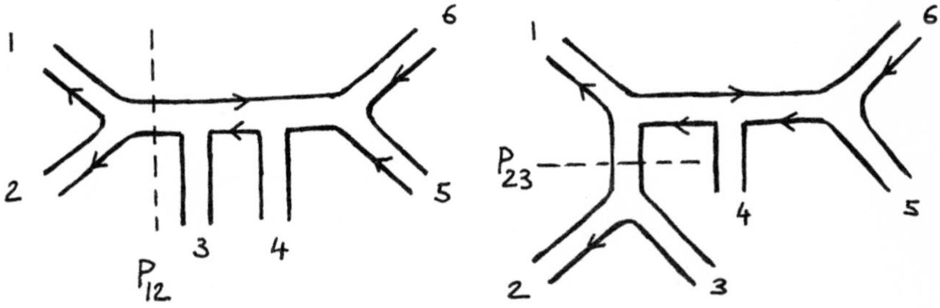

Figure 7-7. Two dual channels connected by a duality transformation.

The complete expression with $\frac{1}{2}(N-1)!$ terms such as (7-4) will then have the complete singularity structure with the correct duality relation among poles in different channels. Each B_N term being constructed to be cyclically symmetric, the sum of all B_N's will be crossing symmetric. We do not, however, have a realistic production amplitude. We have a bootstrap solution for a dual model world of zero width resonances located on linear Regge trajectories. This is furthermore the outcome of a specific construction procedure which could well be non-unique.

B. The Bardakci Ruegg Formula

The generalization of the B_4 amplitude (6-7), with identical particles of mass m and a single exchange degenerate trajectory with negative intercept on which the external particle is located is the following

$$B_N = \prod_{i=2}^{N-2} \int_0^1 du_i \, u_i^{-\alpha(s_i)-1} (1-u_i)^{a-1} \times$$

$$\times \prod_{j \geq i \geq 2}^{N-1} (1 - x_{i,j})^{-2\alpha'(p_i \cdot p_j)} \quad . \tag{7-5}$$

The trajectory function is written as

$$\alpha(s_{ij}) = a + \alpha' \left(\sum_{k=i}^{j} p_k \right)^2 \quad , \qquad a = -\alpha' m^2 .$$

The $N-3$ independent variables u_i which have been selected correspond to a maximal tree graph. u_i corresponds to the partition $1i$ and we should write it explicitly as u_{1i} to keep it on the same footing as the other variables which are not associated with the selected tree graph. We have defined

$$x_{i,j} = u_i u_{i+1} \cdots u_{j-1} \qquad \text{with} \quad x_{i,i} = 0 \quad . \tag{7-6}$$

This explicitly gives the volume element of (7-4).

In order to see how (7-5) is obtained we start with the

B_5 generalization of B_4 [122,123]. There are five different channels once an ordering has been defined and the expression should contain all the associated variables raised to the proper power: $u_{ij}^{-\alpha(s_{ij})-1}$ However there are only two independent variables which are selected as u_{12} and u_{13} (maximal tree graph) and the three others should be expressed in terms of them. A sufficient condition is to write

$$u_P = 1 - \prod_{\bar{P}} u_{\bar{P}} \quad , \tag{7-7}$$

where $\bar{P}$ extends over all partitions dual to the one referred to by P. This obviously forbids dual variables to get to zero at the same time. This gives 5 relations instead of 3 but they are not all independent

$$u_{23} = 1 - u_{34}u_{12} \qquad u_{24} = 1 - u_{12}u_{13}$$

$$u_{13} = 1 - u_{24}u_{34}$$

$$u_{34} \doteq 1 - u_{13}u_{23} \qquad u_{12} = 1 - u_{23}u_{24} \tag{7-8}$$

This is solved as:

$$u_{23} = \frac{1-u_{12}}{1-u_{12}u_{13}} \qquad u_{34} = \frac{1-u_{13}}{1-u_{12}u_{13}} \qquad u_{24} = 1 - u_{12}u_{13} \tag{7-9}$$

B_5 meets all the desired requirements written as

$$B_5 = \int_0^1 \cdots \int_0^1 du_{12}\,du_{13}\,du_{23}\,du_{24}\,du_{34} \frac{u_{12}^{-\alpha(s_{12})}\,u_{13}^{-\alpha(s_{13})}\,u_{23}^{-\alpha(s_{23})}\,u_{24}^{-\alpha(s_{24})}\,u_{34}^{-\alpha(s_{34})}}{u_{12}\,u_{13}\,u_{23}\,u_{24}\,u_{34}} \times$$

$$\times\,\delta(u_{23} + u_{12}u_{34} - 1)\;\delta(u_{24} + u_{12}u_{13} - 1)\;\delta(u_{34} + u_{13}u_{23} - 1)\,. \tag{7-10}$$

Integrating over u_{23}, u_{24}, and u_{34} using the δ - functions gives

$$B_5 = \int_0^1 du_{12}\, du_{13}\, u_{12}^{-\alpha(s_{12})-1}\, u_{13}^{-\alpha(s_{13})-1}\, (1-u_{12})^{-\alpha(s_{23})-1} \times$$

$$\times\, (1-u_{13})^{-\alpha(s_{34})-1}\, (1-u_{12}u_{13})^{-\alpha(s_{24})+\alpha(s_{23})+\alpha(s_{34})}\,.$$

(7-11)

The exponent of the last factor reads

$$a + \alpha'\left(-(p_2+p_3+p_4)^2 + (p_2+p_3)^2 + (p_3+p_4)^2\right) =$$

$$= a + \alpha' m^2 - 2\alpha'(p_2\cdot p_4) = -2\alpha'(p_2\cdot p_4)\,.$$

We can write:

$$-\alpha(s_{23}) = -a - 2\alpha' m^2 - 2\alpha'(p_2\cdot p_3) = a - 2\alpha'(p_2\cdot p_3),$$

$$-\alpha(s_{34}) = a - 2\alpha'(p_3\cdot p_4)\,.$$

Used together with (7-10) this gives the Bardakci-Ruegg formula for B_5, (7-5),

$$B_5 = \prod_{i=2}^{3} \int_0^1 du_i\, u_i^{-\alpha(s_i)-1}\, (1-u_i)^{a-1} \prod_{j\geq i\geq 2}^{4} (1-x_{i,j})^{-2\alpha'(p_i\cdot p_j)},$$

which we can also write directly from (7-10) as originally given, keeping still the trajectories different.

$$B_5 = \int_0^1 u_{12}^{-\alpha(s_{12})-1}\, u_{13}^{-\alpha(s_{13})-1}\, dV^{(2)} \quad \text{with}$$

$$dV^{(2)} = \frac{du_{12}\, du_{13}}{1-u_{12}u_{13}}\, u_{23}^{-\alpha(s_{23})-1}\, u_{24}^{-\alpha(s_{24})-1}\, u_{34}^{-\alpha(s_{34})-1}\,.$$

(7-12)

The integral is defined when all α's are negative. In order to exhibit the poles associated with the end point singularities, one expands the integrand of (7-11) as a double Taylor series in u_{12} and u_{13} and integrates term by term, keeping $\alpha(s_{12})$ and $\alpha(s_{13})$ negative. This gives

$$\sum_{l,m,n=0}^{\infty} \frac{(-1)^{l+m}}{l!\,m!\,n!} \cdot \frac{1}{\big(l+n-\alpha(s_{12})\big)\big(m+n-\alpha(s_{13})\big)} \times$$

$$\times\Big((-\alpha(s_{23})-1)\cdots(-\alpha(s_{23})-l)\Big)\Big((-\alpha(s_{34})-1)\cdots(-\alpha(s_{34})-m)\Big)\times$$

$$\times\Big(\big(\alpha(s_{23})+\alpha(s_{34})-\alpha(s_{24})\big)\cdots\big(\alpha(s_{23})+\alpha(s_{34})-\alpha(s_{24})-n+1\big)\Big). \tag{7-13}$$

Continued to positive values of $\alpha(s_{12})$ and $\alpha(s_{13})$ it displays the double pole structure. The residues are polynomials in the dual variables. If $\alpha(s_{24})$ is taken k times in the last product $\alpha(s_{23})$ and $\alpha(s_{34})$ are both taken $n-k$ times. The residue at the double pole term corresponding to $\alpha(s_{12}) = J$ and $\alpha(s_{13}) = J'$ is then proportional to

$$\sum_{k=0}^{\min(J,J')} (s_{24})^{k}\,(s_{23})^{J-k}\,(s_{34})^{J'-k}\,, \tag{7-14}$$

as required since the channels 24 and 23 (34) are dual to 12 (13).

The amplitude is cyclic symmetric. This is obvious in (7-10) but not obvious in (7-12). When combined with the other eleven amplitudes it yields a crossing symmetric result. The generalization to an arbitrary number of external particles has been worked out by Chan and Tsou , Goebel and Sakita and Koba and Nielsen [122]. It corresponds to (7-10) written for all the $\frac{1}{2}N(N-3)$ channels selected by one of the $\frac{1}{2}(N-1)!$ orderings

$$B_N = \int_0^1 \cdots \int_0^1 \prod_P u_P^{-\alpha(s_P)-1} du_P \prod_{P'} \delta(u_{P'} - 1 - \prod_{\bar{P}'} u_{\bar{P}'}) ,$$

where among all the partitions P only $N-3$ of them are associated with the maximal tree graph. The partitions P' are all the others and the partitions $\bar{P}'$ are all those dual to P'. A partition is defined by the two extreme particle lines which it selects, i and j. u_P is the same as u_{ij}.

The solution of all the non-independent linear relations imposed by the δ- functions is then found to be

$$u_{i,j} = \frac{(1-x_{i,j})(1-x_{i-1,j+1})}{(1-x_{i-1,j})(1-x_{i,j+1})} , \qquad (7\text{-}15)$$

where the $x_{i,j}$ are defined by (7-6) with $u_{11} = u_{1\,N-1} = 0$ by definition.

Integration over all the u variables using the δ function can then be done, after a simple regrouping of terms :

$$B_N = \prod_{i=1}^{N-2} \int_0^1 du_i \, u_i^{-\alpha(s_{ii})-1} (1-u_i)^{-\alpha(s_{i\,i+1})-1} \times$$

$$\times \prod_{i=2}^{N-3} \prod_{j=i+2}^{N-1} (1-x_{i,j})^{-\gamma_{ij}} \qquad (7\text{-}16)$$

with

$$\gamma_{ij} = \alpha(s_{ij}) + \alpha(s_{i+1,\,j-1}) - \alpha(s_{i,j-1}) - \alpha(s_{i+1,j})$$

written of course only for $j \geq i+2$.

This does not yet assume that all intercepts are equal. If

this is the case

$$\gamma_{ij} = \alpha'\left((p_i + p_{i+1} + \cdots + p_j)^2 + (p_{i+1} + \cdots + p_{j-1})^2 + \right.$$

$$\left. - (p_i + \cdots + p_{j-1})^2 - (p_{i+1} + \cdots + p_j)^2 \right) = 2\alpha' (p_i \cdot p_j)$$

and

$$\alpha(s_{i,i+1}) = -a + 2\alpha'(p_i \cdot p_{i+1})$$

we can then rewrite the expression arrived at as

$$B_N = \prod_{i=2}^{N-2} \int_0^1 du_i \, u_i^{-\alpha(s_i)-1} (1-u_i)^{a-1} \prod_{j \geq i \geq 2}^{N-1} (1 - x_{i,j})^{-2\alpha'(p_i \cdot p_j)}$$

with s_i written for s_{1i} .

This is the Bardakci Ruegg formula already quoted. Its pole structure can be obtained as we did for B_5 and the residue possesses the needed properties.

Selecting $N-3$ variables connected to the maximal tree graph, integration over all other variables can then be performed using the duality relations (7-15) . The cyclic symmetry of the initial construction is however not apparent any more. Koba and Nielsen write the u_{ij} as anharmonic ratios of four complex numbers of modulus unity, thereby imposing automatically the duality relation [123]. To see that one uses the invariance of the anharmonic ratio under homographic transformations. This leads to an explicitly symmetric expression which is extremely useful in the formal analysis of the model and which is described in detail by Mandelstam [1]. We shall follow here the standard choice of variables better suited to practical applications.

If the intercepts were not equal we would have

$$\gamma_{ij} = 2\alpha'(p_i \cdot p_j) + \epsilon_{ij}$$

with

$$\epsilon_{ij} = a_{ij} + a_{i+1,j-1} - a_{i,j-1} - a_{i+1,j}$$

Imposing factorization, this may be formally written as [124]

$$\epsilon_{ij} = 2\alpha' \sum_{m=1}^{M} q_i^m \cdot q_j^m \qquad (7\text{-}17)$$

where the q's would be extra momentum components formally attached to the external particles and M does not depend on the number of external lines. One can then keep (7-5), with extra (and generally non-conserved) components. We would for instance use (7-17) with $M=1$ for trajectory intercepts depending quadratically on the total number of quarks and antiquarks [125] and have five-component external momenta.

The B_N amplitude has Regge behaviour, indeed it has multi-Regge behaviour. Multi-Regge behaviour involves for instance all s_{ij} with $i,j = 2, \cdots, N-1$ going to infinity with all s_{ii} $(i = 2, \cdots, N-2)$ kept fixed as well as the ratios $\kappa_i = -\frac{(s_{i,i+1})(s_{i+1,i+2})}{s_{i,i+2}}$ [126]. Such a configuration may be easily represented by a multiperipheral graph as shown in Fig. 7-8. Regge behaviour holds except when the variables are real positive.

In order to show how Regge behaviour results, we sketch here the main steps of the proof of Bardakci and Ruegg [127]. The typical factors of B_N read $(1 - u_i u_{i+1} \cdots u_{j-1})^{-2\alpha' p_i \cdot p_j}$.

With the change of variables $1-u_i = \exp \frac{z_i}{\alpha(s_{i,i+1})}$ made for $i = 2, \cdots, N-2$ when all α's are negative, the asymptotic behaviour as $|\alpha| \to \infty$ is

$$\left(1 + (-1)^{j-i} \frac{z_i \cdots z_{j-1}}{\alpha(s_{i,i+1}) \cdots \alpha(s_{j-1,j})}\right)^{-\alpha(s_{ij})}$$

keeping only the lowest order term in α^{-1}.

One then chooses the multi-Regge configuration of Fig. 7-8 with $s_{12}, \cdots, s_{1,N-2}$ kept fixed and requires that

$$s_{i,i+1}\, s_{i+1,i+2} \cdots s_{j-1,j} = (-K_i) \cdots (-K_{j-2})\, s_{ij}$$

with all K_i fixed. In this limit the selected factor behaves as

$$\exp\left(- \frac{z_i \cdots z_{j-1}}{K_i \cdots K_{j-2}}\right)$$

One then finds

$$B_N \sim \prod_{i=2}^{N-2} \left(-\alpha(s_{i,i+1})\right)^{\alpha(s_{1,i})} G_N ,$$

$$G_N = \prod_{i=2}^{N-2} \int_0^\infty dz_i\, (z_i)^{-\alpha(s_{1i})-1} \exp\left(-z_i - \sum_{j=i+1}^{N-1} \frac{z_i \cdots z_{j-1}}{K_i \cdots K_{j-2}}\right). \tag{7-18}$$

The integral is defined when all α's are negative, and has multi-Regge behaviour. The proof goes through for complex α's but not for real positive values of the α's for which B_N is by construction singular.

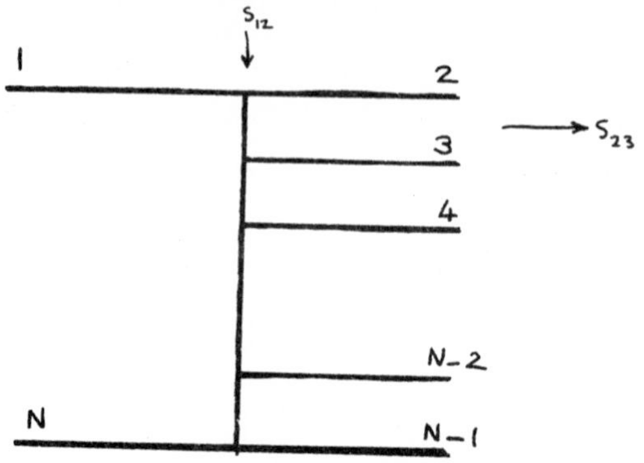

Figure 7-8. Multi-Regge diagram.

One can remove from G_N $\Gamma(-\alpha(s_{ii}))$ factors, (which one may call Reggeon propagators) from a remainder which remains finite for integer or zero values of $\alpha(s_{i1})$. The remainder can be written as the product of Reggeon-Reggeon particle vertices, thus exhibiting a very interesting factorization property [127]. It implies "long range ordering" defined as explicit relations between the successive Treiman Yang angles. This goes against simplifying assumptions so far used in multi-Regge dynamics [128].

Summarizing, the B_N functions have very interesting features which generalize those of B_4 to production processes. They are not realistic production amplitudes but bootstrap solutions in a model world of zero width resonances. They nevertheless summarize many properties which one may then introduce in phenomenological production models.

As bootstrap solutions their pole residues can be checked to factorize. The factors can be expressed in terms of B_M with $M \leq N-1$ and generalizations with satellites. It should be stressed once again that the conditions which have been imposed are sufficient . It is yet unknown if they are necessary. The procedure followed gives no hint toward its generalization to particles with spin or unnatural parities. The case of integer spin particles with natural parity is however now included. The pertinent amplitudes can be obtained as pole residues of B_N amplitudes.

An ambiguity remains with respect to satellite terms. As discussed by Gross [129] they can be introduced if one includes a regular and cyclically symmetric function of the u's in the integrand. The same singularity pattern obviously remains but with different residues. As was the case for B_4 any modification can thus be obtained at the daughter levels. However, an important point is now the inescapable increase of the degeneracy of these levels beyond the exponential behaviour obtained by Fubini and Veneziano [100] which will be discussed in the next section. There is therefore some rationale to prefer B_N to a similarily constructed expression with satellites.

C. B_5 Phenomenology

Even though B_N is not a production amplitude, the connection which it provides between resonance widths and multi-Regge behaviour is a fundamental property which one desires to keep in a phenomenological model for production amplitudes. If one is satisfied with leading Regge behaviour and leading direct channel resonances and is ready to disregard the detailed information about daughters also included in the dual model, the introduction of branch points in the trajectory function with imaginary parts matched to the averaged resonance widths will provide reasonably satisfactory phenomenological models.

For abvious computational reasons applications have been so far limited to B_5 .

Any actual reaction involves however different types of particles, in particular baryons, and it furthermore requires exchanged trajectories with positive intercepts.

The term explicitly written for B_5 in (7-12) does not specify yet that all trajectories are the same and so it can therefore be used with different intercepts. We have twelve different terms to consider but several will have an exotic channel once quantum numbers are attached to the external particles and so should be dropped. For the remaining terms, the leading trajectory which can be included in each channel of (7-12) is selected. If B_5 is not suitable to describe the whole production amplitude where the first particle on the exchanged trajectories are P- wave resonances it might be suitable to describe one of the invariant amplitudes. This is the case if one neglects the baryon spins altogether and writes the meson production amplitude

$$M(q) + B(p) \rightarrow M'(q') + M''(q'') + B' \qquad (7\text{-}19)$$

as:

$$\epsilon_{\mu\nu\sigma\rho}\, q^{\mu} q'^{\nu} q''^{\sigma} p^{\rho}\, B_5 , \qquad (7\text{-}20)$$

where M, M', and M'' stand for pseudoscalar mesons with momenta q, q', and q'' and B and B' for baryons with momenta p and $p+q-q'-q''$. The poles of (7-20) will have the residues which are expected for particles located on a trajectory with a P- wave resonance as the first particle.

B_5 uses as input exchange degenerate trajectories. As discussed previously, one should therefore not have a priori much confidence in the model for reactions dominated by pion exchange. In other words dual models give a singularity structure which may be associated with the imaginary part of the amplitude but resonance dominance does not apply to the real part, which has to be obtained from dispersion relations. A pion exchange amplitude which is almost real where it may be dominant is therefore not controlled by a dual model. It is also obvious that since diffraction scattering is not included in the dual model, one can only use dual models for reactions where Pomeranchon exchange is either absent or where diffraction effects can be separated in a reliable way. There are still many production reactions where the exchange of the leading trajectories plays a dominant role and for which dual models provide much insight. This is in particular the case for the set of reactions [130]

$$K^+ p \rightarrow K^0 \pi^+ p \,, \quad K^- p \rightarrow \bar{K}^0 \pi^- p \,, \quad \pi^- p \rightarrow K^0 K^- p \tag{7-21}$$

related through crossing symmetry and for the reaction

$$K^- p \rightarrow \pi^+ \pi^- \Lambda \tag{7-22}$$

In both cases many of the twelve terms which one can a priori write keeping only one trajectory in each channel have an exotic channel and can be eliminated. Such a term is represented on Fig. 7-9. Special properties associated with the lower mass resonances which one wishes the model to reproduce (for instance the absence of an isospin $\frac{3}{2}$ $\frac{5}{2}^-$ resonance which would naturally appear once the Δ trajectory is introduced in the πp channel) can then be used in order to relate the numerical coefficients which one puts in front of each of the remaining B_5 terms. As discussed by Chan et al for (7-21) and by Peterson and Tornqvist for (7-22) this is enough to fix the relative values of all of them and one

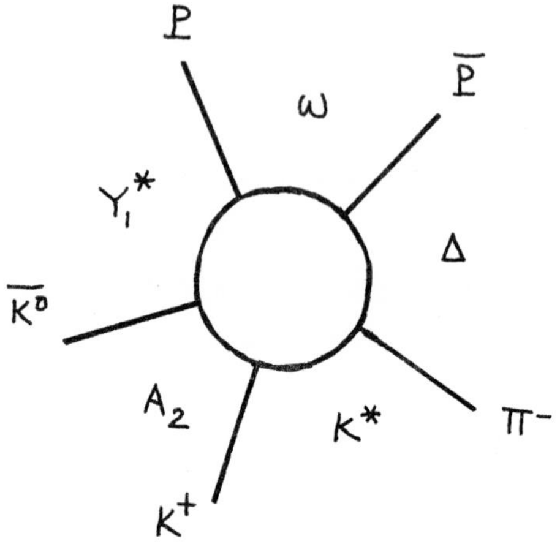

Figure 7-9. One of the twelve terms relevant to the reaction $K^+p \to K^0 \pi^+ p$. All particles are considered as ingoing. The leading trajectories selected in each channel are indicated.

arrives at an expression for the production amplitude (or production amplitudes in the case of (7-21)) which, although it involves many assumptions - which can of course be considered as known properties of Regge trajectories and prominent resonances - has only one parameter. As illustrated on Fig. (7-10) by but a sample of results taken from reference 130, it gives a good global description of the production process, which is thus described in a very satisfactory way from say 3 to 12 Gev. This is a remarkable success for the simple dual model considered. The energy behaviour in all channels is reproduced correctly. The mass distributions and the angular distributions associated with the resonances are also well reproduced.

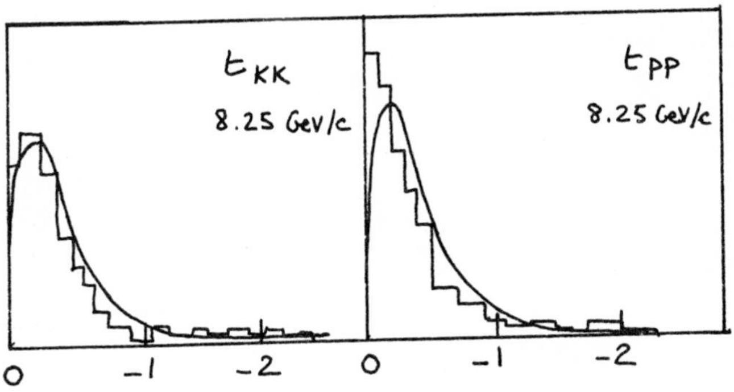

Figure 7-10 (a). Momentum transfer distributions.

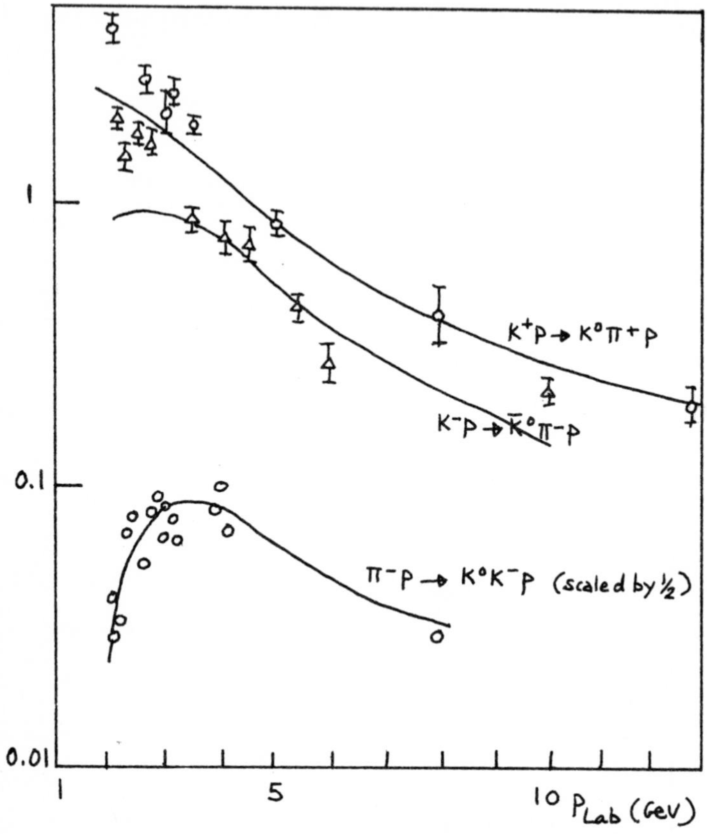

Figure 7-10 (b). Energy behavior.

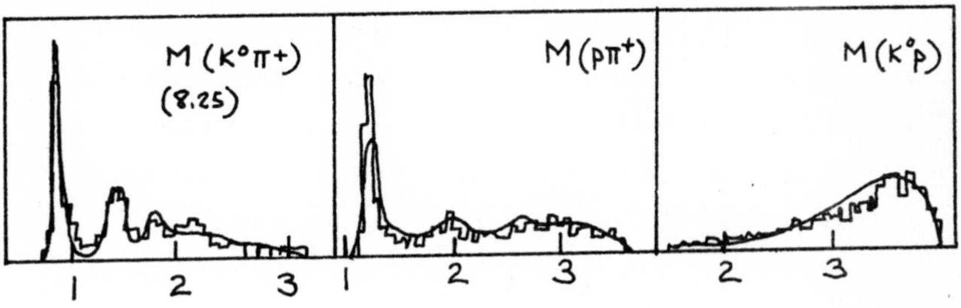

Figure 7-10 (c) . Mass distributions.

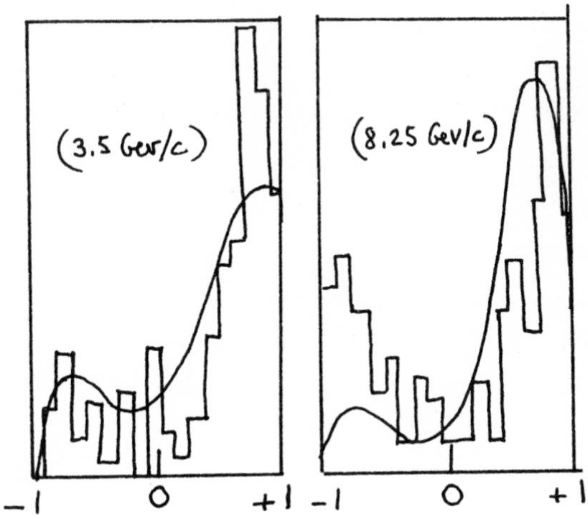

Figure 7-10 (d). Angular distributions $K^*(1420)$ decay.

It should be stressed though that these specific features were actually introduced in the first place by the selection of leading Regge trajectories and proper emphasis of some resonances. Nevertheless it is remarkable that such a simple expression as B_5 (or several B_5 terms but with a single scaling constant) can reproduce them all. This example exhibits duality dramatically; Regge behaviour extends down to energies as low as 2 GeV overlapping with a resonance description. It should be stressed that no detailed test of the dual model is obtained. Only the leading trajectory effects can be followed in a reliable way. Dual models as presently used are very good at handling a large amount of data with poor accuracy (10%). A more subtle treatment of unitarity and spin is necessary to cope with more accurate results. Such models answer long-standing questions associated with resonance overlaps and resonance hunting through mass structures. Cross sections for double resonance production (such as $\rho\Delta$) can be suggestive but have a limited applicability. In general one should give the strength of the dual contribution where both resonances appear in non dual channels. The separation of a resonance contribution from a "background" (ρ production) is interesting to try. This procedure is however limited to special cases. The dual models give us an example where the so-called "background" is really resonance contributions in other channels. It becomes therefore meaningless, at least in general, to speak of resonance peaks as opposed to Regge exchange background.

Beside their interest as summarizers of much information in a simple form, dual Regge models have a deeper theoretical meaning because of their crossing properties. Once the model has been specified according to a particular set of reactions it should also give all amplitudes related by crossing any number of external lines. The last two reactions of (7-21) are related by crossing of two meson lines. Indeed the model adjusted to the first two reproduced the basic features of the third one but for an overall scale factor of 2. This should however be compared to the ratio of the order of 20 experimentally observed between the π and K induced reactions in the energy range studied. This situation is to be associated with the failure of line reversal invariance already discussed in Section 3, and probably results from our neglect of cut contributions and lower trajectory contributions. The situation becomes much worse when one crosses a baryon line to reach the annihilation reaction

$$p\bar{p} \rightarrow K\bar{K}\pi \ . \qquad (7\text{-}23)$$

Here we probably encounter the result of our casual treatment of spin. A proper treatment of spin would involve kinematical terms multiplying dual invariant terms, and the kinematical terms are expected to change in different ways when going from one reaction to the other. A proper treatment of spin would also have a decisive impact on the angular distribution predicted for baryon resonances. Such a treatment is necessary if one wishes to go beyond the P- wave decay mode forced on the lowest spin states. [131]

Models have been considered [132] but no satisfactory answer has been yet proposed for pseudoscalar meson production in pseudoscalar meson-nucleon collisions. Much remains to be done.

Practical applications use a recurrence relation for B_N which has been written by Hopkinson and Plahte [133]. We merely quote it for B_5 written as

$$B_5 = \sum_{k=0}^{\infty} (-1)^k \binom{C_{24}}{k} B_4(y_{34}, y_{13}+k) B_4(y_{12}+k, y_{23})$$

$$\text{with} \quad y_{ij} = -\alpha(s_{ij})$$

$$\text{and} \quad C_{ij} = -\alpha(s_{ij}) + \alpha(s_{i,j-1}) + \alpha(s_{i+1,j}) . \tag{7-24}$$

We get an expansion useful for numerical calculations in terms of beta functions where the arguments correspond to the combination of two external lines into a single one. Writing explicitly B_4 in (7-24) we obtain

$$B_5 = \sum_{k=0}^{\infty} (-1)^k \binom{C_{24}}{k} \int_0^1 dx_{12}\, x_{12}^{-\alpha(s_{12})+k-1} (1-x_{12})^{-\alpha(s_{23})-1} \times$$

$$\times \int_0^1 dx_{13}\, x_{13}^{-\alpha(s_{13})+k-1} (1-x_{13})^{-\alpha(s_{34})-1}$$

with

$$\sum_{k=0}^{\infty} (-1)^k \binom{-\alpha(s_{24}) + \alpha(s_{23}) + \alpha(s_{34})}{k} (x_{12}x_{13})^k =$$

$$= (1 - x_{12}x_{13})^{-\alpha(s_{24}) + \alpha(s_{23}) + \alpha(s_{34})}$$

One gets back (7-11).

8. FROM DUAL MODELS TO A DUAL THEORY

A. The Operator Formalism

Despite many drawbacks dual Regge models are already powerful phenomenological tools. Further exploration of their properties will now lead to very interesting results. At present these results are most easily obtained using the operator formalism introduced by Fubini, Gordon and Veneziano [134, 135]. The new expression for the Bardakci-Ruegg formula which is thus obtained will furthermore make it easier to introduce higher order terms, added to dual models in such a way as to eventually meet unitarity constraints and constructed so as to keep duality relations at each stage of the calculation [7].

B_N, written as

$$B_N = \prod_{i=2}^{N-2} \int_0^1 du_i \, u_i^{-\alpha(s_{1i})-1} (1-u_i)^{a-1} \prod_{j \geq i \geq 2}^{N-1} (1 - x_{i,j})^{-2\alpha' p_i \cdot p_j} , \tag{7-5}$$

has equally spaced levels in all channels associated with a planar connected duality diagram and in particular in the $N-3$ channels selected by the associated maximal tree graph. Indeed this infinite set of equally spaced resonances contains all the other singularities obtained through duality transformations as illustrated in Fig. 7-7. This suggests the introduction of harmonic oscillators the level spacings of which are matched with all possible resonance spin spacings. Spin spacings and mass-squared spacings are of course connected by the unique value of α'. One therefore introduces an infinite set of harmonic oscillators with level spacings $1, 2, \cdots, n, \cdots$ and this for each of the four (space and time) coordinates. The creation (annihilation) operator associated with one oscillator is written as $a_n^{\mu +}$ (a_n^{μ}) where n specifies the level spacing and μ the direction $(\mu = 0, 1, 2, 3)$. The commutation relations are the following:

$$[a_m^{\mu}, a_n^{\nu +}] = - g_{\mu\nu} \delta_{mn} \tag{8-1}$$

The metric tensor $g_{\mu\nu}$ is written as

$$g_{00} = +1 \quad , \quad g_{ii} = -1 \ \text{for} \ i = 1, 2, 3 .$$

The states defined through the action of a_n^{i+} on the vacuum state have therefore positive energy and positive metric whereas the states defined through the action of a_n^{0+} on the vacuum have positive energy and negative metric. They are ghost states, which as we already mentioned, appear as an unavoidable feature of dual models. The energy operator is defined as

$$H = - \sum_{n=1}^{\infty} n\, a_n^{\mu +} a_n^{\mu} \quad , \tag{8-2}$$

where a summation over the Lorentz indices is implied. The vacuum state is defined such that

$$a_n^{\mu} |0\rangle = 0 \ ,$$

and B_N is finally written as the vacuum expectation value of a product of operators

$$B_N = \langle 0 | V(p_{N-1}) D(\alpha(s_{1,N-2})) \cdots V(p_3) D(\alpha(s_{12})) V(p_2) |0\rangle . \tag{8-3}$$

The different variables have been defined according to the maximal tree graph of Fig. 8-1 and the following notations have been introduced

$$V(p) = \exp\left(- \sum_{n=1}^{\infty} \sqrt{\frac{2\alpha'}{n}}\, p \cdot a_n^{+}\right) \exp\left(+ \sum_{n=1}^{\infty} \sqrt{\frac{2\alpha'}{n}}\, p \cdot a_n\right), \tag{8-4}$$

$$D(\alpha(s_{1i})) = \int_0^1 du_i\, u_i^{-\alpha(s_{1i}) - 1 + H} (1 - u_i)^{a-1} . \tag{8-5}$$

$V(p)$ is a "vertex function". Such an operator is associated with each external particle i with momentum p_i, excluding the two extreme ones on the tree $(i = 2, \cdots, N-1)$.

$D(\alpha(s_{1i}))$ is a "propagation function" associated with the set of states exchanged between the vertices defined by particles i and $i+1$. $p \cdot a$ is the scalar product of two Lorentz vectors.

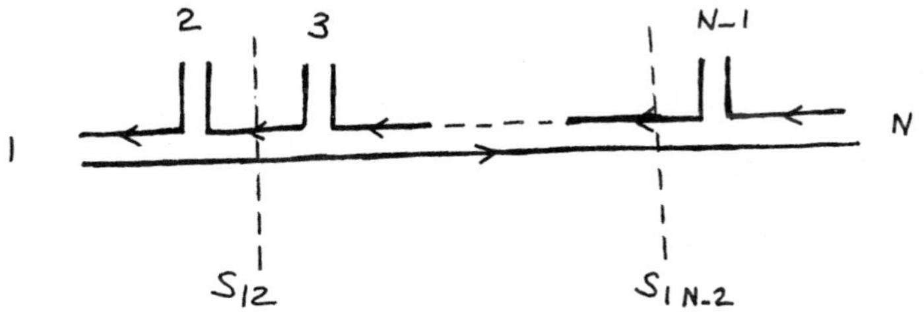

Figure 8-1. Maximal tree graph.

D has pole terms which can be associated with all the intermediate states exchanged between two vertices. They correspond to the states present on a family of exchange degenerate linear trajectories common to all channels. We take the expectation value with an eigenstate of H with eigenvalue $\mathcal{H} \geq 0$. We expand the last factor and integrate term by term keeping $\alpha(s) - \mathcal{H} < 0$. This gives

$$\sum_{n=0}^{\infty} \int_0^1 u^{-\alpha(s)-1+\mathcal{H}+n} (-1)^n \binom{a-1}{n} du =$$

$$= \sum_{n=0}^{\infty} (-1)^n \binom{a-1}{n} \frac{1}{\mathcal{H}+n-\alpha(s)} = \sum_{n=0}^{\infty} \binom{n-a}{n} \frac{1}{\mathcal{H}+n-\alpha(s)} . \tag{8-6}$$

The expansion can then be extended to positive values of $\mathcal{H}-\alpha(s)$, with poles for $\alpha(s) = 0, 1, \cdots$ with $\mathcal{H}$ taking the values $0, 1, \cdots$

We now prove the identity between (7-5) and (8-3). To this end we isolate the operator

$$V(p_{N-1})\, u_{N-2}^{H}\, V(p_{N-2}) \cdots u_2^{H}\, V(p_2) \qquad (8\text{-}7)$$

from a c-number which reads:

$$\prod_{i=2}^{N-2} \int_0^1 du_i\, u_i^{-\alpha(s_{1i})-1} (1-u_i)^{a-1} \qquad (8\text{-}8)$$

In order to get the Bardakci-Ruegg formula we have to show that the vacuum expectation value of (8-7) is equal to

$$\prod_{j \geq i \geq 2}^{N-1} (1 - x_{i,j})^{-2\alpha'(p_i \cdot p_j)}$$

We use the well known relations

$$f(a_n)\, u^H = u^H f(u^n a_n)$$

$$u^H f(a_n^+) = f(u^n a_n^+)\, u^H$$

to bring (7-7) into an ordered product form. We commute the second term through the third one to get

$$\exp\left(\sum_n \sqrt{\tfrac{2\alpha'}{n}}\, p_{N-1} \cdot a_n\right) u_{N-2}^{H} = u_{N-2}^{H} \exp\left(\sum_n \sqrt{\tfrac{2\alpha'}{n}}\, u_{N-2}^{n}\, p_{N-1} \cdot a_n\right)$$

Continuing through the following term, we use the relation $e^A e^B = e^B e^A\, e^{[A,B]}$ since $[A,B]$ is a c-number.

With R_n an arbitrary c-number we therefore find

$$\exp\Big(\sum_n R_n \sqrt{\tfrac{2\alpha'}{n}}\, p_j \cdot a_n\Big) \exp\Big(-\sum_m \sqrt{\tfrac{2\alpha'}{m}}\, p_i \cdot a_m^+\Big) =$$

$$= \exp\Big(-\sum_m \sqrt{\tfrac{2\alpha'}{m}}\, p_i \cdot a_m^+\Big) \exp\Big(\sum_n R_n \sqrt{\tfrac{2\alpha'}{n}}\, p_j \cdot a_n\Big) \exp\Big(\sum_n \tfrac{R_n}{n}\, 2\alpha'\, p_i \cdot p_j\Big).$$

When the right factor associated with p_j has been moved all the way to the left factor associated with p_i its commutation through $\mu_{j-1}^{H}\, \mu_{j-2}^{H} \cdots \mu_{i}^{H}$ has produced a factor R_n equal to

$$R_n = \left(\mu_i\, \mu_{i+1} \cdots \mu_{j-1}\right)^n = \left(x_{i,j}\right)^n .$$

Now commuting the two factors one simply gets an extra term

$$\exp\Big(\sum_n \frac{(x_{i,j})^n}{n}\, 2\alpha' (p_i \cdot p_j)\Big) = \exp\big(-2\alpha'(p_i \cdot p_j) \log(1 - x_{i,j})\big) =$$

$$= \left(1 - x_{i,j}\right)^{-2\alpha'(p_i \cdot p_j)} .$$

The lowest value of i is 2 and the highest value of j is $N-1$. We therefore obtain:

$$\prod_{2 \le i < j}^{N-1} \left(1 - x_{i,j}\right)^{-2\alpha'(p_i \cdot p_j)} \qquad (8\text{-}9)$$

When all the creation operators have been moved to the left and all the annihilation operators have been moved to the right we are obviously left with the operator

$$\exp\Big(-\sum_n \sqrt{\tfrac{2\alpha'}{n}}\, a_n^+ \cdot \big(p_{N-1} + (x_{N-2,N-1})^n p_{N-2} + \cdots + (x_{2,N-1})^n p_2\big)\Big) \times$$

$$\times\, \mu_{N-2}^{H} \cdots \mu_2^{H} \exp\Big(+\sum_n \sqrt{\tfrac{2\alpha'}{n}}\, a_n \cdot \big(p_2 + (x_{2,3})^n p_3 + \cdots + (x_{2,N-1})^n p_{N-1}\big)\Big)$$

Its vacuum expectation value is readily found to be 1. Together with that we have a factored out c-number, which is easily obtained multiplying (8-8) by (8-9). This is the Bardakci-Ruegg formula.

We have then with (8-3) an expression which is equivalent to B_N. It is more suitable than B_N to reveal several interesting properties. The cyclic symmetry of the initial construction is apparently lost and the emphasis is to the contrary put on a maximal tree diagram with specific intermediate states.

B_N simplifies a great deal when $a=1$ since the term $\prod_{i=2}^{N-2}(1-u_i)^{a-1}$ disappears. This has been exploited in several recent analyses [136] in order to write down simpler expressions without too much loss of generality. The price which one has to pay is the introduction of an imaginary mass state at $\alpha=0$. This product can on the other hand be associated with an extra scalar harmonic oscillator mode [137]. The new Hamiltonian is written as

$$H' = a_0^+ a_0 + H \quad . \tag{8-10}$$

One then multiplies each vertex function by an operator V_0 involving a_0 and a_0^+ thus introducing an extra factor as a vacuum expectation value

$$\langle 0 | V_0 \, u_{N-2}^{a_0^+ a_0} \, V_0 \cdots u_2^{a_0^+ a_0} \, V_0 | 0 \rangle \quad . \tag{8-11}$$

If we choose

$$V_0 = \left(1 + \sum_l \frac{\sqrt{(1-a)\cdots(l-a)}}{l!} a_0^{+l}\right) Q_0 \left(1 + \sum_l \frac{\sqrt{(1-a)\cdots(l-a)}}{l!} a_0^l\right) ,$$

where Q_0 is the projection operator on the vacuum state we shall find (8-11) to be equal to

$$\prod_{i=2}^{N-2} (1-u_i)^{a-1} .$$

Introducing new vertex functions $V'(p) = V(p)\, V_0$, the integration over the u_i is now readily done with a new and more compact expression for B_N:

$$B_N = \langle 0|\, V'(p_{N-1}) \frac{1}{H' - \alpha(s_{1,N-2})} \cdots \frac{1}{H' - \alpha(s_{12})} V'(p_2)\, |0\rangle . \qquad (8\text{-}12)$$

V_0 has been defined in such a way that

$$V_0|\ell\rangle = \left(1 + \sum_{\ell'} \frac{\sqrt{(1-a)\cdots(\ell'-a)}}{\ell'!}\, a_0^{+\ell'}\right)|0\rangle \sqrt{\frac{(1-a)\cdots(\ell-a)}{\ell!}} =$$

$$= \sqrt{\frac{(1-a)\cdots(\ell-a)}{\ell!}}\; V_0|0\rangle ,$$

where $|\ell\rangle$ is an occupation number state: $a_0^+ a_0|\ell\rangle = \ell|\ell\rangle$. Sandwiching each $u^{a_0^+ a_0}$ term in (8-11) by a complete set of occupation number states one gets:

$$\prod_{i=2}^{N-2} \left(1 + \sum_{\ell} \langle 0|V_0|\ell\rangle\, u_i^{\ell}\right) =$$

$$= \prod_{i=2}^{N-2} \left(1 + \sum_{\ell} \frac{(1-a)\cdots(\ell-a)}{\ell!}\, u_i^{\ell}\right) = \prod_{i=2}^{N-2} (1-u_i)^{1-a} .$$

The denominator in (8-12) can be formally considered to be propagators with H' the Hamiltonian. The eigenstates of H'

can be labelled according to the occupation number of the different oscillators.

The frequencies are multiples of a fundamental so the system corresponds to a quantized violin string which can vibrate with frequencies $\nu, 2\nu, \cdots, n\nu, \cdots$. If we wish to obtain the state with maximum spin at the squared mass value m_N^2 with $\alpha(m_N^2) = N\nu$, we have to have N quanta on the spatial mode $(\mu = 1, 2, 3)$ with lowest frequency ν. There is only one way to build it and we reach spin $N\nu$. We have of course used ν for the fundamental "frequency." For lower spin values we may consider various constructions involving time modes and modes with higher frequencies. We then obtain a high degeneracy of the levels. This is illustrated on Fig. 8-2.

B. Factorization Property and Spectrum of States

The construction of B_N requires poles in all partition channels for $\alpha = 0, 1, \cdots$. We however do not know how many states there are until factorization is imposed. If as just seen there is only one level with the highest spin (parent trajectory) we expect several levels on each daughter trajectory. The high degeneracy of levels is indeed one of the basic features of the model [100]. This is of course not uncommon in physics [11], but it is new in hadron physics.

The B_N propagator, written as (8-6), reads:

$$D(\alpha(s)) = \sum_{n=0}^{\infty} \binom{n-a}{n} \frac{1}{H + n - \alpha(s)} \qquad (8\text{-}13)$$

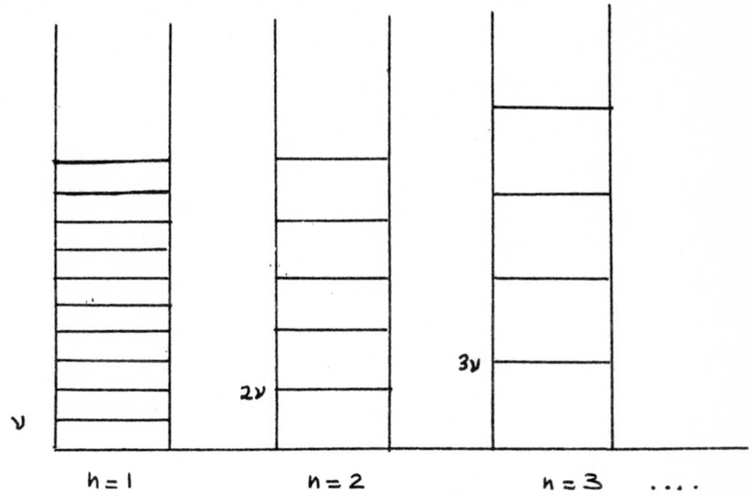

Figure 8-2. Harmonic oscillator levels.

It gives back (8-6) when sandwiched between occupation number states. Isolating such a propagator in B_N one writes

$$B_N = \sum_{n=0}^{\infty} \binom{n-\alpha}{n} \langle p_{i+1} | \frac{1}{H + n - \alpha(s)} | p_i \rangle , \qquad (8\text{-}14)$$

with

$$|p_i\rangle = V(p_i)\, D(\alpha(s_{1,i-1})) \cdots V(p_2)\, |0\rangle$$

$$\langle p_{i+1}| = \langle 0|\, V(p_{N-1})\, D(\alpha(s_{1,N-2})) \cdots V(p_{i+1}) \qquad (8\text{-}15)$$

s in (8-14) stands for s_{1i}.

Introducing a complete set of occupation number states

we get, with $H|\mathcal{H}\rangle = \mathcal{H}|\mathcal{H}\rangle$,

$$B_N = \sum_{n=0}^{\infty} \sum_{\mathcal{H}} \binom{n-a}{n} \langle p_{i+1}|\mathcal{H}\rangle \frac{1}{n+\mathcal{H}-\alpha(s)} \langle \mathcal{H}|p_i\rangle .$$

The residue at the pole $\alpha(s) = J$ is obtained by taking all the integers n and all the occupation number configurations $\mathcal{H}$ such that

$$n + \mathcal{H} = J \quad , \qquad n, \mathcal{H} \text{ positive} . \tag{8-16}$$

The residue

$$\sum_{\substack{n=0 \\ n+\mathcal{H}=J}} \sum_{\mathcal{H}} \binom{n-a}{n} \langle p_{i+1}|\mathcal{H}\rangle\langle \mathcal{H}|p_i\rangle , \tag{8-17}$$

where the sums are extended only over those integers which satisfy (8-16), is written in an obviously factorized form and the degeneracy of the level does not depend on the number of external particles [100]. Indeed the degeneracy of the level corresponds to all possible ways one can form an integer J with positive integers n and n_r such that

$$J = n + \sum_r r n_r \quad , \quad n \geq 0 \; , \; n_r \geq 0 \; .$$

n, since it is associated to a scalar mode, cannot contribute to the angular momentum, so the maximum spin is given by

$$J_{max} = \sum_r n_r \; ,$$

the sum being now extended over all spatial modes. We therefore find that only the oscillator mode with lowest frequency

$(r=1)$ should be excited. When J becomes large the level degeneracy is given by the Hardy-Ramanujan formula [138]

$$d(J) \simeq e^{\pi\sqrt{\frac{8}{3}J}} \qquad (8\text{-}18)$$

This is a very important property of the dual model thus constructed [100]. The level density increases exponentially with the meson mass. Dual terms with satellites give still a higher degeneracy [124, 129].

Such an exponential increase in level density was required by Hagedorn in building up a thermodynamical model [139]. The fixed transverse momentum distribution of secondaries produced even in very high energy collisions can be interpreted as due to the fact that the temperature of the "boiling" matter is maintained fixed as the excitation energy is increased. This temperature limit requires an exponential increase in the level density. Connections between the two approaches remain unclear. As remarked by Lovelace [11] it is however a striking coincidence that the two ultimate temperatures defined by

$$d(J) \sim e^{\frac{\sqrt{s}}{kT}}$$

are comparable.

The residue (8-17) can be factorized into tensor products using the explicit dependence on the external momenta obtained from (8-4). One obtains the expressions originally written by Fubini and Veneziano and Bardakci and Mandelstam.
As readily follows from the definitions (8-1) and (8-4), the factorization involves Lorentz tensors and gives ghosts associated with the time components. From our previous discussion

of the degeneracy they can actually occur only at the first daughter level. However, as shown by Fubini and Veneziano [100] there are Ward-like identities which eliminate them also from the first daughter levels.

These identities express the cyclic and reflection symmetries of B_N which is not apparent any longer in (8-3). This is discussed by Mandelstam [1] and we just recall a few basic points here.

We saw that through a twist of two quark lines we could obtain a new planar graph attached to a different ordering of the external particles. If however we make two twists between the same external line we obviously get back the same ordering. Such a double twisted graph is represented on Fig. 8-3, and the double twist invariance of B_N, which is thus implied, is written as:

$$B_N = \langle p_{i+1} | D(\alpha(s_{1i})) | p_i \rangle = \langle p_{i+1} | \Omega^+(q_i) D(\alpha(s_{1i})) \Omega(q_i) | p_i \rangle . \tag{8-19}$$

The twist operator Ω is known and has been studied in detail [140]. This will not be developed here [135]. It should be mentioned though that double twist invariance holds as a matrix element relation between $|p_i\rangle$ and $|p_{i+1}\rangle$ which are both states built out of external mesons. It does not hold however as an operator equation; instead

$$\Omega^+ D \Omega \neq D .$$

The physical states $|p_i\rangle$ do not form a complete basis for the operators Ω and D. There are "spurious" states necessary in the theory but which are orthogonal to all physical states. They become relevant when one isolates a pole term associated to a well-defined spin J which alone does not correspond to one of the states $|p_i\rangle$. The Ward-like relations can be obtained through the introduction of an operator W which annihilates any physical state [135] and

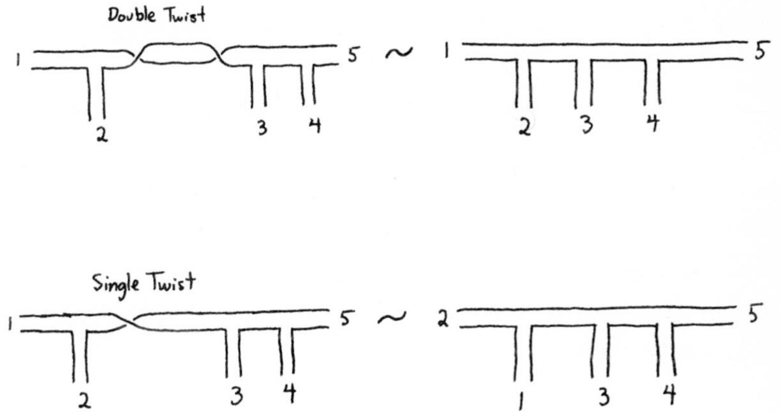

Figure 8-3.

therefore easily isolates the spurious states. They read:

$$\langle \mathcal{H} | W | p \rangle = 0 , \qquad (8\text{-}20)$$

where $|\mathcal{H}\rangle$ is as before an occupation number state and $|p\rangle$ a physical state as defined through (8-15). The coupling constants of the theory as defined in (8-17) are obtained in terms of the components $\langle \mathcal{H} | p \rangle$. Since (8-20) involves an operator W which does not commute with H [1, 135] it gives relations among couplings with different spin values

(differing indeed by one unit). As shown by Fubini and Veneziano [100] each first daughter ghost is associated with a normal level (parent of a new family) with proportional couplings in such a way that they can be combined into a unique level with positive coupling. This cancellation procedure does not hold however at the second daughter level where ghosts do occur. Summarizing, the dual model gives us a unique parent trajectory with normal factorizable couplings, a degenerate first daughter trajectory with normal couplings and an infinite sequence of highly degenerate daughters with ghosts. This degeneracy is a most important and interesting property of the model. The presence of ghosts is however extremely embarrassing if we look at $\mathcal{B}_N$ as a solution of a bootstrap scheme involving an arbitrary number of particles [141].

C. The Dual Loop

This difficulty might not be relevant in a wider scheme in which dual models would be but "Born terms" in a dual theory, a theory being defined by specific prescriptions to construct finite "higher order terms". Such a term can be obtained from Fig. 8-1 by matching together the quark lines of two external particles. In building up such a loop we can refrain from twisting the quark lines or twist them an arbitrary number of times. One twist gives a Mobius strip, two twists may give the possibility of separating the loop into two disconnected terms and may lead to a contribution to diffraction scattering [142] as discussed in Section 5. This is illustrated in Fig. 8-4 where, for simplicity, we have kept only four external particles.

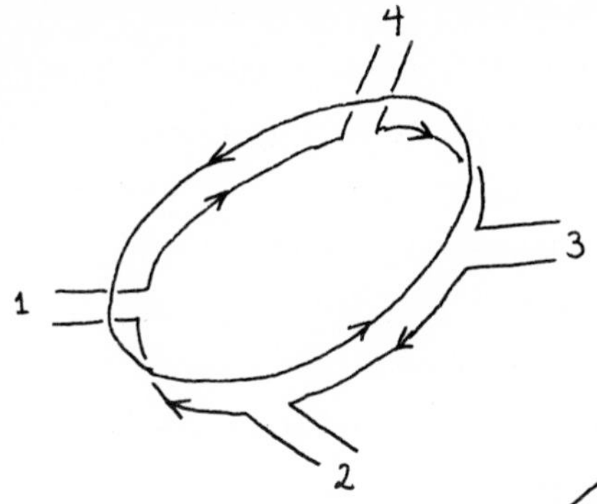

(a) Planar loop

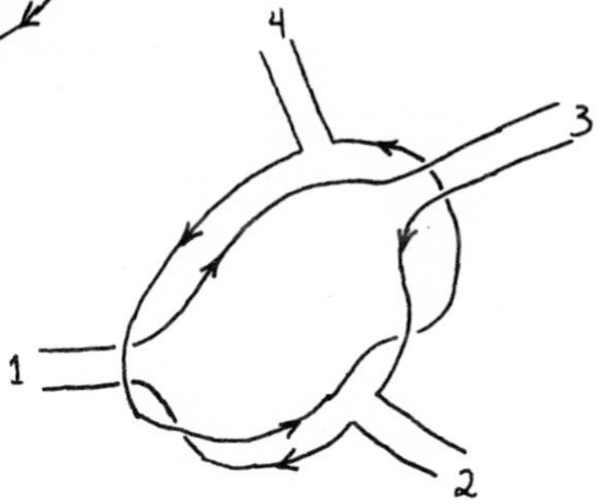

(b) Single-twisted loop

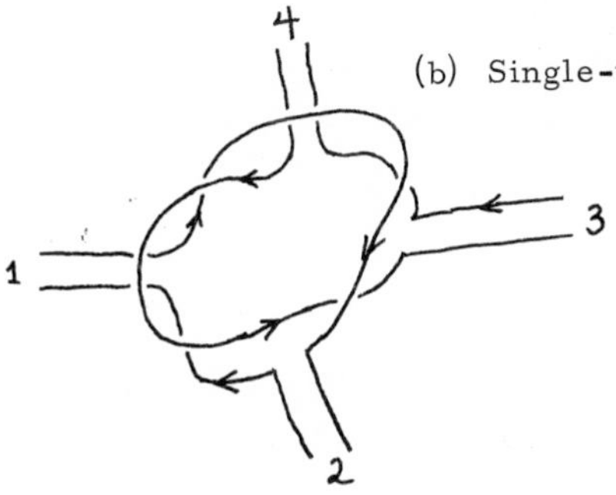

(c) Double-twisted loop

Figure 8-4.

The loops are constructed from B_N in such a way that, cutting any non-twisted part, one finds a quark anti-quark pair associated with an infinite set of resonance states. As a result the loops are invariant under duality transformations which when considered on duality diagrams correspond to all possible deformations maintaining the same topological structure. Some possible deformations of the planar loop of Fig. 8-4a are shown on Fig. 8-5. In each case it consists in writing an infinite set of resonance contributions in one channel as an infinite set of resonance contributions in another dual channel. Such graphs could be respectively associated with a "box diagram", a "vertex term" a "self energy term" and finally a "tadpole term". These contributions should not be considered separately. They are all contained in the planar loop constructed with four external particles constructed from B_6 and merely correspond to our focussing on the possibility for some of the μ's to vanish at the same time. As in Fig. 8-4a a quark loop is kept disconnected from the external quark antiquark lines.

Powerful methods have been devised in order to calculate these loop contributions [137, 143]. We merely sketch here the main step of the calculation of the single planar loop, simply to show how serious divergence problems arise. At present much effort is spent on the calculation of many loop contributions. It is clear that this is the way to eventually meet unitarity constraints, keeping duality at all steps in the calculation, and thus building up a dual theory of hadron interactions in which correction terms to the linear trajectory could be obtained.

The planar loop, as so far defined, is obtained from (8-3) or (8-12) by taking the trace of a series of operators $V(p)$ and $D(\alpha(s))$ associated with the external particles and the intermediate resonance states, as it results for each internal momentum k and then integrating over k, namely

$$F(p_1 \cdots p_N) = \int d^4k \, \mathrm{Tr} \left\{ V'(p_N) \frac{1}{H' - \alpha(s_N)} V'(p_{N-1}) \cdots V'(p_1) \frac{1}{H' - \alpha(s_1)} \right\} . \tag{8-20}$$

Such an expression has by construction the singularity structure imposed by duality.

We write $s_1 = q^2$, $s_{i+1} = (q + p_1 + p_2 + \cdots + p_i)^2$.

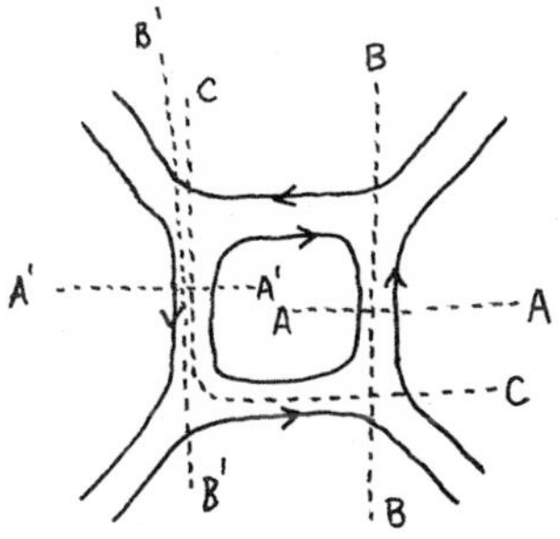

(a) Planar loop.

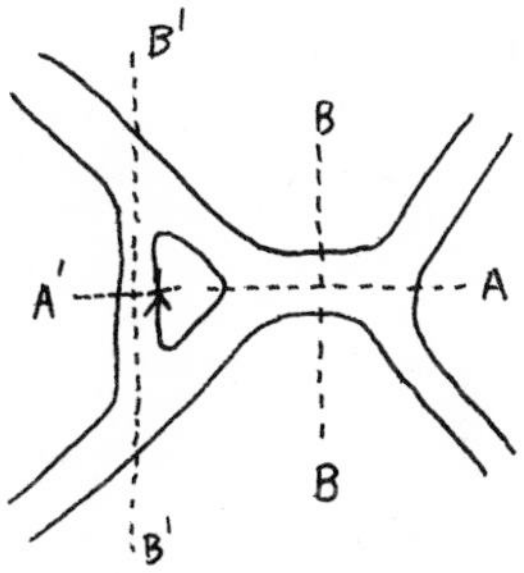

(b) After duality transform on A and B.

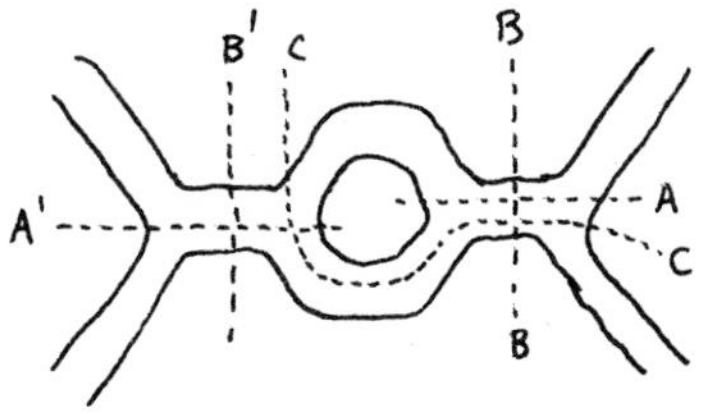

(c) After duality transform on A' and B'.

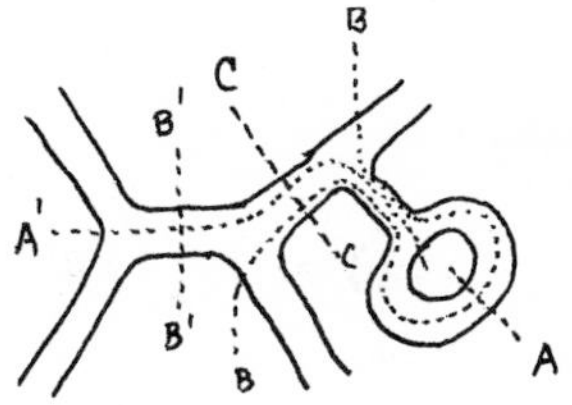

(d) Tadpole obtained after still another duality transformation; Ç, A vs. B.

Figure 8-5.

Using the definition of V' and H' we can rewrite F as a product of traces referring to the different oscillator modes .

$$F = \int d^4k \int_0^1 \cdots \int_0^1 du_1 \cdots du_N \; \mathrm{Tr}\left\{ V_0 \, u_N^{H_0} V_0 \cdots u_1^{H_0} \right\} \times$$

$$\times \prod_n \mathrm{Tr}\left\{ V(p_N) \, u_N^{H_m} \, V(p_{N-1}) \cdots u_1^{H_m} \right\} \prod_{i=1}^{N} u_i^{-\alpha(s_i)-1} ,$$

writing

$$H' = H_0 + \sum_{n=1}^{\infty} H_m .$$

The first trace has been already calculated and gives

$$\left((1-u_1) \cdots (1-u_N)\right)^{a-1} .$$

In order to calculate the other traces it is convenient to use a coherent state basis [135, 137]. The coherent states are defined as

$$|z\rangle = \exp(z a^+) |0\rangle$$

so as to obey the property $a|z\rangle = z|z\rangle$. One further finds $u^{a^+a}|z\rangle = |uz\rangle$, $\langle z'|z\rangle = \exp(z'^* z)$,

and $\exp(z'a^+)|z\rangle = |z'+z\rangle$.

The trace of a product of operators R is then given by

$$\frac{1}{\pi} \int d\,\mathrm{Re}\,z \; d\,\mathrm{Im}\,z \; \exp(-|z|^2) \langle z | R | z \rangle .$$

Using these properties the traces are calculated in a lengthy but straightforward way with the result [137, 143]

$$F = \int d^4k \int_0^1 du_1 \cdots \int_0^1 du_N \prod_{i=1}^{N} u_i^{-\alpha(s_i)-1} (1-u_i)^{a-1} \times$$

$$\times \prod_{n=1}^{\infty} (1-\omega^n)^{-4} \prod_{n=0}^{\infty} \prod_{i,j} (1-\omega^n x_{i+1,j+1})^{-2\alpha'(p_i \cdot p_j)} , \tag{8-21}$$

where
$$\omega = \prod_{i=1}^{N} u_i . \tag{8-22}$$

The $x_{i,j}$ are defined as previously, namely

$$x_{i,j} = u_i u_{i+1} \cdots u_{j-1} .$$

The exponent 4 on the $(1-\omega^n)$ term comes from the fact that a separate trace calculation has to be made for each spatial and time direction of each harmonic oscillator mode.

The expression (8-21) is divergent and the divergence clearly comes from the point $\omega = 1$ reached when all the independent u's become 1 at the same time. In this limit only the first infinite product is relevant to the divergence since momentum conservation $\left(\sum_i p_i = 0\right)$ implies a finite limit for the second one. The behaviour of the first infinite product when $\omega \to 1$ is known to be

$$\prod_{n=1}^{\infty} (1-\omega^n)^{-4} \underset{\omega \to 1-\epsilon}{\sim} \exp \frac{2\pi^2}{3(1-\omega)} . \tag{8-23}$$

This factor is closely related to the one we already met in discussing the degeneracy of levels and the divergence is of course connected with the exponential level degeneracy. It is however due not to the number of levels proper but also to the fact that their individual couplings do not decrease fast enough. The integral over k, neglecting this infinite factor, is on the other hand well behaved and straightforward.

An important point is that none of the singular terms itemized on Fig. 8-5 corresponds to $\omega = 1$. In each of them four μ's are zero but at least one of the μ's selected by the tree diagram sawn into a loop is zero. As a result if we modify the integrand, introducing a cut off at $\omega = 1 - \epsilon$ in order to exclude from the integration the neighborhood of $\omega = 1$, we do not change the singularity pattern of the whole term. We simply change the residues and, as a result, lose the asymptotic Regge behaviour.

It might be that one should not consider the limit $\omega = 1$ in the integrand but expand the integral obtained for $\omega < 1$ in ω [144]. It is also possible to define a dual counter term in such a way that the sum of the two terms has a limit when $\omega \to 1$ [145]. In the procedure proposed by Neveu and Sherk the behaviour of the cut off loop integral, as a function of $(1-\omega)$, can be closely studied by writing it in terms of elliptic functions which have interesting pertinent properties. In so doing a parameter $R = \exp\left(\frac{2\pi^2}{\log \omega}\right)$ is introduced. It goes to zero when $\omega \to 1$ from below and the integrand is found to behave as $R^{-1/3}$ times a convergent power series in R^2.

It is then possible to subtract a dual term with the same divergent behavior but differing from the initial one by terms of order R^2. The difference will now behave as $R^{4/3}$ and be convergent when $\omega \to 1$. [135]

As an example it is now possible to perform a fourth order calculation of the two body amplitudes. One subtracts a dual counterterm from the loop so that the remainder is finite as just sketched. The dual fourth order counterterm is then added to the second order term (a standard dual term) and its leading Regge contribution is used so as to redefine, when combined with the leading Regge contribution of the second order term, the new and physical Regge trajectory. This corresponds to the subtraction of an infinite constant from the initial Regge trajectory as one would do for mass renormalization. The second order term of this dual theory is therefore completely different from the dual term dressed with physical trajectories already considered, but the analogy with renormalization theory is striking. If the procedure can be defined in an unambiguous way and generalized to an arbitrary number of loops this may lead to a dual theory of hadron interactions. Calculation procedures have been developed for the dual loop with twists [143, 140] (non-planar loops). Divergence problems occur in both cases. If resolved, unitarity could be introduced at each order in a dual theory. Many questions are however yet to be answered.

D. Conclusion

Putting it in a nutshell, duality supposes that all forces significant enough to be relevant in the bootstrap scheme originate from resonance exchange. This is most simply conceived in a model world of zero width resonances where the resonances in one channel (infinite in number) generate the resonances of all the dual channels. Unitarity, and in particular shadow scattering, have a prominent role in determining scattering amplitudes but are not considered so relevant to the bootstrap construction of the hadron world. We have seen how such ideas can be abstracted from present experimental information and used to build models which live in a hadronic world of linear trajectories, but which might still have many properties in common with the actual particles we deal with in experiments. This is where the approach now stands, but the duality idea may yet be the guideline to a dual theory of strong interactions, which presents many fascinating aspects and many difficult problems. The purpose of these lectures has been to show how duality can be carried from phenomenology to the present mathematical models.[146]

REFERENCES AND FOOTNOTES

[1] S. Mandelstam, 1970 Brandeis Summer Institute in Theoretical Physics, Vol. I.

[2] What is assumed here is that the high-energy behaviour of the amplitude at fixed transfer can be expressed as the superposition of a few powers $s^{\alpha(t)}$. This is implied in a Regge pole model and, in any case, well met by experiment. For reviews see:

D.R.O. Morrison, Review of inelastic two-body reactions, Stony Brook (1966);

A. Bialas, Topical Conference on High-Energy Collisions of Hadrons, CERN (1968);

V.D. Barger and A.B. Cline, Phenomenological theories of high-energy scattering, (W.A. Benjamin Inc. New York) (1969).

[3] Non-exotic mesons (baryons) are such that their quantum numbers can be considered as those of a quark-antiquark (three quark) system. All hadronic states so far ascertained can be labelled as non-exotic. For a review see:

R.H. Dalitz, Les Houches lecture notes (1965);

H. Harari, Rapporteur's talk, Vienna Conference (1968).

[4] Figures 1-3 are not drawn to hide a most interesting feature which is somewhat troublesome in the framework of our present Regge models. After an "understood" drop off the K^-p total cross-section seems to level off in the energy domain recently explored at Serpukov. Nevertheless, we shall still focus on the 6-20 GeV range where all our experience with Regge models has been gathered;

J.V. Allaby et al., Phys. Letters 30 B, 500 (1969).

[5] Interpretation of the Serpukov results has so far followed seven different lines. Such variety can be found in the following papers quoted as a sample, not as an exhaustive list;

a) Simple Regge poles, see M. Restignoli and G. Violini, Phys. Letters 31 B, 533 (1970);

b) Regge poles and cuts, see V. Barger and R.J.N. Phillips, Wisconsin (1969);

c) Oscillatory component of the total cross-section due to complex Regge poles, see G. F. Chew and D.R. Snider, Phys. Letters 31 B, 75 (1970);

d) Electromagnetic effects important, see R. Wit, Nuovo Cimento (to be published);

e) The observed phenomenon associated with an "ionization" energy, see D. Horn, Phys. Letters 31 B, 30 (1970);

f) Violation of the Pomeranchuk's conditions through Regge dipoles, see V. Barger and R.J.N. Phillips,

Phys. Letters 31 B, 643 (1970); see also R. Arnowitt and P. Rotelli (to be published); and

g) Violation of the Pomeranchuk's conditions in standard Regge theory, see V.N. Gribov, I. Yu Kolsarev, V.D. Mur, L.B. Okun and V.S. Popov, Phys. Letters 32 B, 129 (1970); one should also consult R. Eden, Phys. Rev. (to be published), and B. de Fontenelle, L'histoire des oracles.

[6] G.F. Chew, Invited talk, Irvine Conference on Regge poles (1969);
N.F. Bali, G.F. Chew and A. Pignotti, Phys. Rev. 163 1572 (1967);
L. Caneschi and A. Pignotti, Phys.Rev. 184, 1915 (1969);
F.E. Low, Brookhaven lecture notes (1970).

[7] K. Kikkawa, B. Sakita and M.A. Virasoro, Phys. Rev. 184, 1701 (1969);
K. Bardakci and M.B. Halpern, Phys Rev. 183, 1456 (1969).

[8] B. Sakita and M.A. Virasoro, Phys. Rev. Letters 24, 1146 (1970);
C.S. Hsue, B. Sakita and M. Virasoro, Wisconsin (1970);
H.B. Nielsen and P. Olesen, Phys. Letters 32 B, 203 (1970).

[9] The aim of these lectures is to carry the duality approach from phenomenology up to the point where it is developed into a mathematical construction in Ref. 1.

[10] H. Harari, Brookhaven lecture notes (1969).

[11] C. Lovelace, Veneziano theory, Irvine Conference on Regge Poles, CERN TH 1123 (1969).

[12] M. Jacob, Schladming lecture notes CERN TH 1010 (1969).

[13] M. Jacob, Rapporteur's talk, Lund Conference (1969), CERN TH 1052.

[14] P.D.B. Collins and E. Squires, Regge poles in particle physics, Springer tracts in modern physics 45 (1968);
S.C.Frautschi, Regge poles and S matrix theory (W.A. Benjamin Inc., New York) (1963);
F.E. Low, Brandeis lecture notes (1967);
Ch. Chiu, Lecture notes on Regge theory, Caltech. (1970);
P.D.B. Collins, Phys. Reports (to be published).

[15] L. Van Hove, CERN lecture notes (1968);
L. Durand III, Boulder lecture notes (1969);
R.J.N. Phillips, Erice lecture notes (1966);
G.E. Hite, Rev. Mod. Phys. 41, 669 (1969);
B.E.Y. Svensson, CERN Summer School, lecture notes (1967).

[16] J.D. Jackson, Invited talk, Lund Conference (1969) and Rev. Mod. Phys. 42, 12 (1970).

[17] It is well known, however, that the polarization effect observed in nucleon charge exchange provides evidence for a Regge cut. It seems, however, that the flip amplitude can be described by a single Regge pole and this to a very good approximation. The observed polarization results mainly from the interference between the flip pole contribution and the non-flip cut contribution.

[18] K. Igi and S. Matsuda, Phys. Rev. Letters 18, 625 (1967);
A. Logunov, L.D. Soloviev and A.N. Tarkelidze, Phys. Letters 24 B, 181 (1967).

[19] R. Dolen, D. Horn and C. Schmid, Phys. Rev. 166, 1768 (1968);
D. Horn, Schladming lecture notes (1969);
G.F. Chew, Comments Nucl. Particle Physics 1, 121 (1967).

[20] C. Schmid, What is duality? Royal Society Meeting (1969), CERN TH 1128.

[21] FESR turn out to be very useful phenomenological tools. Besides the integer moment sum rules here derived, continuous moment sum rules can also be written. For a review see Refs. 16 and 22.

[22] H.M. Chan, Rapporteur's talk, Vienna Conference (1968).

[23] However, as we shall stress later, the Regge pole parameters may summarize in an approximate way the contributions of a Regge pole and a Regge cut. This is undoubtedly the case for the non-flip amplitude A' now considered.

[24] A. Donnachie, R.G. Kirsopp and C. Lovelace, Phys. Letters 26 B, 161 (1968).

[25] A' is the π-nucleon amplitude with no helicity flip in the t-channel. For the sake of completeness we recall that the π-nucleon Feynman amplitude is usually written as

$$\bar{u}(p')\left[-A+\tfrac{1}{2}\,i\gamma(q+q')B\right]u(p)$$

where $q\,(q')$ and $p\,(p')$ respectively stand for the momenta of the incoming (outgoing) meson and nucleon. One then defines [26]

$$A' = A + \frac{4m^2\nu}{4m^2-t}\,B \quad , \quad \nu = \frac{s-u}{4m}$$

so that A' and B are respectively proportional to the t-channel ($\pi\pi \to N\bar{N}$) helicity non-flip and flip amplitudes. For the charge exchange amplitude, odd with respect to meson charge exchange, A (and A') and B are, respectively, odd and even under s-u crossing ($\nu \leftrightarrow -\nu$). The π-nucleon total cross-section is equal to

$$\sigma_{TOT} = \frac{1}{k}\,\mathrm{Im}\,A'(s,t=0)$$

where k is the laboratory pion momentum.

[26] V. Singh, Phys. Rev. 129, 1889 (1963).

[27] There is nothing like resonance dominance for the real part of the amplitude. As opposed to the imaginary part it gets s - contributions from distant resonances and no contribution from a particular resonance at the resonance energy.

[28] N. Barash-Schmidt, G. Conforto, A. Barbaro-Galtieri, L.R. Price, M. Roos, A.H. Rosenfeld, P. Söding and C.G. Wohl, Review of particle properties, Phys. Letters 33 B, 1 (1970).

[29] The charge exchange amplitudes B and A' being respectively even and odd under s-u crossing, one should consider odd moment sum rules for B and even moment sum rules for A'.

[30] For a review of expected properties of Regge residues, see L. Bertocchi, Invited talk, Heidelberg Conference (1968) and J.J.J. Kokkedee, Belgium-Dutch School lecture notes, Nijmegen (1970).

[31] P. Sonderegger et al., Phys. Letters 20, 75 (1966); L. Di Lella et al., Phys. Letters 24 B, 77 (1967).

[32] G. Cohen-Tannoudji, A. Morel and H. Navelet, Nuovo Cimento 48 A, 820 (1967).
C.B. Chiu and J. Finkelstein, Nuovo Cimento 46 A, 820 (1967);
R.C. Arnold and R.K. Logan, ANL 68-15 (1968);
J. Finkelstein and M. Jacob, Nuovo Cimento 36 A, 681 (1968).

[33] V.D. Barger and D.B. Cline, Phys. Rev. Letters 16, 913 (1966);
V.D. Barger and M. Olson, Phys. Rev. 151, 1123 (1966).

[34] A. Donnachie and R.G. Krisopp, Glasgow (1969);
V.A. Alessandrini, D. Amati and E.J. Squires, Phys. Letters 27 B, 463 (1968).

[35] H. Harari and Y. Zarmi, Phys. Rev. 187, 2230 (1969).

[36] R. Jengo, Phys. Letters 28 B, 261, 606 (1969).

[37] L. Van Hove, Phys. Letters 24 B, 183 (1967);
L. Durand III, Phys. Rev. 161, 1610 (1967).

[38] C. Schmid, Phys. Rev. Letters 20, 628 (1968) and Phys. Letters 28 B, 348 (1969).

[39] G. F. Chew, S matrix theory (W. A. Benjamin Inc., New York) (1960).

[40] We shall not consider here the splitting of the A_2. The 2^+ A_2, whatever its detailed structure is, can be considered as a unique particle with an effective coupling. As a whole it is exchange degenerate with the ρ since both A_2 peaks couple to the $K\bar{K}$ channel.

[41] B. Maglic, Rapporteur's talk, Lund Conference (1969);
H. Benz et al., Phys. Letters 28 B, 233 (1968);
W. Kienzle, Philadelphia Conference on Meson Spectroscopy (1968).

[42] For a review of the properties of Regge trajectories see Ref. 15 and V. Shirkov, Dubna, p. 2-4726 (1969).

[43] H. Harari, Phys. Rev. Letters 20, 1385 (1968).

[44] P.G.O. Freund, Phys. Rev. Letters 20, 235 (1968).

[45] H. Harari and Y. Zarmi, Phys. Letters 32 B, 291 (1970).

[46] F. J. Gilman, J. Pumplin, A. Schwimmer and L. Stodolsky, Phys. Letters 31 B, 387 (1970).

[47] The parametrizations used for the Pomeranchon contribution are still unsettled. Former Regge fits such as those of Phillips and Rarita [Phys. Rev. 139 B, 1336 (1965)] have a small slope for the leading vacuum contribution, definitely smaller than the one used for the ρ or P' trajectories. Hybrid models such as those of H.I. Abarbanel, S.D. Drell and F.J. Gilman (Phys. Rev. Letters 20, 280 (1968) or Ch. Chiu and J. Finkelstein [Nuovo Cimento 57 A, 649 (1968)]) take a zero slope. The Serpukov data seem to indicate a

slope of the order of 0.4 ; dual models may yield a slope of the same order of magnitude.

[48] As discussed in details later, yet uncertain resonances such as the Z^* are not relevant to our analysis. The Z^* is discussed in S. Anderson et al., Lund Conference (1969) and J.G. Asbury et al., Phys. Rev. Letters 23, 194 (1969).

[49] This of course neglects cut contributions. If one generates these Regge cuts through a Regge pole approximation of the eikonal [C. Arnold, Phys. Rev. 153, 1523 (1967)] one has important non-linear effects and the asymptotic cross-section may then be approached from below. This is the case in the Gribov theory [V.N. Gribov, I. Ya. Pomeranchuk and K.A. Ter Martirosyan, Phys. Rev. 139, B184 (1965).]

[50] D. Morisson, Rapporteur's talk, Lund Conference (1969).

[51] M. Gell-Mann, Phys. Letters 8, 214 (1964); G. Zweig (unpublished) (1964).

[52] For a review see:
R.H. Dalitz, Les H ouches, lecture notes (1965).

[53] Ch. Chiu and J. Finkelstein, Phys. Letters 27 B, 510 (1968).

[54] J. Mandula et al., Phys. Rev. Letters 22, 1147 (1969); and Phys. Rev. Letters 23, 266 (1969);
G.V. Dass et al., Nuovo Cimento 67, 429 (1970);
The H meson could have the mass of the A_1, as required by the degeneracy of slopes [M. Ademollo et al., Phys. Rev. Letters 22, 83 (1969)] or the mass of the B as demanded by the absence of exotic KK^* states.

[55] As remarked by J. Mandula et al., in Ref. 54, the octet mass formula is very well satisfied by the A_1, the D, and the K_A.

[56] A. Schwimmers, as reported in H. Lipkin, Invited talk, Lund Conference (1969).

[57] M. Toller, Nuovo Cimento 53 A, 671 (1968);
D. Freedman and J. Wang, Phys. Rev. 153, 1596 (1967).

[58] C. Schmid, Lettere al Nuovo Cimento 1, 165 (1969).

[59] With resonance dominance, we have a polynomial behaviour in t and extrapolation of the imaginary part is possible to positive values of t.

[60] K.W. Laie and J. Louie, Irvine Conference on Regge Poles (1969); and Nuclear Physics B 19, 205 (1970).

[61] P.R. Auvil, F. Halzen, C. Michael and J. Weyers, Phys. Letters 31 B, 303 (1970).

[62] For a review of spin effects in two-body collisions see for instance M. Jacob, Invited talk, Proceedings of the Saclay Conference on Polarized Targets and Ion Sources (1965).

[63] J. Finkelstein, Phys. Rev. Letters 22, 362 (1969).

[64] V.N. Gribov and I.Ya. Pomeranchuk, Phys. Letters 2, 232 (1962);
S. Mandelstam and L.L. Wang, Phys. Rev. 160, 1490 (1967).

[65] Ch. Chiu and A. Kotanski, Nucl. Phys. B 7, 615 (1968); and Nucl. Phys. B 8, 553 (1968).

[66] Such an association is unambiguous only if the non-resonance background is relatively small. This is what is consistently assumed here. Another attitude (see Ref. 34) calls for a relatively large background.

[67] P.D.B. Collins, R. C. Johnson and E. J. Squires, Phys. Letters 27 B, 23 (1968);
V.A. Alessandrini, P. G. O. Freund, R. Oehme and E. J. Squires, Phys. Letters 27 B, 456 (1968);

[68] V.A. Alessandrini and E.J. Squires, Phys. Letters 27 B, 300 (1968);
C. Schmid, Nuovo Cimento 61, 515 (1969);
M. Jacob, Herceg Novi lecture notes (1968).

[69] C. Schmid, Phys. Rev Letters 20, 689 (1969).

[70] M. Jacob and J. Mandelbrojt, Nuovo Cimento 63 A, 279 (1969); S. Mandelbrojt, Dirichlet series, Rice Institute (1944).

[71] There is now good evidence for a broad s- wave isospin 0 resonance at the mass of the ρ [E. Malamund and P.E. Schlein, Phys. Rev. Letters 18, 1056 (1967), C. Lovelace, R.M. Heinz and A. Donnachie, Phys. Letters 22, 322 (1966)]. The K s-wave resonance at the mass of the K^* may come out as expected once the proper isospin $3/2$ contribution has been included in the analysis. The $\eta\pi$ channel is yet unknown and the ρ', expected at the mass of the f^0, is still hiding.

[72] R. Capps, Phys. Rev. Letters 22, 1215 (1969); V. Barger and C. Michael, Phys. Rev. 186, 1592 (1969).

[73] J. Mandula, J. Weyers and G. Zweig, Ann. Rev. Nucl. Sci. (to be published) (1970).

[74] C. Rebbi and R. Slansky, Rev. Mod. Physics 42, 68 (1970).

[75] Resonance dominance is here assumed to hold as long as there are not too many inelastic channels open. The πN channel is then more reliable than the $\rho\rho$ channel. High above threshold, exotic contributions might also be no longer negligible, thus also contradicting the resonance dominance assumption followed here.

[76] J. Rosner, Phys. Rev. Letters 21, 950 (1968); P.G.O. Freund, J. Rosner and R. Waltz, Nuclear Physics B 13, 237 (1969).

[77] H.J. Lipkin, Phys. Letters 32 B, 301 (1970).

[78] H. Harari, Phys. Rev. Letters 22, 562 (1969).

[79] J. Rosner, Phys. Rev. Letters 22, 689 (1969).

[80] V. Barger and M. Olson, Phys. Rev. 146, 980 (1966).

[81] M. Jacob and J. Weyers, Nuovo Cimento (to be published); If the exotic couplings are as needed to save duality in $B\bar{B}$ scattering, backward production of exotic mesons should be as important as was observed for ρ or A_2 backward

production. They should also show up in $p\bar{p}$ annihilation in flight.

[82] G. Veneziano, Nuovo Cimento 57 A, 190 (1968).

[83] The actual number of states cannot be inferred until the factorization of the pole residue is imposed. This will be discussed later.

[84] C. Lovelace, Phys. Letters 28 B, 265 (1968).

[85] J. Shapiro and J. Yellin, UCRL 18500 (1968).
J. Yellin, UCRL 18637 (1968); and 18664 (1969).

[86] D. Sivers and J. Yellin, UCRL 19418 (1969).

[87] P. Di Vecchia, F. Drago and S. Ferrara, Phys. Letters 29 B, 115 (1969);
F. Scheck, Nuovo Cimento 63, 1074 (1969);
V. Alessandrini and D. Amati, Phys. Letters 29 B, 193 (1969).

[88] The only candidate presently available as the ρ' is the four-pion bump seen in photoproduction at SLAC at 1550 MeV. No bump closer to the f^0 mass has been observed. Nevertheless, the analysis of $\bar{p}n$ annihilation at rest and in flight seems to need a ρ' at the f^0 mass. As is well known semi-local duality already requires daughters. The first t zero in the $\pi\pi$ amplitude at the ρ mass would be much too close to 0 (0.2 instead of 0.6) if there were no ϵ. Similarly a zero at 0.6 at the mass of the f^0 requires a ρ'.

[89] As proposed by M. Virasoro [Phys. Rev. 177, 2309 (1969)] it is possible to write dual terms with poles in all three channels. Such model amplitudes are, however, not suitable to describe reactions with one exotic channel unless proper cancellations between different terms hold. As shown by Mandelstam [Phys. Rev. 183, 1374 (1969)] the Veneziano and the Virasoro models can be considered as two limiting cases of a more general model amplitude. In these notes we shall stick to our present building up of dual amplitudes out of terms written when one channel has no pole.

This is both simple and readily generalizable to production processes.

[90] We mean here further an identification of the linear approximation to the Veneziano amplitude with the Weinberg amplitude which further requires the absence of isospin two σ terms. S. Weinberg, Phys. Rev. Letters 17, 336 (1966). For a detailed discussion see for instance M. Jacob. Algebre des courants, Saclay lecture notes (1967).

[91] We define f_π such that the pion decay amplitude reads

$$\langle 0 | A_\mu^i | \pi^j(q) \rangle = i f_\pi q_\mu \delta_{ij}$$

[92] C. Lovelace, Invited talk, Argonne Conference on $\pi\pi$ and $K\bar{K}$ interactions (1969) CERN TH 1041;
R.G. Roberts and F. Wagner, Phys. Letters 29 B, 368 (1969);
F. Wagner, Nuovo Cimento 64, 206 (1969); and Nuovo Cimento 64 A, 189 (1969).

[93] K. Kawarabayashi, S. Kitakado and H. Yabuki, Phys. Letters 28 B, 432 (1968);
C. Lovelace (unpublished) (1968).

[94] For a detailed discussion see Ref. 12.

[95] As shown by Mandelstam [Phys. Rev. Letters 21, 1724 (1968)] different slopes would result in exponential asymptotic behaviour at fixed angle. We also saw how this is imposed by duality as soon as we impose local relations which have to be satisfied in the Veneziano model.

[96] C. Canning, Nucl. Physics B 14, 319 (1969).

[97] J. Baacke et al., Nuovo Cimento 62 A, 332 (1969).

[98] E. Predazzi and D.B. Lichtenberg, Phys. Rev. Letters 22, 215 (1969).

[99] F. Henyey, Michigan (1969);
For a critical discussion see N:N. Khuri, Phys. Rev. 185, 1876 (1969).

[100] S. Fubini and G. Veneziano, Nuovo Cimento 56 A, 1027 (1968); K. Bardakci and S. Mandelstam, Phys. Rev. 184, 1640 (1969).

[101] F. Wagner, Nuovo Cimento 63, 393 (1969).

[102] A. Martin, Nuovo Cimento 47 A, 265 (1967); and 38 A, 303 (1968);

A.K. Common, Nuovo Cimento 53 A, 946 (1968).

[103] It is possible to write the denominator in such a way that crossing symmetry is still true asymptotically. However, the interest of (6-32) is to provide a low energy amplitude.

[104] M. Baker and R. Blankenbecler, Phys. Rev. 128, 415 (1962); M. Jacob and S. Pokorsky, Nuovo Cimento 61 A, 233 (1969).

[105] J. Roberts and F. Wagner, CERN Conference on Weak Interactions (1969).

[106] G.F. Chew and S. Mandelstam, Phys. Rev. 119, 467 (1960).

[107] G. Altarelli and H.R. Rubinstein, Phys. Rev. 183, 1469 (1969);

J. Hopkinson and R. G. Roberts, Nuovo Cimento 59 A, 181 (1969);

H.R. Rubinstein, E.J. Squires and M. Chaichian, Phys. Letters 30 B, 189 (1969);

H.R. Rubinstein, Phys. Letters 32 B, 370 (1970).

[108] For a detailed discussion see C. Itzykson, M. Jacob and G. Mahoux, Nuovo Cimento Suppl. 5, 978 (1967).

[109] L.L. Loos, N. H. Fuchs, L.J. Gutay and J.H. Scharenguivel Phys. Letters 31 B, 470 (1970).

The σ model Lagrangian also summarizes many of these properties and it can be used through Padé approximants to obtain a meson-meson scattering amplitude also very suitable to describe low energy results. References can be found in Ref. 13. For a review see J. Zinn Justin, Physics Reports (to be published).

[111] The ρ' should be produced at the energy available at Frascati. As already stressed it might be more easily seen in the four π than in the two π channel.

[112] The scattering amplitude is usually written as

$$M_{\mu\nu} = A P_\mu P_\nu + B(P_\mu Q_\nu + P_\nu Q_\mu) + C_1 Q_\mu Q_\nu + C_2 g_{\mu\nu}$$

where P and Q respectively, stand for half sums of the pion and vector meson incoming and outgoing momenta. The four invariant amplitudes A, B, C_1, and C_2 satisfy the Mandelstam representation and can be written as dual amplitudes. They respectively have $s^{\alpha-2}$, $s^{\alpha-1}$, s^{α} and s^{α} asymptotic behaviour in s at fixed t but the same t asymptotic behaviour in t at fixed s. See V. de Alfaro et al., Phys. Letters 21, 576 (1966).

[113] P.G.O. Freund and E. Schonberg, Phys. Letters 28 B, 600 (1969).

[114] C. Goebel, M. Blackmon and K. Wali, Phys. Rev 182, 1487 (1969);
P. Carruthers and E. Lasley, Phys. Rev. D 1, 1204 (1970);
P. Carruthers and F. Cooper, Phys. Rev D 1, 1223 (1970);
A. Capella et al., Lettere al Nuovo Cimento 1, 655 (1969);
S.P. De Alwis, D.A. Nutbrown, P. Brooker and J.M. Kosterlitz, Phys. Letters 29 B, 362 (1969).

[115] H.R. Rubinstein and V. Rittenberg, Weizmann Institute (1970).

[116] S. MacDowell, Phys. Rev. 116, 1774 (1959).

[117] R. Carlitz and M. Kislinger, Phys. Rev. Letters 24, 186 (1970).

[118] M.A. Virasoro, Phys. Rev. 184, 1621 (1969);
K. Igi, Phys. Letters 28 B, 330 (1969);
E.L. Berger and G. Fox, Phys. Rev. 188, 2120 (1970).

[119] C. Lovelace, Regge fits with Veneziano residues, CERN (1969).

[120] M. Ademollo and E. Del Guidice, Nuovo Cimento 63 A, 639 (1969);
R.C. Brower and J.H. Weis, Phys. Rev. 188, 2486, 2495 (1970);
F. Cooper, Phys. Rev. D 1, 1140 (1970);
H.J. Schnitzer, Phys. Rev. Letters 22, 1154 (1969);
R. Arnowitt, M.H. Friedman and P. Nath, Phys. Rev. D 1, 1813 (1970);
F. Csikor, Phys. Letters 31 B, 141 (1970);
J. Ellis and H. Osborn, Phys. Letters 31 B, 580 (1970).

[121] Since the constructed term will be cyclic symmetric, the choice of a particular point on the dual diagram of Figs. 7-3, in order to define N-3 non-dual partitions is irrelevant. Any choice will give the same result.

[122] The original extension of B_4 to B_5 which has been followed up to B_N is to be found in
K. Bardakci and D. Ruegg, Phys. Letters 28 B, 342 (1968);
M.A. Virasoro, Phys. Rev Letters 22, 37 (1969);
The extension to B_N is worked out in
H.M. Chan, Phys. Letters 28 B, 425 (1969);
H.M. Chan and S.T. Tsou, Phys. Letters 28 B, 489 (1969);
K. Bardakci and D. Ruegg, Phys. Rev. 181, 1884 (1969);
Ch. Goebel and B. Sakita, Phys. Rev. Letters 22, 257 (1969);
Z. Koba and H.B. Nielsen, Nuclear Physics B 10, 633 (1969);
The paper of B and R introduces the expression for B_N here used. The second paper of K and N proposes a new choice of variables to replace the u which exhibits explicitly the duality property and which is of very important theoretical interest (see Ref. 1); for a short review see Ref. 13 and for

a detailed review see H.M. Chan, Royal Society Meeting (1969).

[123] Z. Koba and H.B. Nielsen, Nuclear Physics B 12, 517 (1969).

[124] P. Olesen, Nuclear Physics B 18, 473 (1970).

[125] S. Mandelstam, Phys. Rev. D 1, 1720, 1734, 1745 (1970).

[126] For a review of multi-Regge properties see H.M. Chan, Rapporteur's talk, Vienna Conference (1968);
For a review of production processes see L. Van Hove, Physics Reports (to be published).

[127] K. Bardakci and D. Reugg, Phys. Rev. 181, 1884 (1969).

[128] G.F. Chew, F.E. Low and M.L. Goldberger, Phys. Rev. Letters 22, 208 (1969);
For a review see G.F. Chew, Invited talk, Irvine Conference on Regge Poles (1969) (Ref. 6).

[129] D.J. Gross, Nuclear Physics B 13, 467 (1969).

[130] H.M. Chan et al., Nuclear Physics B 19, 173 (1970);
B. Peterson and N.A. Tornquist, Nuclear Physics B 13, 629 (1969);
For a review of B_5 phenomenology see H.M. Chan, Herceg Novi lecture notes (1969); and C. Lovelace, Irvine Conference on Regge Poles (1969).

[131] A correct treatment of spin would probably affect the angular distributions at resonance poles which are reproduced correctly only if specific relations among the B_5 terms are imposed; they may go against the indication which would be given by our writing of duality diagrams. See in particular the analysis of Peterson and Tornquist of the previous reference.

[132] M.B. Green and R.H. Heimann, Phys. Letters 30 B, 642 (1970);
L. Montvay, Phys. Letters 30 B, 653 (1969).

[133] J.F.L. Hopkinson and E. Plathe, Phys. Letters 28 B, 489 (1968).

[134] S. Fubini, D. Gordon and G. Veneziano, Phys. Letters 29 B, 679 (1969);
S. Fubini and G. Veneziano, MIT preprint (1970).

[135] For a review see V. Alessandrini, D. Amati, M. Le Bellac and D. Olive, Phys. Reports (to be published), CERN TH 1160, and Ref. 11.

[136] M. Virasoro, Wisconsin (1969); see also Ref. 8.
The intercept is taken as unity and the zero spin particle has as a result a negative mass squared.

[137] D. Amati, C. Bouchiat and J.L. Gervais, Lettere al Nuovo Cimento 2, 399 (1969).

[138] A. Erdelyi, W. Magnus, F. Oberhettinger and F.G. Tricomi, Vol. 3, p. 179 (1955).

[139] R. Hagedorn, Nuovo Cimento Suppl. 3, 147 (1965); and 6, 311 (1968); Nuovo Cimento 56 A, 1027 (1968).

[140] L. Caneshi, A. Schwimmer and G. Veneziano, Phys. Letters 30 B, 351 (1969);
V. Alessandrini, D. Amati, M. Le Bellac, Phys. Letters 32 B, 285 (1970);
A. Alessandrini, D. Amati, M. Le Bellac and D. Olive, Nuovo Cimento 66 A, 831 (1970);
For a detailed and up to date review see Ref. 135.

[141] One may, however, perhaps only attach a statistical significance to the many couplings thus found. See L.N. Chang, P.G.O. Freund and Y. Nambu, Chicago (1969).

[142] P.G.O. Freund and R.J. Rivers, Phys. Letters 29 B, 510 (1969);
G. Frye and L. Susskind, Phys. Letters 31 B, 589 (1970);
G. Frye, Phys. Letters 31 B, 26 (1970).

[143] K. Bardakci, M.B. Halpern and J. Shapiro, Phys. Rev. 185, 1910 (1969);
M. Kaku and L.P. Yu, Phys. Letters (to be published);

M. Kaku and C.B. Thorn (to be published).

[144] This point is emphasized by Mandelstam (private communication).

[145] A. Neveu and J. Sherk, Phys. Rev. (to be published);
D.J. Gross, A. Neveu, J. Sherk and J.H. Schwartz, Phys. Letters 31 B, 592 (1970);
G. Frye and L. Susskind, Phys. Letters 31 B, 537 (1970);
J.G. Gallardo, E. Galli and L. Susskind, Phys. Rev D 1, 1189 (1970);
C.B. Thorn (to be published);
T.H. Burnett, D.J. Gross, A. Neveu and J. Sherk, Phys. Letters 32 B, 115 (1970).

[146] I wish to thank Profs. M. Grisaru and H. Pendleton for their help with these notes.

Weak Interactions

Henry Primakoff
Department of Physics
University of Pennsylvania
Philadelphia, Pennsylvania

CONTENTS

1. Introduction

Of the four well-defined interactions recognized among the elementary particles (gravitational, weak, electromagnetic, and strong)[1] the weak interaction has been rather fully elaborated from a dynamical as well as a kinematic point of view.[2] Naturally, in the time available, we shall be able to deal only with certain very limited aspects of the subject and, in particular, we shall focus our attention on the purely leptonic weak processes and to some extent on the semileptonic weak processes where the complications arising from the presumed composite structure of the hadrons are either entirely absent or can be essentially subsumed in a small number of form factors. We shall also concern ourselves with the question of whether the weak interactions are transmitted by an as yet undiscovered "intermediate boson" ($W^{\pm}$) in the same sense that electromagnetic interactions are transmitted by the photon. Since practically all of our discussion will deal with processes in which leptons are created or destroyed we shall of necessity consider in some detail the principle of lepton conservation which experimentally is on a less certain footing than the analogous principles of baryon conservation and charge conservation. Again, since the electron-neutrino (ν_e) and the muon-neutrino (ν_μ) are leptons <u>par excellence</u>

(ν_μ and ν_e are the only elementary particles without primitive electromagnetic and strong interactions) we shall naturally pay considerable attention to the properties (mass, magnetic moment, charge radius, and helicity) of these particles. Finally, in our discussion of the semileptonic weak processes, we shall pay especial attention to the question of time-reversal (T) invariance and shall show that no unambiguously interpretable experimental evidence is now available in favor of such T invariance.

We first establish our terminology, notation, etc. The elementary particles are:

$$\begin{array}{l} \text{leptons: } \nu_e,\ e,\ \nu_\mu,\ \mu \\ \text{photon: } \gamma \\ \text{(intermediate boson: } W\ ?\ ) \\ \text{quarks: } q_p,\ q_n,\ q_\Lambda \end{array} \qquad (1)$$

where, on a verbal-qualitative level and without any commitment whatsoever regarding a realistic structural description, we suppose all hadrons with baryon number zero or one (i.e. $\pi, \eta, K, \ldots;\ p, n, \Lambda,\ \ldots$) to be composites of two or three fractionally charged quarks (i.e. $\pi^+ = [q_p \bar{q}_n]$, $K^+ = [q_p \bar{q}_\Lambda], \ldots;\ p = [q_p q_p q_n], \Lambda = [q_p q_n q_\Lambda], \ldots$)

The basic weak processes are then the purely leptonic:

$$\nu_e + e^- \rightarrow \nu_e + e^- \quad (2)$$

$$\nu_\mu + \mu^- \rightarrow \nu_\mu + \mu^- \quad (3)$$

$$\nu_\mu + e^- \rightarrow \nu_e + \mu^- \quad (4)$$

$$\mu^- \rightarrow \nu_\mu + \bar{\nu}_e + e^- \quad (5)$$

$$\nu_\mu + (Z,A) \rightarrow \nu_\mu + \mu^+ + \mu^- + (Z,A) \quad (6)$$

$$\nu_\mu + (Z,A) \rightarrow \nu_e + e^+ + \mu^- + (Z,A) \quad (7)$$

the semileptonic:

$$q_n \rightarrow q_p + e^- + \bar{\nu}_e \quad (8a)$$

$$q_n + \bar{q}_p \rightarrow e^- + \bar{\nu}_e \quad (8b)$$

$$q_p \rightarrow q_n + e^+ + \nu_e \quad (9a)$$

$$q_p + \bar{q}_n \rightarrow e^+ + \nu_e \quad (9b)$$

$$q_\Lambda \rightarrow q_p + e^- + \bar{\nu}_e \quad (10a)$$

$$q_\Lambda + \bar{q}_p \rightarrow e^- + \bar{\nu}_e \quad (10b)$$

$$q_p \rightarrow q_\Lambda + e^+ + \nu_e \quad (11a)$$

$$q_p + \bar{q}_\Lambda \rightarrow e^+ + \nu_e \quad (11b)$$

and the nonleptonic:

$$q_\Lambda + q_p \rightarrow q_p + q_n \quad (12a)$$

$$q_\Lambda \rightarrow q_p + q_n + \bar{q}_p \quad (12b)$$

$$q_n + q_p \rightarrow q_n + q_p \quad (13a)$$

$$q_n \rightarrow q_p + q_n + \bar{q}_p \quad (13b)$$

Various comments can be made regarding the processes in Eqs. (2) - (13b), viz:

(a) At the present moment, muon decay (Eq. (5)) is the only one of the purely leptonic processes which has been extensively studied experimentally and whose rate has been precisely measured. The "inelastic scattering" process of Eq. (4) is essentially the inverse of muon decay.

(b) The μ^+, μ^-; e^+, μ^- "pair-production" processes of Eqs. (6), (7) (the nucleus (Z, A) provides a Coulomb field whose interaction with the final μ^+, μ^- or e^+, μ^- is required for momentum conservation) involve the same weak vertices as the "elastic scattering" processes of Eqs. (3) and (4).[3] If the weak interactions are in fact transmitted by an intermediate boson, then, for neutrino energies above a threshold determined by the boson mass,

$$\nu_\mu + (Z, A) \longrightarrow W^+ + \mu^- + (Z, A) \qquad (14)$$

$$W^+ \begin{cases} \nearrow \nu_\mu + \mu^+ & (15a) \\ \searrow \nu_e + e^+ \quad , & (15b) \end{cases}$$

where the W^+ lifetime is expected to be so short that the final states in Eqs. (14), (15a), (15b) will be distinguishable from the final states in Eqs. (6), (7) only on the basis of the energy distributions of μ^+ and μ^-, or e^+ and μ^- [4].

(c) Typical semileptonic or nonleptonic decays,

e.g.,

$$\{[q_{bp}\bar{q}_{bn}] = \pi^+\} \longrightarrow e^+ + \nu_e \tag{16}$$

$$\{[\bar{q}_{bp} q_{b\Lambda}] = K^-\} \longrightarrow \{[\bar{q}_{bp} q_{bn}] = \pi^-\} + \{[q_{bp}\bar{q}_{bp}] = \pi^0\} \tag{17}$$

are viewed as a manifestation of the semileptonic or non-leptonic quark decays in Eqs. (9b) and (12b), the $\bar{q}_{bp}$ in Eq. (17) participating only as a "spectator."

(d) The semileptonic decay in Eq. (16) defines operationally the electron-neutrino, i.e. ν_e is that "neutrino" which is emitted together with e^+ in $\pi^+ \to e^+ + \nu_e$ (or is emitted upon the absorbtion of the e^- in $e^- + p \to n + \nu_e$). Similarly, the operational definition of the electron-antineutrino $\bar{\nu}_e$, the muon-neutrino, ν_μ and the muon-antineutrino $\bar{\nu}_\mu$ is based on the reactions

$$\pi^- \longrightarrow e^- + \bar{\nu}_e \tag{18}$$

$$\pi^+ \longrightarrow \mu^+ + \nu_\mu \tag{19}$$

$$\pi^- \longrightarrow \mu^- + \bar{\nu}_\mu \tag{20}$$

One should stress that, to the best of our present knowledge, the only physical distinction between ν_e and $\bar{\nu}_e$ other than in this circumstance of their emission lies in the experimentally established fact that the helicity of the ν_e is very close to $-\frac{1}{2}$ and that of the $\bar{\nu}_e$ very close to $+\frac{1}{2}$; precisely the same remark can be made regarding the physical distinction between ν_μ and $\bar{\nu}_\mu$.

(e) With the conventional dynamical description of the weak interactions given below the purely leptonic processes

$$\nu_\mu + e^- \rightarrow \nu_\mu + e^- \qquad (21)$$

$$\nu_\mu + (Z, A) \rightarrow \nu_\mu + e^+ + e^- + (Z, A) \qquad (22)$$

are "2nd-order weak" while the purely leptonic processes in Eqs. (2)-(7), the semileptonic processes in Eqs. (8a)-(11b) and the nonleptonic processes in Eqs. (12a)-(13b) are all "1st-order weak." Thus, we may suppose that, at least in a limited energy range of the incident ν_μ , the cross section for the process of Eq. (22) should be appreciably smaller than the cross section for the processes of Eqs. (6), (7). Experimental verification or contradiction of this supposition is of paramount importance for the theory of weak interactions as is indeed the whole question of the relative magnitude of the higher-order weak effects. In general, it is reasonable to imagine that all questions regarding the higher-order weak interactions are more easily answered if one confines one's attention to the "simpler" purely leptonic and semileptonic weak processes and this offers another justification for our emphasis on these processes.

(f) It is worth mentioning that while γ has no primitive or direct weak interaction and ν_e, ν_μ no primitive or direct electromagnetic interaction, γ interacts with $e^\mp$ and $\mu^\mp$ as do ν_e and ν_μ, respectively. As a consequence, γ acquires an induced or indirect weak interaction — see Fig. 1 — while the ν_e or ν_μ

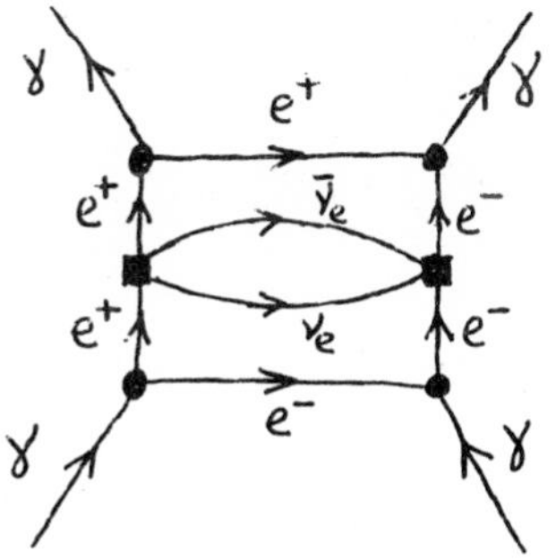

Fig. 1

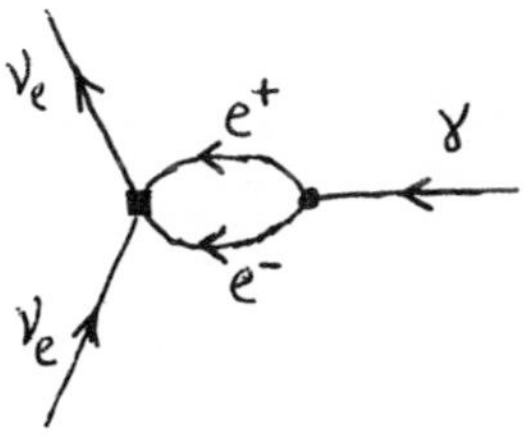

Fig. 2

acquires an induced or indirect electromagnetic interaction - see Fig. 2.[5] The induced electromagnetic interaction of ν_e, ν_μ corresponds to a small but nonvanishing charge radius while the induced weak interaction of γ slightly modifies the purely electrodynamic expression for the amplitude of photon-photon scattering.

We proceed to record the conventional weak-interaction Hamiltonian which describes quantitatively all the 1st-order leptonic and semileptonic weak processes. If no intermediate bosons exist so that the weak currents are directly coupled to each other, we have

$$H_{WEAK} = \int d^3x\, \mathcal{H}_{WEAK}(x) \qquad (23)$$

$$\mathcal{H}_{WEAK}(x) = \frac{G}{\sqrt{2}} J^{\dagger}_{\alpha}(x)\, J_{\alpha}(x)$$

$$J_{\alpha}(x) = l_{\alpha}(x) + h_{\alpha}(x) = \left(l^{(e\nu_e)}_{\alpha}(x) + l^{(\mu\nu_\mu)}_{\alpha}(x)\right) +$$

$$+ \left(\cos\theta_C\, h^{(q_n q_p)}_{\alpha}(x) + \sin\theta_C\, h^{(q_\lambda q_p)}_{\alpha}(x)\right)$$

$$l^{(e\nu_e)}_{\alpha}(x) = \overline{\psi}_e(x)\gamma_\alpha(1+\gamma_5)\psi_{\nu_e}(x) \,, \quad l^{(\mu\nu_\mu)}_{\alpha} = \overline{\psi}_\mu(x)\gamma_\alpha(1+\gamma_5)\psi_{\nu_\mu}(x) ,$$

$$h^{(q_n q_p)}_{\alpha}(x) = \overline{\psi}_{q_n}(x)\gamma_\alpha(1+\gamma_5)\psi_{q_p}(x) \,,$$

$$h^{(q_\lambda q_p)}_{\alpha} = \overline{\psi}_{q_\lambda}(x)\gamma_\alpha(1+\gamma_5)\psi_{q_p}(x)$$

where:

(a) $\mathcal{H}_{WEAK}(x)$ is the weak-interaction Hamiltonian density at the space-time point $x = [\vec{x}, it]$ corresponding to H_{WEAK}.[6]

(b) G is the universal weak coupling constant of dimensions (energy) x (volume); numerically, $G = \left(\frac{1.026 \times 10^{-5}}{m_p^{\,3}}\right) m_p$

as obtained by comparison of the experimental value for the rate of muon decay with the theoretical expression for this rate [2]

$$\Gamma(\mu^+ \to e^+ + \nu_e + \bar{\nu}_\mu) = \frac{G^2 m_\mu^5}{192\,\pi^3}\left(1 - \alpha\left(\frac{\pi}{2} - \frac{25}{8\pi}\right)\right) = \frac{G^2 m_\mu^5}{192\,\pi^3}(0.996)$$

(c) $J_\alpha(x)$ is the weak current with $J_\alpha^\dagger(x) \equiv$ (Hermitian conjugate of $J_\alpha(x)$) x $(1-2\delta_{\alpha 4})$

(d) $l_\alpha(x)$ is the total lepton weak current, $h_\alpha(x)$ is the total hadron weak current, $l_\alpha^{(e\nu_e)}(x)$ and $l_\alpha^{(\mu\nu_\mu)}(x)$ are, respectively, the (e, ν_e) and (μ, ν_μ) lepton weak currents, while $h_\alpha^{(q_n q_p)}(x)$ and $h_\alpha^{(q_\Lambda q_p)}(x)$ are, respectively, the (q_n, q_p) strangeness-nonchanging and the (q_Λ, q_p) strangeness-changing hadron weak currents.

(e) θ_c is the Cabibbo angle; numerically, $\theta_c = 0.2$ as obtained by comparison of the experimental value for the rate of $K^+ \to \pi^0 + e^+ + \nu_e$ with the theoretical expression for this rate [2]

$$\Gamma(K^+ \to \pi^0 + e^+ + \nu_e) = \frac{G^2 (\sin\theta_c)^2 m_K^5}{1536\,\pi^3}\left(1 - 8\left(\frac{m_\pi}{m_K}\right)^2 - 24\left(\frac{m_\pi}{m_K}\right)^4 \log\left(\frac{m_\pi}{m_K}\right) + 8\left(\frac{m_\pi}{m_K}\right)^6 - \left(\frac{m_\pi}{m_K}\right)^8\right) ; \; f_+([p_K - p_\pi]^2) \cong f_+(0) \cong 1$$

(f) $\psi_e(x)$, $\psi_{\nu_e}(x)$, $\psi_\mu(x)$, $\psi_{\nu_\mu}(x)$, $\psi_{q_n}(x)$, $\psi_{q_p}(x)$, $\psi_{q_\Lambda}(x)$ are field operators for the indicated (fermion-type) particles and γ_α, γ_5 are the anticommuting Dirac matrices.[7]

On the other hand, if intermediate bosons exist so that the weak currents are directly coupled not to each other but to the intermediate boson field, we have

$$H_{\text{SEMIWEAK}} = \int d^3x\, \mathcal{H}_{\text{SEMIWEAK}}(x) \qquad (24a)$$

$$\mathcal{H}_{\text{SEMIWEAK}}(x) = g\, W_\alpha^\dagger(x)\, J_\alpha(x) + \text{herm. conj.}$$

where g is the universal "semiweak" coupling constant (see below) and $W_\alpha(x)$ is the field operator for the intermediate boson, and where, since the $J_\alpha(x)$ are polar-vector and axial-vector charged currents, the intermediate bosons introduced have spin one and are charged: W^+ and W^-; $W_\alpha(x)$ annihilates W^+ and creates W^-. Using perturbation theory to eliminate the explicit appearance of the boson field operators, we see that H_{SEMIWEAK} gives rise in 2nd order in g to an effective $H^{(W)}_{\text{WEAK}}$ specified by

$$H^{(W)}_{\text{WEAK}} = \int d^3x\, \mathcal{H}^{(W)}_{\text{WEAK}}(x) \qquad (24b)$$

$$\mathcal{H}^{(W)}_{\text{WEAK}}(x) = g^2 \int d^4x'\, \Delta_{\beta\alpha}(x-x')\, T(J_\beta^\dagger(x'), J_\alpha(x))$$

$$\Delta_{\beta\alpha}(x-x') = \int d^4q \left(\frac{\delta_{\beta\alpha} + \frac{q_\beta q_\alpha}{m_W^2}}{m_W^2 + q^2} \right) e^{iq\cdot(x-x')}$$

where $T(J_\beta^\dagger(x'), J_\alpha(x))$ represents the time-ordered product of $J_\alpha(x)$ with $J_\beta^\dagger(x')$ and $\Delta_{\beta\alpha}(x-x')$ is the propagator of the intermediate boson. A typical 1st-order weak process, e.g. $\nu_e + e^- \rightarrow \nu_e + e^-$ (Eq. (2)) then proceeds in lowest order (i.e. 2nd order) in g by means of the exchange of an intermediate boson between the two leptons - see Fig. 3.[8]

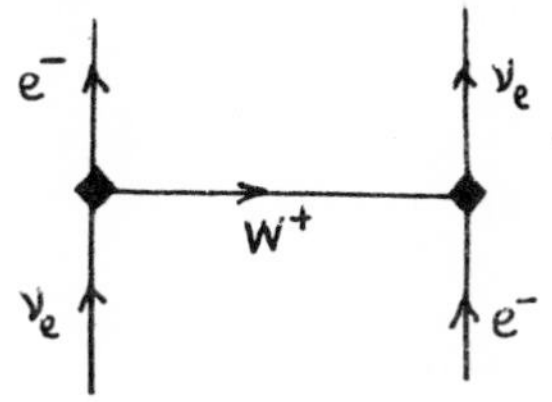

Fig. 3

Comparison of the $H_{WEAK}^{(w)}$ of Eq. (24b) with the H_{WEAK} of Eq. (23) shows that these two Hamiltonians will coincide in the limit of small four-momentum transfer at the vertices where the $J_\alpha(x)$ act provided that

$$\frac{g^2}{m_w^2} = \frac{G}{\sqrt{2}} = \frac{1}{\sqrt{2}}\left(\frac{10^{-5}}{m_p^2}\right) \qquad (25)$$

In connection with Eq. (25) we recall that the lower limit set by available experiments on (m_w/m_p) is $\cong 2$.

We now make some preliminary comments about the magnitude of higher order (e.g. 2nd order) weak interaction effects. Thus, using the H_{WEAK} of Eq. (23) a typical 2nd-order weak amplitude, e.g. that for

$\nu_\mu + e^- \longrightarrow \nu_\mu + e^-$ (Eq. (21)), is smaller than the corresponding 1st-order weak amplitude: $\nu_\mu + e^- \rightarrow \mu^- + \nu_e$ by a factor which on dimensional grounds is $\approx G\xi^2$ — here ξ is an energy relevant to the process, e.g. in Figure 4: $\xi = \xi_\mu + \xi_\nu = \xi_\mu + \sqrt{\xi_\mu^2 - m_\mu^2}$ with ξ_μ the maximum possible energy of the virtual μ^- in the c. of m. frame. There is, however, no natural limit imposed on ξ_μ by H_{WEAK} so that $G\xi^2 \cong G(4\xi_\mu^2)$ instead of being comfortably less than unity is actually divergent.

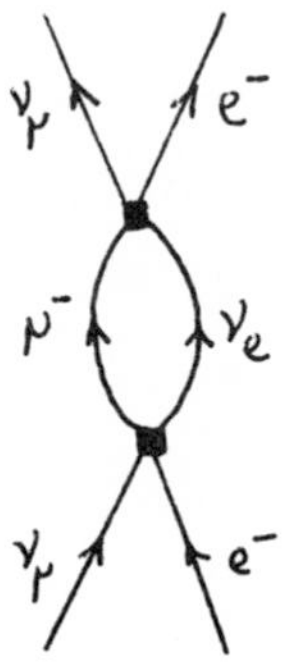

Fig. 4

An analogous divergence also occurs in the 4th-order semiweak amplitude for $\nu_\mu + e^- \rightarrow \nu_\mu + e^-$ associated with the $H_{SEMIWEAK}$ of Eq. (24a) - see Fig. 5; both of these divergences are particularly striking examples of the mathematical pathology disfiguring our (unrenormalizable) H_{WEAK} and $H_{SEMIWEAK}$. It is nevertheless reassuring that such

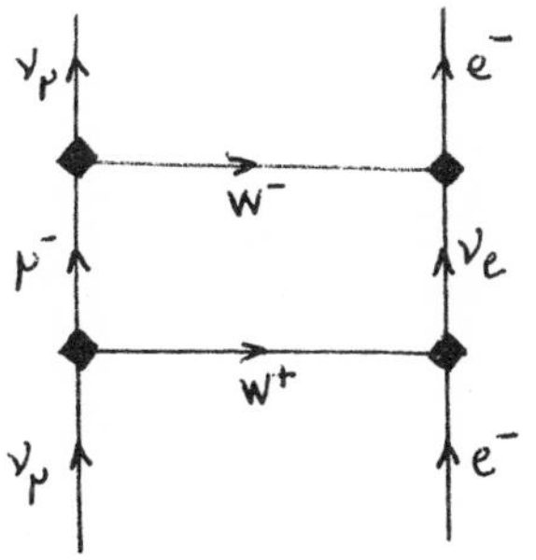

Fig. 5

2nd-order weak or 4th-order semiweak divergences can be removed by a variety of more or less plausible physical modifications of the underlying H_{WEAK} or $H_{SEMIWEAK}$. Thus, form factors which decrease with increasing momentum transfer and which describe spatially extended leptons might be introduced at the $\nu_\mu \mu^- e^- \nu_e$ vertex (Fig. 4) or at the $\nu_\mu \mu^- W$ and $\nu_e e^- W$ vertices (Fig. 5).

In this case ξ would be an inverse length specifying the "size" of the lepton and so would be an entirely new parameter introduced into particle theory. Alternatively, $H_{SEMIWEAK}$ could be modified in such a way $\left(W_\alpha^\dagger(x)\, J_\alpha(x) \rightarrow \left(W_\alpha^\dagger(x)\, J_\alpha(x)\right) \Big/ \left(1 + \frac{a}{m_W^2} W_\beta^\dagger(x) W_\beta(x)\right)^b ;\ a, b\right.$ numerical constants$\Big)$ that the 4th-order semiweak amplitude of Fig. 5 is finite, the corresponding value of ξ being m_W. Assuming that some definite finite values can be assigned to the $G\xi^2$ for the various 2nd-order weak or 4th-order semiweak amplitudes we note that, unless these values are all appreciably less than unity, various predictions made on the basis of H_{WEAK} in 1st-order or $H_{SEMIWEAK}$ in 2nd-order will be nullified. Thus, if the $G\xi^2$ associated with Fig. 5 is ≈ 1, the exchange of a W^+ and a W^- between the ν_μ and the e^- leads to an amplitude comparable to that arising from the exchange of a W^+ between a ν_e and an e^-. But, the exchange of a W^+ and a W^- is equivalent, as far as charge transfer is concerned, to the exchange of a W^0 (neutral intermediate boson) so that with $G\xi^2 \approx 1$ the $H_{SEMIWEAK}$ containing no W^0 and no neutral lepton weak currents nevertheless simulates in 4th-order semiweak processes the effects that would arise in 2nd-order semiweak processes on the basis of an $H_{SEMIWEAK}$ which contained a relatively important term coupling a neutral intermediate boson to a neutral lepton weak current. Further, any non-negligible amplitudes involving the exchange of three W's (e.g. W^-, W^+, W^-) between a lepton and a quark (as in the process of Eq. (8a) :

$$e^- + q_p \to \nu_e + W^- + q_p \to \nu_e + q_n \to e^- + W^+ + q_n \to$$

$$\to e^- + q_p \longrightarrow \nu_e + W^- + q_p \to \nu_e + q_n \;)$$

would be associated with an operator possessing transformation properties under rotations in isospace different from the transformation properties of the operator associated with the exchange of just a single W $(e^- + q_p \to \nu_e + W^- + q_p \to \nu_e + q_n)$.

This conclusion follows since $h_\alpha^{(q_n q_p)}(x) =$

$$= \bar{\psi}_{q_n}(x) \gamma_\alpha (1 + \gamma_5) \psi_{q_p}(x) = \bar{\psi}_{q_N}(x) \gamma_\alpha (1 + \gamma_5) \tau^{(-)} \psi_{q_N}(x)$$

is an isovector (see Eq. (23) et seq.)[9] and the $W^- W^+ W^-$ exchange amplitude is trilinear in $h_\alpha^{(q_n q_p)}(x)$ while the W^- exchange amplitude is linear in $h_\alpha^{(q_n q_p)}(x)$. In general, various experimentally established conservation laws and selection rules characterizing the lowest order amplitudes associated with H_{WEAK} or $H_{SEMIWEAK}$ will become inoperative if higher order corrections to these amplitudes $(\approx G\xi^2$ or $(G\xi^2)^2$ or ...) are important so that it behooves us to construct the theory so that these higher order corrections are always small.

We conclude this introductory section by presenting a preliminary discussion of the transformation properties of the weak current $J_\alpha(x)$ under time reversal ($\mathcal{T}$) and of the transformation properties of the strangeness-nonchanging hadron part of this current $h_\alpha^{(q_n q_p)}(x)$ under charge symmetry ($e^{i\pi I^{(2)}}$ = rotation through 180° about second axis in isospace). We note from the explicit expression for the various parts of $J_\alpha(x)$ in Eq. (23) that $l_\alpha(x)$ (the lepton part of $J_\alpha(x)$) as well as $h_\alpha^{(q_n q_p)}(x)$ and $h_\alpha^{(q_\lambda q_p)}(x)$ (the strangeness-changing hadron part) are all "normal" ($\mathbf{n}$), i.e., odd, under $\mathcal{T}$.

$$\begin{aligned} \mathcal{T}\, l_\alpha(x)\, \mathcal{T}^{-1} &= -\left(l_\alpha(x)\right)^\dagger \\ \mathcal{T}\, h_\alpha^{(q_n q_p)}(x)\, \mathcal{T}^{-1} &= -\left(h_\alpha^{(q_n q_p)}(x)\right)^\dagger \\ \mathcal{T}\, h_\alpha^{(q_\lambda q_p)}(x)\, \mathcal{T}^{-1} &= -\left(h_\alpha^{(q_\lambda q_p)}(x)\right)^\dagger \end{aligned} \qquad (26)$$

while $h_\alpha^{(q_n q_p)}(x)$ is also "regular" ($\mathbf{r}$), i.e., odd, under $e^{i\pi I^{(2)}}$

$$e^{i\pi I^{(2)}}\, h_\alpha^{(q_n q_p)}(x)\, e^{-i\pi I^{(2)}} = -\left(h_\alpha^{(q_n q_p)}(x)\right)^\dagger . \qquad (27)$$

It is however possible that one should augment the axial-vector part of the $h_\alpha^{(q_n q_p)}(x)$ given in Eq. (23):

$$\overline{\psi}_{q_n} \gamma_\alpha \gamma_5 \psi_{q_p} = \overline{\psi}_{q_r} \gamma_\alpha \gamma_5 \tau^{(-)} \psi_{q_r} ,$$

by terms which are "abnormal" (a), i.e., even, under T and/or terms which are "irregular" (i), i.e., even, under $e^{i\pi I^{(2)}}$. In this case we write [10]

$$\left(h_\alpha^{(q_n q_p)}(x) \right)_{\text{AXIAL PART}} \equiv A_\alpha(x) = A_\alpha^{(n)(r)}(x) + A_\alpha^{(a)(r)}(x) + A_\alpha^{(n)(i)}(x) + A_\alpha^{(a)(i)}(x) \tag{28}$$

with

$$T A_\alpha^{(x)(y)} T^{-1} = -a_{xy} \left(A_\alpha^{(x)(y)} \right)^\dagger, \tag{29a}$$

$$a_{nr} = -a_{ar} = a_{ni} = -a_{ai} = 1 ;$$

$$e^{i\pi I^{(2)}} A_\alpha^{(x)(y)} e^{-i\pi I^{(2)}} = -b_{xy} \left(A_\alpha^{(x)(y)} \right)^\dagger, \tag{29b}$$

$$b_{nr} = b_{ar} = -b_{ni} = -b_{ai} = 1 ;$$

and

$$A_\alpha^{(n)(r)} = \overline{\psi}_{q_r} \gamma_\alpha \gamma_5 \tau^{(-)} \psi_{q_r} \tag{30}$$

but do not necessarily attempt to set down explicit expressions for $A_\alpha^{(a)(r)}(x)$, $A_\alpha^{(n)(i)}(x)$, and $A_\alpha^{(a)(i)}(x)$ in terms of $\overline{\psi}_{q_r}(x)$ and $\psi_{q_r}(x)$ [11]. We must also mention that it is quite reasonable that the polar-vector part of the $h_\alpha^{(q_n q_p)}(x)$ given in Eq. (23):

$$\bar{\Psi}_{q_n} \gamma_\alpha \psi_{q_p} = \bar{\Psi}_{q_r} \gamma_\alpha \tau^{(-)} \psi_{q_r} = I_\alpha^{(-)} = \{\alpha\text{-component}$$

of the total isospin current of the hadron world in our quark model} is not to be augmented either by $\mathcal{T}$-abnormal terms or by $e^{i\pi I^{(2)}}$-irregular terms; this follows because of the impressive experimental evidence in favor of the conserved polar-vector current hypothesis (CVC) which identifies in a completely model-independent way the polar-vector part of the strangeness-nonchanging hadron weak current with the $\mathcal{T}$-normal, $e^{i\pi I^{(2)}}$-regular total isospin current $I_\alpha^{(-)}$ [12]. We also refrain at this stage from augmenting the expression for the $\ell_\alpha(x)$ in Eq. (23) by $\mathcal{T}$-abnormal terms. For the sake of completeness we note that $A_\alpha^{(n)(r)}$ and $A_\alpha^{(a)(i)}$ are odd under $G = C e^{i\pi I^{(2)}}$ and are called "first-class" (axial strangeness-nonchanging hadron weak) currents while $A_\alpha^{(a)(r)}$ and $A_\alpha^{(n)(i)}$ are even under G and are called "second-class" (axial strangeness-nonchanging hadron weak) currents. [13]

We now emphasize that the available unambiguously interpretable experimental evidence on the semileptonic strangeness-nonchanging weak processes rules out only the first-class current $A_\alpha^{(a)(i)}$ and is as yet incapable of making any very definite statements about the presence of the second-class currents $A_\alpha^{(n)(i)}$ and $A_\alpha^{(a)(r)}$ - this situation will be discussed below and we shall there describe several experiments whose outcomes should decide whether $A_\alpha^{(a)(r)}$ and $A_\alpha^{(n)(i)}$ are in fact contained in A_α. In particular, if the second-class $\mathcal{T}$-abnormal current $A_\alpha^{(a)(r)}$ is in fact contained in A_α, the H_{WEAK} of Eqs. (23) and (28) violates $\mathcal{T}$ invariance since this H_{WEAK} then contains a term of the form: $\ell_\alpha^\dagger (\overset{(-)}{I_\alpha} + A_\alpha^{(n)(r)})$ which is even under $\mathcal{T}$ together with a term of the form: $\ell_\alpha^\dagger A_\alpha^{(a)(r)}$ which is odd under $\mathcal{T}$. Clearly, settlement of this issue is of great importance.

2. Lepton Conservation and the Implications of a Possible Lepton Nonconservation

We proceed to discuss the experimental evidence for lepton conservation and to specify the experimentally available upper limits on a possible lepton nonconservation. Assuming further that lepton nonconservation consistent with these limits actually occurs and that this lepton nonconservation is associated with the properties of the neutrino, we discuss the consequences that follow and point out some directions for future investigation. In general, we may argue that the validity of lepton conservation is to be anticipated on the basis of a presumed analogy between leptons and baryons or leptons and electrically charged particles, together with the experimentally well founded belief in the validity of baryon conservation and charge conservation.

We recall the "operational definition" of the electron-neutrino ν_e, the electron-antineutrino $\bar{\nu}_e$, the muon-neutrino ν_μ, and the muon-antineutrino $\bar{\nu}_\mu$ as referring respectively to the "neutrinos" emitted together with e^+, e^-, μ^+, and μ^- in $\pi^\pm$ decay (Eqs. (16), (18)-(20) and associated discussion) and also repeat that, to the best of our present knowledge, the only physical distinction between ν_e and $\bar{\nu}_e$ (ν_μ and $\bar{\nu}_\mu$) other than in this circumstance of their emission lies in the experimentally established fact that the helicity of the ν_e (ν_μ) lies very close to $-\frac{1}{2}$ while the helicity of the $\bar{\nu}_e$ ($\bar{\nu}_\mu$) lies very close to $+\frac{1}{2}$.

This whole situation naturally suggests the assignment of electron lepton number L_e and muon lepton number L_μ according to:

$$
\begin{array}{ll}
L_e = 1 : e^-, \nu_e & L_\mu = 1 : \mu^-, \nu_\mu \\
L_e = -1 : e^+, \bar{\nu}_e & L_\mu = -1 : \mu^+, \bar{\nu}_\mu \\
L_e = 0 : \mu, \nu_\mu ; \gamma ; W ; q_n, q_p, q_\Lambda & \\
L_\mu = 0 : e, \nu_e ; \gamma ; W ; q_n, q_p, q_\Lambda &
\end{array}
\qquad (31)
$$

$$\left(\left\{(L_e)_{TOTAL}\right\}_{INITIAL} = -1,\ \left\{(L_e)_{TOTAL}\right\}_{FINAL} = +1\right) \quad ;$$

equivalently the succession of the processes in Eq. (33) and Eq. (32b) results in the overall process

$$n + n \to e^- + e^- + p + p \tag{35}$$

which obviously violates the conservation of $(L_e)_{TOTAL}$.

If now the lepton-nonconserving process:

$$\bar{\nu}_e + n\,[q_n q_n q_p] \to e^- + p\,[q_p q_n q_p]$$

(i.e., $\bar{\nu}_e + q_n \to e^- + q_p$) is characterized by a coupling constant $G\eta_e$ analogous to the characterization of the lepton-conserving process:

$$\bar{\nu}_e + p\,[q_p q_n q_p] \to e^+ + n\,[q_n q_n q_p] \quad \left(\text{i.e., } \bar{\nu}_e + q_p \to e^+ + q_n\right)$$

by the coupling constant G entering into the H_{WEAK} of Eq. (23) then one can set the upper limit

$$\eta_e < \sqrt{4\%} = 0.2 \tag{36}$$

b) The processes

$$\nu_\mu + p \to \mu^+ + n,\ N^{*0},\ \ldots \tag{37a}$$

and

$$\nu_\mu + n \to \mu^+ + N^{*-},\ \ldots \tag{37b}$$

are not observed. Here the ν_μ originates in π^+ decay

$$\pi^+ \to \mu^+ + \nu_\mu \tag{38}$$

and the upper limit on the cross sections of

$\nu_\mu + p \to \mu^+ + n,\ N^{*0}, \ldots$ and $\nu_\mu + n \to \mu^+ + N^{*-}, \ldots$

together with the rule that the $(L_e)_{TOTAL}$ and $(L_\mu)_{TOTAL}$ of a many particle system are additively composed from the L_e and L_μ of the individual particles and to the hypothesis that $(L_e)_{TOTAL}$ and $(L_\mu)_{TOTAL}$ are conserved in every allowed particle process — e.g. in Eq. (16)

$$\{(L_e)_{TOTAL}\}_{INITIAL} = (L_e)_{\pi^+} = 0 ,$$

$$\{(L_e)_{TOTAL}\}_{FINAL} = (L_e)_{e^+} + (L_e)_{\nu_e} = -1 + 1 = 0 .$$

The experimental evidence in favor of this $\{(L_e)_{TOTAL}, (L_\mu)_{TOTAL}\}$ - conservation principle (briefly, lepton-conservation principle) is as follows:

a) The process

$$\bar{\nu}_e + {}_{17}^{37}Cl_{20} \longrightarrow e^- + {}_{18}^{37}Ar_{19} \qquad (32a)$$

i. e.

$$\bar{\nu}_e + n_{BOUND} \longrightarrow e^- + p_{BOUND} \qquad (32b)$$

is not observed. Here the $\bar{\nu}_e$ originate in the beta decay of neutron-rich fission products

$$n_{BOUND} \longrightarrow p_{BOUND} + e^- + \bar{\nu}_e \qquad (33)$$

and the upper limit on the cross section of

$$\bar{\nu}_e + n_{BOUND} \longrightarrow e^- + p_{BOUND}$$

is 4% of the measured cross section of

$$\bar{\nu}_e + p \longrightarrow e^+ + n \qquad (34)$$

Note that the (unobserved) process $\bar{\nu}_e + n \rightarrow e^- + p$ violates the conservation of $(L_e)_{TOTAL}$

is less than 1/2% of the measured cross sections of

$$\nu_\mu + p \to \mu^- + N^{*++}, \ldots \qquad (39a)$$

and

$$\nu_\mu + n \to \mu^- + p, N^{*+}, \ldots \qquad (39b)$$

In this case the (unobserved) processes $\nu_\mu + p \to \mu^+ + n, N^{*0}, \ldots$ and $\nu_\mu + n \to \mu^+ + N^{*-}, \ldots$ violate the conservation of $(L_\mu)_{TOTAL}$ $\left(\{(L_\mu)_{TOTAL}\}_{INITIAL} = 1, \{(L_\mu)_{TOTAL}\}_{FINAL} = -1\right)$ or, equivalently, the succession of the processes in Eq. (38) and Eqs. (37a), (37b) results in the overall processes

$$\pi^+ + p \longrightarrow \mu^+ + \mu^+ + n, N^{*0}, \ldots \qquad (40a)$$

$$\pi^+ + n \longrightarrow \mu^+ + \mu^+ + N^{*-}, \ldots \qquad (40b)$$

which obviously violates the conservation of $(L_\mu)_{TOTAL}$.

If now the lepton-nonconserving processes:

$$\nu_\mu + p \longrightarrow \mu^+ + n, N^{*0}, \ldots$$

and

$$\nu_\mu + n \longrightarrow \mu^+ + N^{*-}, \ldots$$

(i.e. $\nu_\mu + q_p \to \mu^+ + q_n$)

are characterized by a coupling constant $G\eta_\mu$ analogous to the characterization of the lepton-conserving processes:

$\nu_\mu + p \to \mu^- + N^{*++}, \ldots$

and

$\nu_\mu + n \to \mu^+ + N^{*-}, \ldots$

(i.e. $\nu_\mu + q_n \rightarrow \mu^- + q_p$)
by the coupling constant G entering into the H_{WEAK} of Eq. (23) then one can set the upper limit

$$\eta_\mu < \sqrt{\tfrac{1}{2}\%} = 0.07 \quad . \qquad (41)$$

c) The processes

$$\nu_\mu + p \rightarrow e^- + N^{*++}, \ldots \qquad (42a)$$

$$\nu_\mu + n \rightarrow e^- + p, N^{*+}, \ldots \qquad (42b)$$

are not observed, the upper limit on their cross sections being 1% of the measured cross sections of the corresponding processes of Eqs. (39a) and (39b); the processes of Eqs. (42a) and (42b) (i. e. $\nu_\mu + q_n \rightarrow e^- + q_p$) obviously violate the conservation of both $(L_e)_{TOTAL}$ and $(L_\mu)_{TOTAL}$. Thus, if these lepton-nonconserving processes are characterized by a coupling constant $G\eta_{\mu e}$ analogous to the characterization of the lepton-conserving processes of Eqs. (39a), (39b) by the coupling constant G entering into the H_{WEAK} of Eq. (23) one can set the upper limit

$$\eta_{\mu e} < \sqrt{1\%} = 0.1 \qquad (43)$$

d) Further, a considerably smaller limit on $\eta_{\mu e}$ can be set on the basis of the non-observation of the weak process

$$\mu^+ \to e^+ + \gamma \, . \qquad (44)$$

Here, on the basis of the mechanism: $\mu^+ \to \bar{\nu}_\mu + W^+ \to e^+$ with either the μ^+ or the e^+ or the W^+ emitting the γ, the rate of $\mu^+ \to e^+ + \gamma$ is calculated to be

$$\Gamma(\mu^+ \to e^+ + \gamma) \cong \Gamma(\mu^+ \to e^+ + \nu_e + \bar{\nu}_\mu)\left[\frac{\alpha}{24\pi} \log\left(\frac{\Lambda^2}{m_\mu^2}\right)\right]^2 \eta_{\mu e}^2, \quad (45)$$

where $\Lambda \equiv$ cut-off parameter of the $\nu_\mu W^+$ loop — we take $\Lambda \cong m_W \cong 5 m_p$ — and $g\eta_{\mu e}$ characterizes the lepton-nonconserving process: $\bar{\nu}_\mu + W^+ \to e^+$ in a manner analogous to the characterization of the lepton-nonconserving process

$$\nu_\mu + q_n \to e^- + W^+ + q_n \to e^- + q_p$$

by the coupling constant $G\eta_{\mu e} = (g\eta_{\mu e}) \cdot g$.

Eq. (45), together with the experimental limit on $\Gamma(\mu^+ \to e^+ + \gamma)$, viz.:

$$\{\Gamma(\mu^+ \to e^+ + \gamma) / \Gamma(\mu^+ \to e^+ + \nu_e + \bar{\nu}_\mu)\}_{\text{EXPERIM}} < 10^{-8}$$

yield

$$\eta_{\mu e} < 10^{-2} \, . \qquad (46)$$

e) Finally, we record the smallest available upper limit on η_e, namely that obtained on the basis of the comparison of theory and experiment in the second-order weak processes of nuclear double-beta decay, viz.:[14]

$$\eta_e < 10^{-3} \qquad (47)$$

This upper limit is obtained by comparing the observed lifetime of ${}_{52}Te^{130}_{78} \rightarrow {}_{54}Xe^{130}_{76} + e^- + e^- + (\bar{\nu}_e + \bar{\nu}_e\ ?)$,

namely $10^{21.34 \pm 0.12}$ yr. [15], with the calculated lifetime as follows.

Suppose that in the nuclear double-beta decay process the two $\bar{\nu}_e$ always accompany the two e^- (experimentally, this question is not yet settled since the electron momentum spectrum has not yet been observed, only the daughter ${}_{54}Xe^{130}_{76}$ atoms being detected mass-spectrometrically in rocks containing the parent ${}_{52}Te^{130}_{78}$). Then $(L_e)_{TOTAL}$ is conserved

$$\left(\{(L_e)_{TOTAL}\}_{INITIAL} = 0,\ \{(L_e)_{TOTAL}\}_{FINAL} = 2(L_e)_{e^-} + 2(L_e)_{\bar{\nu}_e} = 2 - 2 = 0 \right)$$

and the double-beta decay essentially involves the single beta decay of two neutrons in the initial nucleus into two protons in the final nucleus:

$$\begin{aligned} (n_1)_{BOUND} &\rightarrow (p_1)_{BOUND} + e_1^- + (\bar{\nu}_e)_1 \\ (n_2)_{BOUND} &\rightarrow (p_2)_{BOUND} + e_2^- + (\bar{\nu}_e)_2 \end{aligned} \qquad (48)$$

or, in terms of the decomposition of the second-order weak double-beta decay amplitude into the "succession" of two first-order weak single-beta decay amplitudes,

$$ {}_{52}Te^{130}_{78} \rightarrow \left({}_{53}I^{130}_{77}\right)_{N^{TH}\ STATE} + e_1^- + (\bar{\nu}_e)_1 \rightarrow {}_{54}Xe^{130}_{76} + e_2^- + (\bar{\nu}_e)_2 + e_1^- + (\bar{\nu}_e)_1 \tag{49} $$

it being remembered on the basis of the known masses of ${}_{52}Te^{130}_{78}$ and $\left({}_{53}I^{130}_{77}\right)_{GROUND\ STATE}$ that all ${}_{52}Te^{130}_{78} \rightarrow \left({}_{53}I^{130}_{77}\right)_{N^{TH}\ STATE} + e_1^- + (\bar{\nu}_e)_1$ are virtual. The mechanism of Eq. (49), together with a semiempirical estimate of the

$$ {}_{52}Te^{130}_{78} \rightarrow \left({}_{53}I^{130}_{77}\right)_{N^{TH}\ STATE} + e_1^- + (\bar{\nu}_e)_1 $$

and

$$ \left({}_{53}I^{130}_{77}\right)_{N^{TH}\ STATE} \rightarrow {}_{54}Xe^{130}_{76} + e_2^- + (\bar{\nu}_e)_2 $$

single-beta decay amplitudes and the summation of these amplitudes for the various $\left({}_{53}I^{130}_{77}\right)$ states by a closure technique, yields a lifetime for ${}_{52}Te^{130}_{78}$ of $10^{22\pm2}$ yr., which is certainly consistent with the observed value. On the other hand, if one assumes that in the nuclear double-beta process the two e^- appear in most decays <u>without</u> the two $\bar{\nu}_e$, $(L_e)_{TOTAL}$ is not conserved

$$ \left(\{(L_e)_{TOTAL}\}_{INITIAL} = 0,\ \{(L_e)_{TOTAL}\}_{FINAL} = 2(L_e)_{e^-} = 2\right) $$

and the double-beta decay involves the single-beta decay of two neutrons in the initial nucleus into two protons in the final nucleus with the $\bar{\nu}_e$ emitted with the first e^- by the first neutron being absorbed upon the emission of the second e^- by the second neutron, viz:

$$(n_1)_{\text{BOUND}} \rightarrow (p_1)_{\text{BOUND}} + e_1^- + \bar{\nu}_e \quad (50)$$

$$\bar{\nu}_e + (n_2)_{\text{BOUND}} \rightarrow (p_2)_{\text{BOUND}} + e_2^- \quad (51)$$

The combination of the two processes in Eqs. (50) and (51) is exactly analogous to the combination of the two processes in Eqs. (32), (33) the difference being that the $\bar{\nu}_e$ is here virtual and there real. Alternatively, in terms of the decomposition of the second-order weak double-beta decay amplitude into the "succession" of two "first-order" weak single beta decay amplitudes, we may write

$${}_{52}\text{Te}^{130}_{78} \rightarrow \left({}_{53}\text{I}^{130}_{77}\right)_{N^{\text{TH}}\ \text{STATE}} + e_1^- + \bar{\nu}_e \rightarrow {}_{54}\text{Xe}^{130}_{76} + e_2^- + e_1^- . \quad (52)$$

The mechanism of Eq. (52), together with a semiempirical estimate of the ${}_{52}\text{Te}^{130}_{78} \rightarrow \left({}_{53}\text{I}^{130}_{77}\right)_{N^{\text{TH}}\ \text{STATE}} + e_1^- + \bar{\nu}_e$ and $\bar{\nu}_e + \left({}_{53}\text{I}^{130}_{77}\right)_{N^{\text{TH}}\ \text{STATE}} \rightarrow {}_{54}\text{Xe}^{130}_{76} + e_2^-$ single-beta decay amplitudes and the summation of these amplitudes for the various virtual $\bar{\nu}_e$ momenta and spin orientations and the various ${}_{53}\text{I}^{130}_{77}$ states by a closure technique, yields a lifetime for ${}_{52}\text{Te}^{130}_{78}$ of $10^{15\pm2}\,\text{yr}/\eta_e^2$. Hence, by comparison with the observed lifetime, we obtain the limit on η_e given in Eq. (47); we note that for reasons to be elucidated below and contrary to experience with second-order weak amplitudes generally, no divergences appear in the calculation of the second-order weak double-beta decay amplitudes either in the lepton-conserving ($e^- e^- \bar{\nu}_e \bar{\nu}_e$) case or in the lepton-nonconserving ($e^- e^-$) case.

We must now mention that some, possibly suspect, experimental evidence exists for the double-beta decay:

$$ {}_{52}Te^{128}_{76} \rightarrow {}_{54}Xe^{128}_{74} + e^- + e^- + (\bar{\nu}_e + \bar{\nu}_e\ ?) $$

with a lifetime of $10^{22.5 \pm 0.5}$ yr. [16] From the above quoted experimental value for the double-beta lifetime of ${}_{52}Te^{130}_{78}$ ($10^{21.3}$ yr.) one expects the double-beta lifetime of ${}_{52}Te^{128}_{76}$ to be

$$ \cong 10^{21.3} \cdot \left(\frac{\text{Energy Release in } Te^{130} \rightarrow Xe^{130}}{\text{Energy Release in } Te^{128} \rightarrow Xe^{128}} \right)^{8.4} \text{yr.} = 10^{21.3} \cdot 3^{8.4} \text{ yr} = $$

$$ = 10^{25.3} \text{ yr.} $$

if lepton conservation holds so that 4 leptons ($e^- e^- \bar{\nu}_e \bar{\nu}_e$) are always emitted, while a double-beta life of ${}_{52}Te^{128}_{76} \cong$

$$ \cong 10^{21.3} \cdot \left(\frac{\text{Energy Release in } Te^{130} \rightarrow Xe^{130}}{\text{Energy Release in } Te^{128} \rightarrow Xe^{128}} \right)^{4.2} \text{yr.} = 10^{21.3} \cdot 3^{4.2} \text{ yr} = $$

$$ = 10^{23.3} \text{ yr.} $$

is anticipated if lepton conservation does not hold so that 2 leptons ($e^- e^-$) are normally emitted. We therefore see that the ${}_{52}Te^{128}_{76}$ double-beta lifetime theoretically estimated on the assumption of lepton nonconservation, agrees better with the (possibly suspect) corresponding measured lifetime than the lifetime theoretically estimated on the assumption of lepton conservation, which disagrees with the measured lifetime by a factor close to 10^3. It is also to be emphasized that these theoretical estimates of the lifetime of ${}_{52}Te^{128}_{76}$, using as they do the measured lifetime of ${}_{52}Te^{130}_{78}$, depend only on the reasonable assumption that the various matrix element ratios

$$\left\{ \frac{\langle (I^{130})_n | (h_\alpha^{(q_n q_p)})^\dagger | Te^{130} \rangle}{\langle (I^{128})_n | (h_\alpha^{(q_n q_p)})^\dagger | Te^{128} \rangle} \right\}$$

and

$$\left\{ \frac{\langle Xe^{130} | (h_\alpha^{(q_n q_p)})^\dagger | (I^{130})_n \rangle}{\langle Xe^{128} | (h_\alpha^{(q_n q_p)})^\dagger | (I^{128})_n \rangle} \right\}$$

are of order unity since the dependence of lifetime on energy release is essentially given by the corresponding phase space available to the leptons (the phase space associated with energy release for massless leptons varies $\sim \epsilon^{3n-1}$ i.e. as $\epsilon^{3\cdot 4-1} = \epsilon^{11}$ for $e^- e^- \bar{\nu}_e \bar{\nu}_e$ and as $\epsilon^{3\cdot 2-1} = \epsilon^5$ for $e^- e^-$). Thus, it would appear that if the double-beta lifetime of ${}_{52}Te^{128}_{76}$ is indeed $10^{22.5}$ yr. lepton conservation is violated to the extent specified by $\eta_e \approx 10^{-3}$ (Eq. (47)).

As a conclusion to our discussion of the experimental evidence for lepton conservation and for a possible lepton non-conservation it is interesting to record a second scheme of assigning lepton numbers to the various leptons different from the $\{(L_e)_{TOTAL}, (L_\mu)_{TOTAL}\}$ – conserved scheme discussed above but again containing two conserved lepton numbers. This second scheme is as yet also consistent with all available experimental evidence and entails the assignment of lepton numbers L and L':

$$\begin{array}{ll} L = 1 & : e^-, \nu_e, \mu^-, \nu_\mu \\ L = -1 & : e^+, \bar{\nu}_e, \mu^+, \bar{\nu}_\mu \\ L' = 1 & : e^-, \nu_e, e^+, \bar{\nu}_e \\ L' = -1 & : \mu^-, \nu_\mu, \mu^+, \bar{\nu}_\mu \\ L = 0,\ L' = 1 & : \gamma, W, q_n, q_p, q_\Lambda \end{array} \qquad (53)$$

together with the rule that the $(L)_{TOTAL}$ and the $(L')_{TOTAL}$ of a many particle system are, respectively, additively and multiplicatively composed from the L and L' of the individual particles and the hypothesis of the conservation of $(L)_{TOTAL}$ and $(L')_{TOTAL}$ in every particle reaction. Thus, in the process of Eq. (32) which is not observed one has:

$$\{(L')_{TOTAL}\}_{INITIAL} = 1, \quad \{(L')_{TOTAL}\}_{FINAL} = 1$$

but

$$\{(L)_{TOTAL}\}_{INITIAL} = -1, \quad \{(L)_{TOTAL}\}_{FINAL} = +1$$

while in the process of Eq. (33) which is observed one has:

$$\{(L')_{TOTAL}\}_{INITIAL} = 1, \quad \{(L')_{TOTAL}\}_{FINAL} = 1$$

and

$$\{(L)_{TOTAL}\}_{INITIAL} = -1, \quad \{(L)_{TOTAL}\}_{FINAL} = -1 .$$

In general the $\{(L_e)_{TOTAL}, (L_\mu)_{TOTAL}\}$ – conserved scheme (Eq. (31)) and the $\{(L)_{TOTAL}, (L')_{TOTAL}\}$ – conserved scheme (Eq. (53)) forbid and allow the same reactions; a difference between them arises however in the case of the muon-decay type processes. Thus, the processes of Eqs. (5) and (7)

$$\mu^- \to \nu_\mu + \bar{\nu}_e + e^- \qquad (54)$$

and

$$\nu_\mu + (Z,A) \to \nu_e + e^+ + \mu^- + (Z,A) \qquad (55)$$

are allowed by both schemes (e.g. for Eq. (54):
$\{(L_e)_{TOTAL}\}_{INITIAL} = 0$, $\{(L_e)_{TOTAL}\}_{FINAL} = 0 - 1 + 1 = 0$;
$\{(L_\mu)_{TOTAL}\}_{INITIAL} = 1$, $\{(L_\mu)_{TOTAL}\}_{FINAL} = 1 + 0 + 0 = 1$ while
$\{(L)_{TOTAL}\}_{INITIAL} = 1$, $\{(L)_{TOTAL}\}_{FINAL} = 1 - 1 + 1 = 1$;
$\{(L')_{TOTAL}\}_{INITIAL} = -1$, $\{(L')_{TOTAL}\}_{FINAL} = (-1)\cdot 1 \cdot 1 = -1$).
On the other hand, the processes

$$\mu^- \to \bar{\nu}_\mu + \nu_e + e^- \qquad (56)$$

$$\nu_\mu + (Z,A) \to \nu_e + e^- + \mu^+ + (Z,A) \qquad (57)$$

are forbidden by the $\{(L_e)_{TOTAL}, (L_\mu)_{TOTAL}\}$ – conserved scheme but are allowed by the $\{(L)_{TOTAL}, (L')_{TOTAL}\}$ – conserved scheme (e.g. for Eq. (56):

$\{(L_e)_{TOTAL}\}_{INITIAL} = 0$, $\{(L_e)_{TOTAL}\}_{FINAL} = 0 + 1 + 1 = 2$;

$\{(L_\mu)_{TOTAL}\}_{INITIAL} = 1$, $\{(L_\mu)_{TOTAL}\}_{FINAL} = -1 + 0 + 0 = -1$

while $\{(L)_{TOTAL}\}_{INITIAL} = 1$, $\{(L)_{TOTAL}\}_{FINAL} = -1 + 1 + 1 = 1$;

$\{(L')_{TOTAL}\}_{INITIAL} = -1$, $\{(L')_{TOTAL}\}_{FINAL} = -1 \cdot 1 \cdot 1 = -1$.)

Thus, if the $\{(L)_{TOTAL}, (L')_{TOTAL}\}$ — conserved scheme is valid both the processes in Eq. (56) and in Eq. (54) are allowed and the muon-"neutrinos" from μ^--decay should, by means of processes allowed on the basis of either the lepton-conservation schemes i. e.

$$\nu_\mu + n \rightarrow \mu^- + p, N^{*+}, \ldots \quad (58a)$$

$$\nu_\mu + p \rightarrow \mu^- + N^{*++}, \ldots \quad (58b)$$

and

$$\bar{\nu}_\mu + n \rightarrow \mu^+ + N^{*-}, \ldots \quad (59a)$$

$$\bar{\nu}_\mu + p \rightarrow \mu^+ + n, N^{*0}, \ldots, \quad (59b)$$

produce <u>both</u> μ^- and μ^+ in collisions with neutrons and protons within nuclei. On the other hand, if the $\{(L_e)_{TOTAL}, (L_\mu)_{TOTAL}\}$ — conserved scheme is valid, only the process in Eq. (54) is allowed and the muon-"neutrinos" from μ^--decay will, by means of the processes of Eqs. (58a) and (58b), produce <u>only</u> μ^- in nuclear collisions. The experiment just described involving muon-"neutrinos" from decay has not yet been performed but the (e^-, μ^+) production process in Eq. (57) allowed by the $\{(L)_{TOTAL}, (L')_{TOTAL}\}$ — conserved scheme but not by the $\{(L_e)_{TOTAL}, (L_\mu)_{TOTAL}\}$ — conserved scheme, has been sought without success. However, the ($e^+ \mu^-$) production process in Eq. (55), allowed by both schemes, was not found either in the same experimental circumstances so that the $\{(L)_{TOTAL}, (L')_{TOTAL}\}$ — conserved scheme cannot yet be ruled out.

We now assume that lepton conservation is basically governed by the $\{(L_e)_{TOTAL}, (L_\mu)_{TOTAL}\}$ — conserved scheme but that the $(L_e)_{TOTAL}$ and $(L_\mu)_{TOTAL}$ conservation laws are not exact but fail to the extent specified by $\eta_e \approx 10^{-3}$ (Eq. (47)) and also $\eta_\mu \approx 10^{-3}$, $\eta_{\mu e} \approx 10^{-3}$ (The $\eta_{\mu e}$-failure will not be discussed further apart from the comment that were it to occur one would predict

$$\{\Gamma(\mu^+ \to e^+ + \gamma) \,/\, \Gamma(\mu^+ \to e^+ + \nu_e + \bar{\nu}_\mu)\} \approx 10^{-10}$$

i.e. a factor 100 less than the present experimental upper limit -- see Eqs. (45), (46) .) We proceed to parameterize the assumed η_e-failure by modifying the expression for the (e, ν_e) lepton weak current $\ell_\alpha^{(e\nu_e)}(x)$ which enters into the weak-interaction Hamiltonian H_{weak} of Eq. (23). (The assumed η_μ-failure is handled analogously.) Our procedure is as follows:

Consider the electron-neutrino field operator $\psi_{\nu_e}(x)$. Take this $\psi_{\nu_e}(x)$ to be a Majorana neutrino field operator

$$\psi_{\nu_e}(x) = \sum_{\substack{\vec{p} \\ s = 1/2, -1/2}} \left(a_{\nu_e}(\vec{p},s)\, u_{\nu_e}(\vec{p},s)\, e^{i\vec{p}\cdot\vec{x}} + a^\dagger_{\nu_e}(\vec{p},s)\, u^*_{\nu_e}(\vec{p},s)\, e^{-i\vec{p}\cdot\vec{x}} \right) \qquad (60)$$

which is hermitian and so invariant under particle-antiparticle conjugation

$$C\,\psi_{\nu_e}(x) = \psi^\dagger_{\nu_e}(x) = \psi_{\nu_e}(x) \qquad (61)$$

and which describes a spin-1/2 particle indistinguishable from the corresponding antiparticle ($a^{\dagger}_{\nu_e}(\vec{p},s)$ is the operator for creation of the particle into the $\vec{p},s$ eigenstate described by the spinor $u_{\nu_e}(\vec{p},s)$ and $a_{\nu_e}(\vec{p},s)$ is the corresponding destruction operator). The physical distinction between ν_e and $\bar{\nu}_e$ based on the fact that

$$(\vec{s}\cdot\hat{p})_{\nu_e} = -\tfrac{1}{2}\;, \quad (\vec{s}\cdot\hat{p})_{\bar{\nu}_e} = +\tfrac{1}{2}$$

is then introduced into Eq. (23) by coupling the $\psi_{\nu_e}(x)$ in the $\ell^{(e\nu_e)}_{\alpha}(x)$ with the $(1+\gamma_5)$ helicity-projection operator

$$(1+\gamma_5)\psi_{\nu_e}(x) = \sum_{\substack{\vec{p}\\ s=1/2,-1/2}} \Big(a_{\nu_e}(\vec{p},s)\big((1+\gamma_5)u_{\nu_e}(\vec{p},s)\big)e^{i\vec{p}\cdot\vec{x}} +$$

$$+ a^{\dagger}_{\nu_e}(\vec{p},s)\big((1-\gamma_5)u_{\nu_e}(\vec{p},s)\big)^{*} e^{-i\vec{p}\cdot\vec{x}}\Big) =$$

$$= \sum_{\substack{\vec{p}\\ s=1/2,-1/2}} \Big(a_{\nu_e}(\vec{p},s)\big((1-2\vec{s}\cdot\hat{p})\, u_{\nu_e}(\vec{p},s)\big)e^{i\vec{p}\cdot\vec{x}} +$$

$$+ a^{\dagger}_{\nu_e}(\vec{p},s)\big((1+2\vec{s}\cdot\hat{p})u_{\nu_e}(\vec{p},s)\big)^{*} e^{-i\vec{p}\cdot\vec{x}} \tag{62}$$

so that the particle destroyed by $a_{\nu_e}(\vec{p},s)$, the ν_e , has helicity = - 1/2 and may be assigned $L_e = +1$ while the particle created by $a^{\dagger}_{\nu_e}(\vec{p},s)$, the $\bar{\nu}_e$, has helicity = + 1/2 and may be assigned $L_e = -1$. If now the $\psi_{\nu_e}(x)$

in the (e, ν_e) lepton weak current is instead coupled with the helicity-projection operator $[(1+\gamma_5)+\eta_e(1-\gamma_5)]$ this (e, ν_e) lepton weak current entering into the H_{weak} of Eq. (23) becomes

$$\ell_\alpha^{(e\nu_e)}(x) = \bar{\psi}_e(x)\gamma_\alpha[(1+\gamma_5)+\eta_e(1-\gamma_5)]\psi_{\nu_e}(x) \quad (63)$$

and it then follows from Eqs. (63), (60) and (23) that the $\bar{\nu}_e$ emitted in, for example, $n \rightarrow p+e^-+\bar{\nu}_e$ (Eq. (33) or Eq. (50)) is in a superposition of helicity states + 1/2 and - 1/2 with relative weights $w(\frac{1}{2}) = \frac{1}{1+\eta_e^2}$ and $w(-\frac{1}{2}) = \frac{\eta_e^2}{1+\eta_e^2}$. The probability of absorbtion of such a $\bar{\nu}_e$ in, for example, $\bar{\nu}_e+n \rightarrow e^-+p$ (Eq. (32) or Eq. (51)) is now proportional to ($p(\pm\frac{1}{2})$ = relative probabilities of absorbtion of helicity components $\pm$ 1/2):

$$w(-\tfrac{1}{2})p(-\tfrac{1}{2}) + w(\tfrac{1}{2})p(\tfrac{1}{2}) = \left(\frac{\eta_e^2}{1+\eta_e^2}\right)\left(\frac{1}{1+\eta_e^2}\right) + \left(\frac{1}{1+\eta_e^2}\right)\left(\frac{\eta_e^2}{1+\eta_e^2}\right) = \frac{2\eta_e^2}{(1+\eta_e^2)^2}, \quad (64)$$

so that lepton conservation is violated to an extent proportional to η_e^2 for small η_e^2. Alternatively, one can say that, for example, the emission of a $\bar{\nu}_e$ in $n \rightarrow p+e^-+\bar{\nu}_e$ in a superposition of helicity states + 1/2 and - 1/2 with the above relative weights $w(\frac{1}{2})$ and $w(-\frac{1}{2})$ corresponds to the emission of this $\bar{\nu}_e$ in a superposition of electron-lepton-number states $L_e = -1$ and $L_e = +1$ with the same relative weights $w(\frac{1}{2})$ and $w(-\frac{1}{2})$ and with the $L_e = +1$ component associated with the nonconservation of $(L_e)_{TOTAL}$ (remember that the emitted e^- also has $L_e = +1$). From the point of view of a connection between the failure of exact $(L_e)_{TOTAL}$ -conservation and the breaking of a symmetry we note that the H_{weak} of Eq. (23) with $\ell_\alpha^{(e\nu_e)}(x)$ given by Eqs. (63) and (60) is not invariant under the transformation $\psi_{\nu_e}(x) \to \gamma_5 \psi_{\nu_e}(x)$, the extent of the noninvariance being proportional to η_e .

Our parameterization of lepton nonconservation according to Eqs. (63), (60) and (23) is, so to speak, complementary to the parameterization of parity nonconservation (in the leptonic and semileptonic weak processes) by these equations; thus $\eta_e = 0$ corresponds to lepton conservation and maximal parity nonconservation while $\eta_e = 1$ corresponds to parity conservation and maximal lepton nonconservation. Thus, our general approach is to associate any lepton nonconservation with a deviation from maximal parity nonconservation in the various leptonic and semileptonic weak processes or, more specifically, with a deviation of the average longitudinal spin polarization of the ν_e and the $\bar{\nu}_e$ from, respectively, -1 and +1 . Correspondingly, the average longitudinal spin polarization of e^- emitted in $n \to p + e^- + \bar{\nu}_e$ is now predicted to be

$$\left(-\frac{v_{el}}{c}\right)\left\{(1)\left(\frac{1}{1+\eta_e^2}\right) + (-1)\left(\frac{\eta_e^2}{1+\eta_e^2}\right)\right\} = \left(-\frac{v_{el}}{c}\right)\left(\frac{1-\eta_e^2}{1+\eta_e^2}\right) \qquad (65)$$

which, by comparison with experiment, can be used to set an upper limit of 1% on η_e^2 and so an upper limit of 0.1 on η_e. We note further that if η_e is complex with $\mathrm{Re}\,\eta_e \approx \mathrm{Im}\,\eta_e$ we can conceivably relate CP or T nonconservation to lepton nonconservation, it being recalled that 2×10^{-3} is the size of the CP-nonconserved amplitude in $K_L^0 \to \pi^+ + \pi^-$ in comparison to the CP-conserved amplitude in $K_s^0 \to \pi^+ + \pi^-$. In fact, the H_{weak} of Eqs. (23), (63), (60) with such a complex η and with an admixture of $\Delta S = -\Delta Q$ hadron weak current consistent with the experimental limits can give a rough account of the observed CP nonconservation in the K_L^0 decays.[17] To round out this general discussion we also comment that one can, of course, always describe any observed lepton nonconservation by simply postulating a "superweak" lepton-nonconserving interaction associated, for example, with the processes:

$$n \to N^{*++} + W^{--} \to N^{*++} + e^- + e^- \qquad (66)$$

with "semi-superweak" coupling constants $g\sqrt{\eta_e}$ and $g\sqrt{\eta_e}$ — such a superweak description of lepton nonconservation would not relate the lepton nonconservation to modifications in the conventionally expected properties of the neutrino and could in principle be distinguished from our description by the experimental discovery of such modifications.

In addition, a superweak description of lepton nonconservation would in general predict, for example, a different e_1^-, e_2^- angular correlation in nuclear double-beta decay than our $\{\eta_e \bar{\psi}_e(x)\gamma_\alpha(1-\gamma_5)\psi_{\nu_e}(x)\}$ admixture-in- $\ell_\alpha^{(e\nu_e)}(x)$ description.

The two most striking modifications in the conventionally expected properties of the electron-neutrino if lepton conservation fails and if this failure is due to the $\{\eta \bar{\psi}_e(x)\gamma_\alpha(1-\gamma_5)\psi_{\nu_e}(x)\}$ admixture into $\ell_\alpha^{(e\nu_e)}(x)$ (Eq. (63)) are the modifications of the expected values of the neutrino mass m_{ν_e} and neutrino magnetic moment μ_{ν_e} from the zero values demanded by the invariance of H_{WEAK} to $\psi_{\nu_e}(x) \rightarrow \gamma_5 \psi_{\nu_e}(x)$ to non-zero values proportional to η_e. From the experimental point of view, upper limits are available on m_{ν_e} and m_{ν_μ} and also μ_{ν_e} and μ_{ν_μ}, viz.:

$$\frac{m_{\nu_e}}{m_e} < 10^{-4} \tag{65a}$$

$$\frac{m_{\nu_\mu}}{m_\mu} < 10^{-2} \tag{65b}$$

$$\frac{\mu_{\nu_e}}{\mu_e} < 10^{-10} \tag{66a}$$

$$\frac{\mu_{\nu_\mu}}{\mu_\mu} < 10^{-6} \tag{66b}$$

These upper limits are obtained as follows:

a) $\frac{m_{\nu_e}}{m_e}$: from the shape of the electron energy spectrum near its high energy end in a low energy-release beta decay; this shape depends on the phase space available and is quite sensitive to a non-zero m_{ν_e} . The method has been applied to the beta decay $H^3 \to He^3 + e^- + \bar{\nu}_e$ and yields the upper limit of Eq. (65a). We note that the value of m_{ν_e} corresponding to this upper limit produces a much smaller γ_5-symmetry breaking in H_{weak} than that produced by $\eta_e = 10^{-3}$ so that m_{ν_e} can be safely set equal to zero in discussions of lepton-nonconserving nuclear double-beta decay ($(1+\gamma_5)u(\vec{p},s) = \left(1 - \frac{2\vec{s}\cdot\vec{p}}{\sqrt{\vec{p}^2 + m_{\nu_e}^2}} + \frac{\gamma_5\gamma_4 m_{\nu_e}}{\sqrt{\vec{p}^2 + m_{\nu_e}^2}}\right)u(\vec{p},s) \cong \left(1 - 2\vec{s}\cdot\hat{p} + \frac{\gamma_5\gamma_4 m_{\nu_e}}{|\vec{p}|}\right)u(\vec{p},s)$ with $\langle m_{\nu_e}/|\vec{p}|\rangle_{AVER.} \approx 10^{-6}$ for the virtual ν_e of Eqs. (50), (51) or Eq. (52)).

b) $\frac{m_{\nu_\mu}}{m_\mu}$: from the energy and momentum balance in $\pi^+ \to \mu^+ + \nu_\mu$ and measurements of m_π , m_μ and the μ^+ kinetic energy.

c) $\frac{\mu_{\nu_e}}{\mu_e}$ and $\frac{\mu_{\nu_\mu}}{\mu_\mu}$: from the experimental upper limits on the cross sections of $\bar{\nu}_e + e^- \to \bar{\nu}_e + e^-$ and $\nu_\mu + e^- \to \nu_\mu + e^-$ (1.3×10^{-46} cm^2 for fission spectrum $\bar{\nu}_e$ and 3×10^{-42} cm^2 for CERN beam ν_μ) assuming these cross sections to arise from the mechanism $\bar{\nu}_e + e^- \to \bar{\nu}_e + \gamma + e^- \to \bar{\nu}_e + e^-$

and

$$\nu_\mu + e^- \rightarrow \nu_\mu + \gamma + e^- \rightarrow \nu_\mu + e^- ,$$

the contribution to $\mathcal{H}_{ELMAG}(x)$ appropriate to the $\nu_\mu \nu_\mu \gamma$ and $\nu_e \nu_e \gamma$ vertices being taken as

$$\mu_{\nu_e} \overline{\psi}_{\nu_e}(x)\, \sigma_{\alpha\beta}\, \psi_{\nu_e}(x)\, \mathcal{F}_{\alpha\beta}(x) + \mu_{\nu_\mu} \overline{\psi}_{\nu_\mu}(x) \sigma_{\alpha\beta}\, \psi_{\nu_\mu}(x) \mathcal{F}_{\alpha\beta}(x). \tag{67}$$

From a theoretical point of view, we can calculate m_{ν_e}/m_e from the H_{weak} of Eq. (23) with the therein contained $\ell_\alpha^{(e\nu_e)}(x)$ given by Eqs. (63), (60) and

$$\psi_e(x) = \sum_{\substack{\vec{p} \\ s=1/2,-1/2}} \left(a_e(\vec{p},s) u_e(\vec{p},s) e^{i\vec{p}\cdot\vec{x}} + b_e^\dagger(\vec{p},s) u_e^*(\vec{p},s) e^{-i\vec{p}\cdot\vec{x}} \right) \tag{68}$$

where $a_e(\vec{p},s)$ is the e^- destruction operator and $b_e^\dagger(\vec{p},s)$ is the e^+ creation operator ($\frac{m_{\nu_\mu}}{m_\mu}$ is calculated analogously). We then have (see Fig. 2 without the γ line)

$$\frac{m_{\nu_e}}{m_e} = \frac{1}{m_e} \left\{ \langle \nu | \frac{G}{\sqrt{2}} \int \left(\ell_\alpha^{(e\nu_e)}(x) \right)^\dagger \ell_\alpha^{(e\nu_e)}(x)\, d^3x\, | \nu \rangle + \right.$$

$$\left. - \langle vac | \frac{G}{\sqrt{2}} \int \left(\ell_\alpha^{(e\nu_e)}(x) \right)^\dagger \ell_\alpha^{(e\nu_e)}(x)\, | vac \rangle \right\} =$$

$$= \frac{G}{\sqrt{2}} \left(\frac{\Lambda^2}{4\pi^2} \right) 16\eta_e \equiv \xi\,(16\eta_e) , \tag{69}$$

where Λ is the cut-off parameter used to render finite the otherwise divergent integral over the e loop. With $m_{\nu e}/m_e < 10^{-4}$ (Eq. (65a)), $\eta_e = 10^{-3}$ (Eq. (44)) and $G = 10^{-5}/m_p^2$ we have

$$\xi < 6 \times 10^{-3}, \quad \Lambda < 180 \text{ GeV}. \qquad (70)$$

These values of Λ and ξ are smaller by a factor of about 2 and about 4, respectively, than the so-called unitarity-limit values of Λ and ξ (see below).

As regards the calculated values of $\frac{\mu_{\nu e}}{\mu_e}$ and $\frac{\mu_{\nu_\mu}}{\mu_\mu}$ we use Fig. 2 and obtain

$$\frac{\mu_{\nu e}}{\mu_e} \cong \eta_e \frac{G m_e^2}{\sqrt{2}\, 4\pi^2} \log \frac{\Lambda^2}{m_e^2} = 10^{-15} \qquad (71)$$

$$\frac{\mu_{\nu_\mu}}{\mu_\mu} \cong \eta_\mu \frac{G m_\mu^2}{\sqrt{2}\, 4\pi^2} \log \frac{\Lambda^2}{m_\mu^2} = 10^{-11} \qquad (72)$$

for $\Lambda = 180 \text{ GeV}$, $\eta_e = 10^{-3}$, and $\eta_\mu = 10^{-3}$. These values appear to be too small to be observed in the foreseeable future.

Finally, we record the values of the charge radius of ν_e calculated from Fig. 2.[18] This is given by (η_e is here taken = 0)

$$\langle \nu_e' | J_\alpha^{ELMAG} | \nu_e \rangle = e \left(\bar{u}_{\nu_e} \gamma_\alpha (1+\gamma_5) u_{\nu_e} \right) F_{\nu_e}^{ELMAG}(q^2)$$

$$q^2 = (p'_{\nu_e} - p_{\nu_e})^2 \; ; \quad F_{\nu_e}^{ELMAG}(q^2) = F_{\nu_e}(0) + q^2 F'_{\nu_e}(0) + \cdots$$

$$= 0 + \frac{1}{6} q^2 r_{\nu_e}^2 + \cdots \tag{73}$$

with

$$r_{\nu_e} \cong \left[6 \left\{ \frac{G}{\sqrt{2}} \frac{1}{4\pi^2} \frac{1}{3} \log\left(\frac{\Lambda^2}{m_e^2}\right) \right\} \right]^{\frac{1}{2}} \tag{74}$$

Thus $r_{\nu_e} = 10^{-16}$ cm for $\Lambda = 180\,GeV$. and this value is so small that, for example, the corresponding scattering cross sections of ν_e in nuclear Coulomb fields are well below available experimental limits. Entirely analogous results are obtained for the charge radius of ν_μ ($m_e \to m_\mu$ in Eq. (74)).

3. First-Order and Second-Order Weak Collision Processes of Neutrinos with Electrons and with Nucleons - General Treatment of Second-Order Weak Processes

We now present a treatment of first-order weak and second-order weak collision processes of neutrinos with electrons and with protons and neutrons and also outline a general treatment of second-order weak processes. This will be done on the assumption that weak interactions are transmitted by intermediate charged spin-one bosons ($W^{\pm}$) so that "first-order weak" $\longleftrightarrow$ "second-order semiweak" and "second-order weak" $\longleftrightarrow$ "fourth-order semiweak."

We begin with a discussion of the elastic scattering of incident electron-neutrinos by target electrons $\nu_e + e^- \rightarrow \nu_e + e^-$ (Eq. (2) - see Fig. 3). The differential cross section for this second-order semiweak process is, using Eqs. (24a), (24b) with

$$J_\alpha(x) = \ell_\alpha^{(e\nu_e)}(x) = \bar{\Psi}_e \gamma_\alpha (1+\gamma_5)\psi_{\nu_e}$$

(so that $\eta_e = 0$),

$$\frac{d\sigma(\nu_e + e^- \rightarrow \nu_e + e^-)}{d(2\pi\cos\theta)} = \frac{p^2}{\pi}\,\frac{d\sigma(\nu_e + e^- \rightarrow \nu_e + e^-)}{d(q^2)} = \frac{\left(\frac{g^2}{4\pi}\right)^2 32p^2}{[m_W^2 + 4p^2(\sin\frac{\theta}{2})^2]^2} =$$

$$= \frac{\left(\frac{g^2}{4\pi}\right)^2 8s}{[m_W^2 + s(\sin\frac{\theta}{2})^2]^2} \qquad (75a)$$

where all lepton masses have been neglected, $\frac{g^2}{m_W^2} = \frac{G}{\sqrt{2}}$ (Eq. (25)), $p \equiv |\vec{p}_{\nu_e}| = |-\vec{p}_e|$ is the magnitude of the three-momentum of the ν_e in the center-of-mass frame, $\cos\theta = (\hat{p}_e)_{FINAL} \cdot (\hat{p}_{\nu e})_{INITIAL}$, and

$$q^2 \equiv \left((p_e)_{FINAL} - (p_{\nu e})_{INITIAL}\right)^2 = 2p^2(1-\cos\theta) = 4p^2\left(\sin\frac{\theta}{2}\right)^2$$

$$s \equiv -\left((p_e)_{INITIAL} + (p_{\nu e})_{INITIAL}\right)^2 = (2p)^2 = 2m_e E_\nu \qquad (75b)$$

where E_ν is the energy of the incident ν_e in the laboratory frame. It is to be noted that the $(q_\beta q_\alpha / m_W^2)$ terms in the W propagator (Eq. (24b)) give here a relatively negligible contribution $\approx m_e^2/m_W^2$ since the $(\nu_e)_{INITIAL}$, $(e^-)_{FINAL}$, $(e^-)_{INITIAL}$, $(\nu_e)_{FINAL}$ to which the W is coupled are all on their mass shell. The total cross section corresponding to the differential cross section of Eq. (75a) is

$$\sigma_{TOTAL}(\nu_e + e^- \to \nu_e + e^-) = \int \frac{d\sigma(\nu_e + e^- \to \nu_e + e^-)}{d(\cos\theta)} d(\cos\theta) = \frac{2}{\pi}\frac{g^4}{m_W^2}\frac{4p^2}{m_W^2 + 4p^2} =$$

$$= \frac{2}{\pi}\frac{g^4}{m_W^2}\frac{s}{m_W^2 + s} \qquad (76a)$$

and approaches a constant value as s becomes large:

$$\lim_{(s/m_W^2)\to\infty} \sigma_{TOTAL}(\nu_e + e^- \to \nu_e + e^-) = \frac{2}{\pi}\frac{g^4}{m_W^2} = \frac{G}{\pi} m_W^2 \qquad (76b)$$

The unitarity limit on this second-order semiweak $\nu_e + e^- \rightarrow \nu_e + e^-$ elastic scattering cross section is obtained by considering the partial wave expansion of the corresponding scattering amplitude:

$$\sqrt{\frac{d\sigma(\nu_e + e^- \rightarrow \nu_e + e^-)}{d(2\pi\cos\theta)}} = \frac{\frac{g^2}{4\pi}\sqrt{32}\,p}{\left[m_W^2 + 4p^2\left(\sin\frac{\theta}{2}\right)^2\right]} = A_{SWAVE}(p)\,P_0(\cos\theta) + A_{PWAVE}(p)\,P_1(\cos\theta) + \cdots \quad (77a)$$

with

$$A_{SWAVE}(p) = \left(\frac{g^2}{4\pi}\right)\frac{\sqrt{2}}{p}\log\left(1 + \frac{4p^2}{m_W^2}\right), \quad A_{PWAVE}(p) = \cdots, \quad (77b)$$

where $P_0(\cos\theta) = 1$, $P_1(\cos\theta) = \cos\theta$, $\cdots$ are Legendre polynomials and where, on the basis of the validity of unitarity for the scattering amplitude:

$$A_{SWAVE}(p) = \left(\frac{g^2}{4\pi}\right)\frac{\sqrt{2}}{p}\log\left(1 + \frac{4p^2}{m_W^2}\right) < \frac{1}{2p}, \cdots \quad (78a)$$

so that

$$p < p_{UNITAR.LIM.} \equiv \left(e^{\frac{\sqrt{2}\pi}{g^2}} - 1\right)^{1/2}\frac{m_W}{2} = \left(e^{\frac{2\pi}{Gm_W^2}} - 1\right)^{\frac{1}{2}}\frac{m_W}{2} \quad (78b)$$

$$\cong 10^3\ \text{GeV} \quad \text{for } m_W = 10\,m_p$$

In the limit $m_W = \infty$ (but with $\frac{g^2}{m_W^2} = \frac{G}{\sqrt{2}}$) which corresponds to the direct coupling of the weak current to its hermitian conjugate (Eq. (23)) these results become

$$\frac{d\sigma(\nu_e+e^- \to \nu_e+e^-)}{d(2\pi\cos\theta)} = \frac{p^2}{\pi}\frac{d\sigma(\nu_e+e^- \to \nu_e+e^-)}{d(q^2)} = \frac{G^2}{\pi^2}p^2 = \frac{G^2 s}{4\pi^2} \tag{79a}$$

so that $\frac{d\sigma(\nu_e+e^- \to \nu_e+e^-)}{d(q^2)}$ is independent of s and of q^2

$$\frac{d\sigma(\nu_e+e^- \to \nu_e+e^-)}{d(q^2)} = \frac{G^2}{\pi} \tag{79b}$$

$$\sigma_{TOTAL}(\nu_e+e^- \to \nu_e+e^-) = \frac{G^2}{\pi}4p^2 = \frac{G^2}{\pi}s =$$

$$= \frac{G^2}{\pi}(2m_e E_\nu) = 1.5\times 10^{-41}\,\text{cm}^2\,(E_\nu^{\text{in GeV}}) \tag{80}$$

$$A_{SWAVE}(p) = \frac{G}{\pi}p < \frac{1}{2p} \tag{81}$$

$$p < p_{UNITAR.LIM.} \equiv \sqrt{\frac{\pi}{2}}\frac{1}{\sqrt{G}} = 350\,\text{GeV} \tag{82}$$

Eq. (80) shows that $\sigma_{TOTAL}(\nu_e+e^- \to \nu_e+e^-)$ calculated in the first-order weak approximation increases linearly with E_ν and attains a value $= 2G = 8\times 10^{-33}\,\text{cm}^2$ for $p = p_{UNITAR.LIM.} = 350$ GeV which corresponds to $E_\nu = 5\times 10^8$ GeV !!! It is also plausible that an appropriate inclusion of 2nd-order weak contributions, third-order weak contributions, etc., to $\sigma_{TOTAL}(\nu_e+e^- \to \nu_e+e^-)$ would result in an expression of the form

$$\sigma_{TOTAL}(\nu_e+e^- \to \nu_e+e^-) = \frac{\frac{G^2}{\pi}4p^2}{1+\frac{1}{4}\left(\frac{G}{\pi}4p^2\right)^2} = \frac{\frac{G^2}{\pi}2m_e E_\nu}{1+\frac{1}{4}\left(\frac{G}{\pi}2m_e E_\nu\right)^2} \tag{83}$$

which never exceeds the unitarity limit on $\sigma_{TOTAL}(\nu_e + e^- \to \nu_e + e^-)$ viz.: $4\pi\left(\frac{1}{2p}\right)^2 = \frac{\pi}{p^2}$. We note in addition that while

$$p_{UNITAR. LIM} \gg \sqrt{\frac{\pi}{2}}\frac{1}{\sqrt{G}} = 350\ \text{GeV}\ (m_W \lesssim 10\, m_p)\ \text{for}\ \nu_e + e^- \to \nu_e + e^-$$

(Eq. (78b)), for second-order semiweak processes where the W is coupled to leptons off their mass shell so that $(q_\alpha q_\beta / m_W^2) \approx p^2/m_W^2$, as e.g. for $e^- + e^+ \longrightarrow W^- + \nu_e + e^+ \longrightarrow W^- + W^+$, one has

$$A_{PWAVE}(p) \approx Gp \tag{84}$$

and $p_{UNITAR. LIM.}$ is $\approx \frac{1}{\sqrt{G}}$ for all m_W.

Thus the $p_{UNITAR. LIM.}$ are not, in general, essentially different for the case of relatively small m_W than for the case of $m_W = \infty$ which corresponds to the directly coupled current-current H_{WEAK} (Eq. (23)).

As regards experiment, a recent study of the first-order weak process

$$\bar{\nu}_e + e^- \to \bar{\nu}_e + e^- \tag{85}$$

whose differential and total cross sections are

$$\frac{d\sigma(\bar{\nu}_e + e^- \to \bar{\nu}_e + e^-)}{d(q^2)} = \frac{1}{3}\frac{d\sigma(\nu_e + e^- \to \nu_e + e^-)}{d(q^2)} = \frac{1}{3}\frac{G^2}{\pi} \tag{86a}$$

and

$$\sigma_{TOTAL}(\bar{\nu}_e + e^- \to \bar{\nu}_e + e^-) = \frac{1}{3}\sigma_{TOTAL}(\nu_e + e^- \to \nu_e + e^-) = \frac{1}{3}\left(\frac{G^2}{\pi} 2m_e E_\nu\right) \tag{86b}$$

has yielded an upper bound of $1.3 \times 10^{-46}\ \mathrm{cm}^2$ for the cross section appropriate to incident fission-spectrum $\bar{\nu}_e$ and with recoil e^- detected in the energy region $3.8\ \mathrm{MeV.} < E_e < 5.0\ \mathrm{MeV.}$ [19] Since the corresponding theoretically calculated cross section is $3.3 \times 10^{-47}\ \mathrm{cm}^2$ (Eq. (86a) integrated over an appropriate range of q^2) we obtain the experimental upper limit

$$\left(G_{\nu_e e \nu_e e}\right)_{\mathrm{EXPER.}} < 2G \quad . \qquad (87)$$

Further, astrophysical evidence regarding the need of non-zero cross sections for first-order weak processes such as

$\gamma + e^- \rightarrow e^- + \nu_e + \bar{\nu}_e$ and $e^+ + e^- \rightarrow \nu_e + \bar{\nu}_e$

to account for the observed evolutionary time scale of certain types of stars indicates that[20]

$$\left(G_{\nu_e e \nu_e e}\right)_{\mathrm{EXPER.}} > \frac{1}{10} G \quad . \qquad (88)$$

These upper and lower limits on $\left(G_{\nu_e e \nu_e e}\right)_{\mathrm{EXPER.}}$ are of particular interest in view of theories[21] which, in order to mitigate higher-order weak divergences arising from the $\left(q_\beta q_\alpha / m_w^2\right)$ terms in the propagators of the intermediate charged spin-one bosons (see below), introduce such gradient-coupled-to $J_\alpha(x)$ intermediate spin-zero bosons as to predict that the effective coupling constant

$G_{\nu_e e \nu_e e}$ for the "diagonal" process $\nu_e + e^- \to \nu_e + e^-$ might well be different from the effective coupling constant $G_{\nu_\mu e \nu_e \mu} = G$ for the basic "nondiagonal" process $\nu_\mu + e^- \to \nu_e + \mu^-$ or $\mu^- \to \nu_\mu + \bar{\nu}_e + e^-$.

We proceed to give a very brief discussion of several representative processes which can produce a $W^\pm$. We have already mentioned (see Eqs. (14), (15a), (15b) and footnote 4) the first-order semiweak, second-order electromagnetic process (incident neutrino beam)

$$\nu_\mu + (Z,A) \to W^+ + \mu^- + (Z,A) \to W^+ + \mu^- + \gamma_{cov} + (Z,A) \to$$
$$\to W^+ + \mu^- + (Z,A) \tag{89}$$
$$W^+ \to \nu_\mu + \mu^+ \tag{90a}$$
$$W^+ \to \nu_e + e^+ \tag{90b}$$

(Fig. 6) which may be crudely estimated to have the cross section:[22]

$$\sigma(\nu_\mu + (Z,A) \to W^+ + \mu^- + \text{all final } (Z,A) \text{ states}) \approx Z \left(\frac{g^2}{4\pi}\right) \alpha^2 \left(\frac{1}{m_W^2}\right) =$$
$$= Z\left(\frac{G}{\sqrt{2}\, 4\pi} \alpha^2\right) = Z\,(1 \times 10^{-38}\,\text{cm}^2) \tag{91}$$

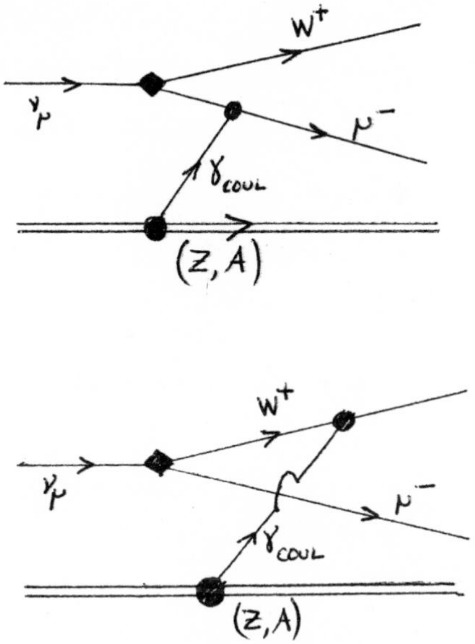

Fig. 6

We have also mentioned (Eq. (84)) the second-order semiweak process (colliding electron-position beams)

$$e^- + e^+ \longrightarrow W^- + \nu_e + e^+ \longrightarrow W^- + W^+ \qquad (92)$$

$$W^\pm \longrightarrow \nu_\mu (\bar{\nu}_\mu) + \mu^\pm \qquad (93a)$$

$$W^\pm \longrightarrow \nu_e (\bar{\nu}_e) + e^\pm \qquad (93b)$$

(Fig. 7a) with the crudely estimated cross section

$$\sigma(e^- + e^+ \rightarrow W^- + W^+) \approx \left(\frac{g^2}{4\pi}\right)^2 \frac{s}{m_W^4} = \frac{G^2}{32\pi^2} s = \qquad (94)$$

$$= 2 \times 10^{-37}\ \mathrm{cm}^2 \text{ for } s = (2 \times 20\ \mathrm{GeV})^2$$

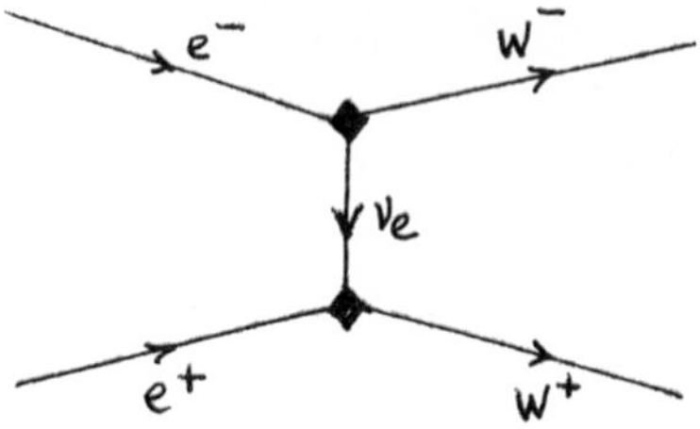

Fig. 7a

Considerably more important is the second-order electromagnetic process

$$e^- + e^+ \to \gamma \to W^- + W^+ \qquad (95)$$

(Fig. 7b) with cross section

$$\sigma(e^- + e^+ \to W^- + W^+) \cong \frac{\pi}{24}\alpha^2 \frac{s}{m_W^4} = \qquad (96)$$

$$= 4\times 10^{-34}\ \text{cm}^2 \text{ for } s = (2\times 20\ \text{GeV.})^2 \text{ and } m_W = 10\, m_p .$$

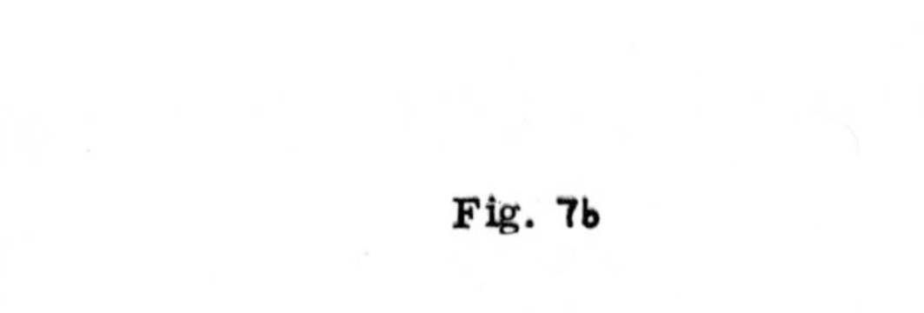

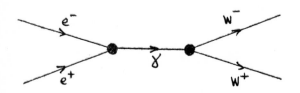

Fig. 7b

As indicated in Eqs. (90a), (90b) or Eqs. (93a), (93b) the dominant decay modes of the $W^{\pm}$ are expected to be leptonic since the hadronic decay modes (e.g. $W^{\pm} \rightarrow \pi^{\pm} + \pi^{0}$) are generally inhibited by strong-interaction type form factors.

Finally, we may mention the first-order semiweak, all-orders strong process (incident proton beam)

$$\pi^{+} + p \rightarrow N^{*++}, \ldots \rightarrow W^{+} + p, N^{*+}, \ldots \tag{97}$$

$$W^{+} \rightarrow \nu_{\mu} + \mu^{+} \tag{98a}$$

$$W^{+} \rightarrow \nu_{e} + e^{+} \tag{98b}$$

(Fig. 8) which is analogous to that in Eq. (89) with the difference that e.g. a q_p in $N^{*++}[q_p q_p q_p]$ rather than the ν_{μ} emits the W^{+} ($N^{*++}[q_p q_p q_p] \rightarrow W^{+} + p$, $N^{*+}, \ldots [q_n q_p q_p]$) i.e. $q_p \rightarrow W^{+} + q_n$

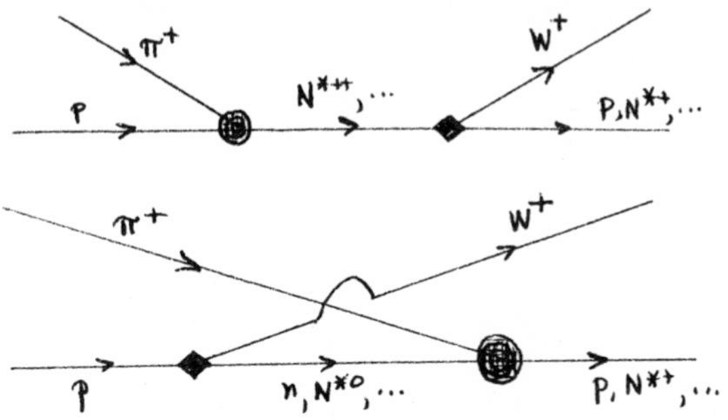

Fig. 8

rather than $\nu_\mu \longrightarrow W^+ + \mu^-$. The cross section for the process of Eq. (97) may be crudely estimated as

$$\sigma(\pi^+ + p \to W^+ + p, N^{*+}, \ldots) \approx$$

$$\approx \left\{ \frac{\frac{g^2}{4\pi}}{\frac{g^2_{\pi n p}}{4\pi}} \right\} \sigma(\pi^+ + p \to \pi^+ + p, N^{*+}, \ldots)$$

$$\cong \left\{ \frac{G m_W^2}{\sqrt{2} \cdot 4\pi \cdot 14} \right\} (3 \times 10^{-26}\,\text{cm}^2)$$

$$= 2 \times 10^{-31}\,\text{cm}^2 \text{ for } m_W = 10 m_p \qquad (99)$$

and is much larger than the cross sections in Eq. (96) or in Eq. (91). However, background problems are correspondingly more severe for $\pi^+ + p \longrightarrow W^+ + p, N^{*+}, \ldots$ than for $e^- + e^+ \longrightarrow W^- + W^+$ or for $\nu_\mu + (Z, A) \to W^+ + \mu^- + (Z, A)$ and it is generally believed that the $\nu_\mu + (Z, A) \longrightarrow W^+ + \mu^- + (Z, A)$ process is the most promising at present.

We now give an account of neutrino-nucleon collision processes and begin with a treatment of 2nd-order semiweak, i.e., 1st-order weak, processes such as

$$\nu_\mu + p[q_p q_p q_n] \to \mu^- + W^+ + p[q_p q_p q_n] \to \mu^- + N^{*++}[q_p q_p q_p] \qquad (100a)$$

i.e. (see Eqs. (8a), (23), (24a), (24b))

$$\nu_\mu + q_n \rightarrow \mu^- + W^+ + q_n \rightarrow \mu^- + q_p \; . \qquad (100b)$$

We note that processes such as $\nu_\mu + q_p \rightarrow \mu^+ + q_n$ are forbidden by conservation of $(L_\mu)_{TOTAL}$ (Eq. (31) et seq.); further, processes such as

$$\nu_\mu + q_p \rightarrow \mu^- + W^+ + q_p \rightarrow \mu^- + q_p + \bar{q}_n + q_p \qquad (100c)$$

while just as allowed as those in Eq. (100b) for neutrinos incident on free quarks, are suppressed in our ultra-crude quark model where quark-antiquark pair terms are neglected in the wavefunctions of the bound states corresponding to

$$p, \; n, \; N^{*++}, \; N^{*+}, \; N^{*0}, \; N^{*-},$$

etc. Thus, in our model, when a muon-neutrino is incident on a proton or a neutron, only the bound q_n acts as a target particle and, similarly, when a muon-antineutrino is incident on a proton or a neutron, only the bound q_p acts as a target particle. As we shall see below, this consequence of our model together with the H_{weak} of Eq. (23) or the $H^{(W)}_{weak}$ of Eq. (24b), (24a) will enable us to deduce several simple and easily testable relations among cross sections for

$$\nu_\mu + p \rightarrow \mu^- + N^{*++}, \cdots \; ; \; \nu_\mu + n \rightarrow \mu^- + p, N^{*+}, \cdots \; ;$$

$$\bar{\nu}_\mu + p \rightarrow \mu^+ + n, \; N^{*0}, \cdots \; ;$$

and $\bar{\nu}_\mu + n \rightarrow \mu^+ + N^{*-}, \cdots$.

We first set down the cross sections for the elastic collisions between incident muon-neutrinos and target free quarks. In view of Eqs. (23), (24b), (24a) and in exact analogy to Eqs. (79a), (79b), (80), we have, in the limit $m_W \to \infty$,

$$\frac{d\sigma(\nu_\mu + q_n \to \mu^- + q_p)}{d(2\pi \cos\theta)} = \frac{d\sigma(\bar{\nu}_\mu + q_p \to \mu^+ + q_n)}{d(2\pi\cos\theta)} =$$

$$= \frac{(G\cos\theta_c)^2}{\pi^2}\, p^2 = \frac{(G\cos\theta_c)^2}{4\pi^2}\, s\,, \qquad (101a)$$

$$\frac{d\sigma(\nu_\mu + q_n \to \mu^- + q_p)}{d(q^2)} = \frac{d\sigma(\bar{\nu}_\mu + q_p \to \mu^+ + q_n)}{d(q^2)} = \frac{(G\cos\theta_c)^2}{\pi}\,, \qquad (101b)$$

$$\sigma_{TOTAL}(\nu_\mu + q_n \to \mu^- + q_p) = \sigma_{TOTAL}(\bar{\nu}_\mu + q_p \to \mu^+ + q_n) = \frac{(G\cos\theta_c)^2}{\pi}\, 4p^2 =$$

$$= \frac{(G\cos\theta_c)^2}{\pi}\, s = \frac{(G\cos\theta_c)^2}{\pi}\, 2m_q E_\nu =$$

$$= (1.5\times 10^{-41}\ \text{cm}^2)\left(\frac{m_q}{m_e}\right) E_\nu^{\text{in GeV}}\,, \qquad (102)$$

with $q^2 \equiv (p_{q_n} - p_{q_p})^2 = (p_\mu - p_{\nu_\mu})^2 \cong 2p^2(1-\cos\theta)$; $s \equiv -(p_{\nu_\mu} + p_{q_n})^2 \cong$

$\cong (2p)^2 \cong 2m_q E_\nu$; $p \equiv |\vec{p}_{\nu_\mu}| = |-\vec{p}_{q_n}| \cong |\vec{p}_\mu| = |-\vec{p}_{q_p}|$;

$$\cos\theta \equiv \hat{p}_{\nu_\mu}\cdot\hat{p}_\mu \qquad (103)$$

and

$$m_q \ll p < p_{\text{UNITAR. LIM.}} = \sqrt{\frac{\pi}{2}}\,\frac{1}{\sqrt{G}} = 350\ \text{GeV}\,., \qquad (104)$$

so that we suppose that m_q, the mass of a free quark, while presumably much larger than m_p $(\cong 9\,m_\mu)$ is less than, say, $20\,m_p$. We then recall the well-known theorem (proof just below) that in a scattering process of a high-energy incident particle (e.g. a neutrino) by a composite system (e.g. a hadron such as a proton) consisting of $\mathcal{N}$ well-defined but otherwise arbitrarily strongly interacting constituents (e.g. quarks), the differential cross section, calculated in impulse approximation for the various final states of the composite system and then summed over the (generally very numerous) kinematically accessible final states, is equal to the number of constituents $\mathcal{N}_{\text{TARGET}} \leq \mathcal{N}$ which act as targets multiplied by the corresponding differential cross section of a single target constituent. Thus, using this theorem and Eq. (101b), we have

$$\frac{d\sigma(\nu_\mu + p[q_p q_p q_n] \to \mu^- + N^{*++}[q_p q_p q_p], \cdots)}{d(q^2)} \equiv$$

$$\equiv \sum_{\substack{f:\ \text{kinematically} \\ \text{accessible } N^{*++}, \ldots}} \frac{d\sigma(\nu_\mu + p \to \mu^- + f)}{d(q^2)_f} \cong$$

$$\cong 1 \cdot \frac{d\sigma(\nu_\mu + q_n \to \mu^- + q_p)}{d(q^2)} = 1 \cdot \frac{(G\cos\theta_c)^2}{\pi}, \quad (105a)$$

$$\frac{d\sigma(\nu_\mu + n[q_p q_n q_n] \to \mu^- + p[q_p q_p q_n], N^{*+}[q_p q_p q_n], \ldots)}{d(q^2)} \equiv$$

$$\equiv \sum_{\substack{f:\text{ kinematically} \\ \text{accessible } p, N^{*+}, \ldots}} \frac{d\sigma(\nu_\mu + n \to \mu^- + f)}{d(q^2)_f} \cong$$

$$\cong 2 \cdot \frac{d\sigma(\nu_\mu + q_n \to \mu^- + q_p)}{d(q^2)} = 2 \cdot \frac{(G\cos\theta_c)^2}{\pi}, \quad (105b)$$

$$\frac{d\sigma(\bar{\nu}_\mu + p[q_p q_p q_n] \to \mu^+ + n[q_p q_n q_n], N^{*0}[q_p q_n q_n], \ldots)}{d(q^2)} \equiv$$

$$\equiv \sum_{\substack{f:\text{ kinematically} \\ \text{accessible } n, N^{*0}, \ldots}} \frac{d\sigma(\bar{\nu}_\mu + p \to \mu^+ + f)}{d(q^2)_f} \cong 2 \cdot \frac{d\sigma(\bar{\nu}_\mu + q_p \to \mu^+ + q_n)}{d(q^2)} = 2 \cdot \frac{(G\cos\theta_c)^2}{\pi}, \quad (105c)$$

$$\frac{d\sigma(\bar{\nu}_\mu + n[q_p q_n q_n] \to \mu^+ + N^{*-}[q_n q_n q_n], \ldots)}{d(q^2)} \equiv \sum_{\substack{f:\text{ Kinematically} \\ \text{accessible } N^{*-}, \ldots}} \frac{d\sigma(\bar{\nu}_\mu + n \to \mu^+ + f)}{d(q^2)_f} \cong$$

$$\cong 1 \cdot \frac{d\sigma(\bar{\nu}_\mu + q_p \to \mu^+ + q_n)}{d(q^2)} = 1 \cdot \frac{(G\cos\theta_c)^2}{\pi}, \quad (105d)$$

where

$$(q^2)_f \equiv (p_i - p_f)^2 = -M_i^2 - M_f^2 + 2M_i E_f = M_i^2 - M_f^2 + 2M_i(E_\nu - (E_\mu)_f) =$$

$$= (p_\mu - p_{\nu_\mu})^2 \cong 2(|\vec{p}_\mu|)_f |\vec{p}_{\nu_\mu}|(1-\cos\theta) \cong 2p^2\left(1 - \frac{M_f^2 - M_i^2}{4p^2}\right)(1-\cos\theta);$$

$$|i\rangle = |p\rangle \text{ or } |n\rangle; \quad p \equiv |\vec{p}_{\nu_\mu}| = |-\vec{p}_i|; \quad (|\vec{p}_\mu|)_f = |-\vec{p}_f| \cong$$

$$\cong \frac{1}{2}\left(p + \sqrt{M_i^2 + p^2}\right)\left(1 - \frac{M_f^2}{\left(p + \sqrt{M_i^2 + p^2}\right)^2}\right) =$$

$$= p\left(1 - \frac{M_f^2 - M_i^2}{4p^2} + \dots\right); \qquad \cos\theta \equiv \hat{p}_{\nu_\mu} \cdot (\hat{p}_\mu)_f, \qquad (106)$$

and where each $d(q^2)_f$ is taken equal to $d(q^2)$.

(Proof of Theorem: For incident neutrino energy large compared to the average energy transferred by the neutrino to the target proton and for this average energy in turn large compared to the mean spacing between excited nucleon states, i.e. for $E_\nu \gg \{E_\nu - (E_\mu)_f\}_{AV} \gg \frac{1}{3} m_p$,

$$\sum_{\substack{f:\text{ kinematically}\\ \text{accessible } N^{*++},\dots}} \frac{d\sigma(\nu_\mu + p \longrightarrow \mu^- + f)}{d(q^2)_f} \cong \sum_{f:\text{ all } N^{*++},\dots} \frac{d\sigma(\nu_\mu + p \rightarrow \mu^- + f)}{d(q^2)_f} =$$

$$= \sum_{f:\text{ all } N^{*++},\dots} 2\pi \rho_f \left| \langle f | \sum_{j=1}^{N} \Gamma_j e^{i\vec{q}_f \cdot \vec{r}_j} + \begin{pmatrix}\text{Corrections to Impulse}\\ \text{Approximation}\end{pmatrix} | p \rangle \right|^2 \equiv$$

$$\cong 2\pi \{\rho_f\}_{AV} \langle p | \sum_{j=1}^{\mathcal{N}} \Gamma_j^{+}\Gamma_j + \sum_{j=1,k=1}^{\mathcal{N}} (1-\delta_{jk}) \Gamma_k^{+}\Gamma_j \, e^{i\vec{q}_f\cdot(\vec{r}_j-\vec{r}_k)} | p \rangle$$

$$\cong \mathcal{N}_{TARGET} \, 2\pi \{\rho_f\}_{AV} \langle p | \Gamma_{q_n}^{+} \Gamma_{q_n} | p \rangle$$

$$\left(\text{for } |\vec{q}_f| = |\vec{q}| \gg |\vec{r}_j - \vec{r}_k|^{-1} \right)$$

$$\cong \mathcal{N}_{TARGET} \frac{d(\nu_\mu + q_n \to \mu^- + q_p)}{d(q^2)} ; \qquad (107)$$

$$\rho_f = \left(\frac{\{(|\vec{p}_\mu|)_f\}^2}{(2\pi)^3 \left((|\vec{\nu}_\mu|)_f + |\vec{\nu}_f| \right)} \Big/ \left(\frac{d(q^2)_f}{d(2\pi\cos\theta)} \right) \right); \quad \text{here } \mathcal{N} = 3, \; \mathcal{N}_{TARGET} = 1.)$$

Therefore, from Eqs. (105a) – (105d) and Eq. (106) (compare also Eq. (102))

$$\sigma_{TOTAL}(\nu_\mu + p \to \mu^- + N^{*++}, \dots) \cong \frac{1}{2} \sigma_{TOTAL}(\nu_\mu + n \to \mu^- + p, N^{*+}, \dots) \cong$$

$$\cong \frac{1}{2} \sigma_{TOTAL}(\bar{\nu}_\mu + p \to \mu^+ + n, N^{*0}, \dots) \cong \sigma_{TOTAL}(\bar{\nu}_\mu + n \to \mu^+ + N^{*-}, \dots) \cong$$

$$\cong \frac{(G\cos\theta_c)^2}{\pi} 2p^2 \left(1 - \frac{\{M_f^2 - M_i^2\}_{AV}}{4p^2} \right) \left[(1-\cos(\pi)) - (1-\cos(0)) \right] \cong$$

$$\cong \frac{(G\cos\theta_c)^2}{\pi} 4p^2 \qquad \cong$$

$$\cong \frac{(G\cos\theta_c)^2}{\pi}\; 2\,(m_q)_{EFFECTIVE}\, E_\nu =$$

$$= \left(1.5\times 10^{-41}\,\text{cm}^2\right)\left(\frac{(m_q)_{EFFECTIVE}}{m_e}\right)\left(E_\nu^{\text{in GeV}}\right), \qquad (108)$$

where $(m_q)_{EFFECTIVE}$ is the dynamically effective mass of a quark bound in $p, n, N^{*++}, N^{*+}, N^{*o}, N^{*-}$, etc. Hence, using Eq. (108), and with $N_i, N_{f_1}, N_{f_2}, \ldots$ referring to a given nucleus of mass number A in the indicated initial and kinematically accessible final states

$$\frac{\sigma_{TOTAL}(\nu_\mu + N_i \to \mu^- + N_{f_1}, N_{f_2}, \ldots)}{A} \cong \frac{1}{2}\Big(\sigma_{TOTAL}(\nu_\mu + p \to \mu^- + N^{*++}, \ldots) +$$

$$+ \sigma_{TOTAL}(\nu_\mu + n \to \mu^- + p, N^{*+}, \ldots)\Big) \cong \frac{3}{2}\,\sigma_{TOTAL}(\nu_\mu + p \to \mu^- + N^{*++}, \ldots) \cong$$

$$\cong \frac{3}{2}\left(1.5\times 10^{-41}\,\text{cm}^2\right)\left(1.84\times 10^3\right)\left(\frac{(m_q)_{EFFECTIVE}}{m_p}\right)\left(E_\nu^{\text{in GeV}}\right) =$$

$$= \left(4.1\times 10^{-38}\,\text{cm}^2\right)\left(\frac{(m_q)_{EFFECTIVE}}{m_p}\right)\left(E_\nu^{\text{in GeV}}\right), \qquad (109)$$

so that if we take, consistent with quark-model studies of the electromagnetic properties of the proton and neutron,[23]

$$(m_q)_{EFFECTIVE} \cong \frac{1}{3} m_p \ll m_q \qquad (110)$$

we obtain the calculated value

$$\left(\frac{\sigma_{TOTAL}(\nu_\mu + N_i \to \mu^- + N_{f_1}, N_{f_2}, \ldots)}{A}\right)_{THEORETICAL} \cong \left(1.4 \times 10^{-38}\,cm^2\right)\left(E_\nu^{\text{in GeV}}\right), \tag{111}$$

which is to be compared with the measured value[24]

$$\left(\frac{\sigma_{TOTAL}(\nu_\mu + N_i \to \mu^- + N_{f_1}, N_{f_2}, \ldots)}{A}\right)_{EXPERIMENTAL} = \left((0.8 \pm 0.2) \times 10^{-38}\,cm^2\right)\left(E_\nu^{\text{in GeV}}\right). \tag{112}$$

Considering the extreme crudity of our quark model, the agreement between the measured and calculated values of

$$\frac{\sigma_{TOTAL}(\nu_\mu + N_i \to \mu^- + N_{f_1}, N_{f_2}, \ldots)}{A}$$

is quite encouraging. It would also be of great interest to test experimentally the predictions of Eq. (108) regarding the ratios of the various σ_{TOTAL}.

We next note on the basis of Eq. (107) and Eqs. (105a), (105b) that

$$\frac{d\sigma(\nu_\mu + N_i \to \mu^- + N_{f_1}, N_{f_2}, \ldots)}{d(q^2)} \cong$$

$$\cong \frac{A}{2}\left\{\frac{d\sigma(\nu_\mu + p \to \mu^- + N^{*++}, \ldots)}{d(q^2)} + \frac{d\sigma(\nu_\mu + n \to \mu^- + p, N^{*+}, \ldots)}{d(q^2)}\right\}$$

$$\cong \frac{3}{2} A \left\{\frac{d\sigma(\nu_\mu + p \to \mu^- + N^{*++}, \ldots)}{d(q^2)}\right.$$ is expected to be approximately independent of q^2 for

$$E_\nu \gg \left\{E_\nu - (E_\mu)_f\right\}_{AV} \gg \tfrac{1}{3} m_p$$

("deep inelastic" differential scattering cross section) while a strong dependence of $\frac{d\sigma(\nu_\mu + p \to \mu^- + N^{*++}, \ldots)}{d(q^2)}$ on q^2

is expected for $E_\nu \gg \{E_\nu - (E_\mu)_f\}_{AV} \approx \frac{1}{3} m_p$

("shallow inelastic" differential scattering cross section)

since in this latter case

$$\sum_{\substack{f:\text{ kinematically} \\ \text{accessible } N^{*++},\ldots}} \frac{d\sigma(\nu_\mu + p \to \mu^- + f)}{d(q^2)}$$

is quite different from

$$\sum_{f:\text{ all } N^{*++}} \frac{d\sigma(\nu_\mu + p \to \mu^- + f)}{d(q^2)}$$

and is expressible in terms of $p \leftrightarrow N^{*++}$ form factors of only a small number of relatively low-lying N^{*++} every one of which favor small q^2. This expectation is well borne out by the experimental data as shown for example by the following.[25]

Number of $\nu_\mu \to \mu^-$ Events	Values of q^2 in $(\text{GeV.})^2$
24	0 - 1
18	1 - 2
10	2 - 3
10	3 - 4

$\{E_\nu - (E_\mu)_f\}_{AV} > 2$ GeV. (1.13a)

which is to be compared with

Number of $\nu_\mu \to \mu^-$ Events	Values of q^2 in $(\mathrm{GeV.})^2$
403	0 - 1
12	1 - 2
0	2 - 3
0	3 - 4

$\{E_\nu - (E_\mu)_f\}_{AV} < 0.5\ \mathrm{GeV.}$ (113 b)

We also remark that Eqs. (105a) - (105d) are consistent, as indeed they must be, with the sum rules[26]

$$\left\{ \frac{d\sigma(\bar{\nu}_\mu + p \to \mu^+ + n, N^{*0}, \ldots)}{d(q)^2} - \frac{d\sigma(\nu_\mu + p \to \mu^- + N^{*++}, \ldots)}{d(q^2)} \right\} = \frac{(G\cos\theta_c)^2}{\pi} ;$$

$$\left\{ \frac{d\sigma(\bar{\nu}_\mu + n \to \mu^+ + N^{*-}, \ldots)}{d(q^2)} - \frac{d\sigma(\nu_\mu + n \to \mu^- + p, N^{*+}, \ldots)}{d(q^2)} \right\} = - \frac{(G\cos\theta_c)^2}{\pi} ;$$

$$E_\nu \gg \{E_\nu - (E_\mu)_f\}_{AV} \gg \tfrac{1}{3} m_p \qquad (114)$$

deduced essentially solely on the basis of the equal-time commutation relations satisfied by the strangeness-non-changing hadron weak currents $h_\alpha^{(q_n q_p)}(x)$ of Eq. (23). These equal-time commutation relations are, in turn, either obtained directly from the anticommutation relations of the $\psi_{q_n}(x)$, $\psi_{q_p}(x)$ or are postulated phenomenologically in case the strangeness-nonchanging hadron weak currents are not explicitly expressed in terms of the quark field operators.[2]

Finally, we recall that rather extensive experimental studies have been made of the differential cross section:

$$\frac{d\sigma(\nu_\mu + n \to \mu^- + p)}{d(q^2)}$$

with the general conclusion[2] that the most important (i.e. axial) form factor associated with

$$\langle p | \{ h_\alpha^{(q_n q_p)}(x) \}^\dagger_{\text{AXIAL PART}} | n \rangle$$

has a q^2-dependence not very different from that of the (polar and weak magnetism) form factors associated with

$$\langle p | \{ h_\alpha^{(q_n q_p)}(x) \}^\dagger_{\text{POLAR PART}} | n \rangle = \langle p | I_\alpha^{(+)}(x) | n \rangle = 2 \langle p | I_\alpha^{(3)}(x) | p \rangle$$

$$= 2 \langle p | \left(J_\alpha^{\text{ELMAG}}(x) \right)_{\text{ISOVECTOR}} | p \rangle$$

where $J_\alpha^{\text{ELMAG}}(x) = \left(J_\alpha^{\text{ELMAG}}(x) \right)_{\text{ISOSCALAR}} + \left(J_\alpha^{\text{ELMAG}}(x) \right)_{\text{ISOVECTOR}} = \frac{1}{2} Y_\alpha(x) + I_\alpha(x)$

is the hadron electromagnetic current. Similar though not so detailed experimental studies have been made of the differential cross section:

$$d\sigma(\nu_\mu + p \to \mu^- + N^{*++}) .$$

It is also worth mentioning that the important $n \leftrightarrow p$ form factors have such a q^2-dependence that a nucleon essentially behaves as a point particle for $\{\frac{1}{6} q^2 \times$ (r. m. s. nucleon radius)$^2\}$ $\cong$ $\cong \frac{1}{6} q^2 \left(\frac{3}{m_p}\right)^2 \lesssim 1$ so that (compare Eq. (101b))

$$\frac{d\sigma(\nu_\mu + n \to \mu^- + p)}{d(q^2)} \cong \frac{(G\cos\theta_c)^2}{\pi} \quad : \quad q^2 \lesssim \tfrac{2}{3} m_p^2$$

$$\cong 0 \quad : \quad q^2 \gtrsim \tfrac{2}{3} m_p^2$$

$$\sigma_{TOTAL}(\nu_\mu + n \to \mu^- + p) \cong \frac{G^2}{\pi} \frac{2}{3} m_p^2 = 0.9 \times 10^{-38} \text{ cm}^2 \qquad (115)$$

which agrees well with the measured value [25]

$$\{\sigma_{TOTAL}(\nu_\mu + n \to \mu^- + p)\}_{EXPERIMENTAL} \cong 0.8 \times 10^{-38} \text{ cm}^2 \text{ for } E_\nu \gtrsim 1 \text{ GeV}.$$

We proceed to discuss 2nd-order weak, i.e. 4th-order semiweak, collision processes and begin with a treatment of the elastic scattering of incident muon-neutrinos by target electrons: see Eq. (21) et seq. and Figs. 9a and 9b. Considering first the current-current H_{weak} of Eq. (23) the matrix element for $\nu_\mu + e^- \to \mu^- + \nu_e \to \nu_\mu + e^-$ is given by (see Fig 9a)

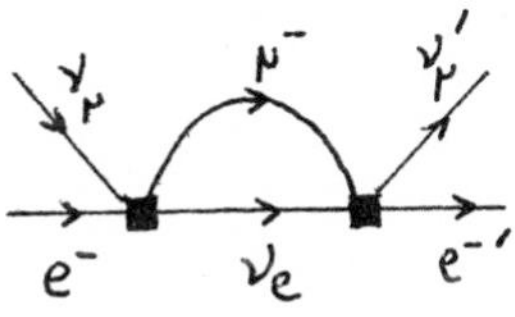

Fig. 9a

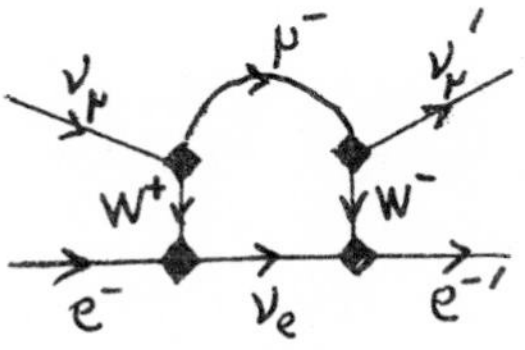

Fig. 9b

$$M(\nu_\mu + e^- \to \nu_\mu + e^-) = \sum_n \langle \nu_\mu' e^{-\prime} | \frac{G}{\sqrt{2}} \int \{ \ell_\beta^{(\mu\nu_\mu)}(y) \}^\dagger \ell_\beta^{(e\nu_e)}(y) \, d^3y \, | \{\nu_e \mu^-\}_n \rangle \times$$

$$\times \langle \{\nu_e \mu^-\}_n | \frac{G}{\sqrt{2}} \int \{ \ell_\alpha^{(e\nu_e)}(x) \}^\dagger \ell_\alpha^{(\mu\nu_\mu)}(x) \, d^3x \, | \nu_\mu e^- \rangle \left[\{(E_{\nu_e} + E_{\mu^-})\}_n - (E_{e^-} + E_{\nu_\mu}) \right]^{-1} =$$

$$= \left(\bar{u}'_{\nu_\mu} \gamma_\beta (1+\gamma_5) u_{\nu_\mu} \right) \left[\left(\frac{G}{\sqrt{2}}\right)^2 \int \frac{d^4 p_\mu}{(2\pi)^4} \mathrm{Trace} \left\{ \left(\frac{1}{\gamma \cdot p_\mu}\right) \gamma_\beta (1+\gamma_5) \left(\frac{1}{\gamma \cdot p_\mu}\right) \gamma_\alpha (1+\gamma_5) \right\} \right] \times$$

$$\times \left(\bar{u}'_e \gamma_\alpha (1+\gamma_5) u_e \right) + M'' \equiv M' + M'' ; \quad \gamma \cdot p_\mu \equiv \vec{\gamma} \cdot \vec{p}_\mu + \gamma_4 (p_\mu)_4 , \tag{116}$$

where the integral over the momentum of the virtual muon in M' diverges quadratically and M'' contains an integral over this momentum with at most a linear divergence. Introducing the cut-off parameter Λ to render the integrals over p_μ finite we have (compare Eq. (69) et seq.)

$$M(\nu_\mu + e^- \to \nu_\mu + e^-) = M' + M'' = \frac{G}{\sqrt{2}}\,\xi\left(\bar{u}'_{\nu_\mu}\gamma_\alpha(1+\gamma_5)u_{\nu_\mu}\right)\times$$

$$\times\left(\bar{u}'_e\gamma_\alpha(1+\gamma_5)u_e\right) + M'' ;$$

$$\xi \equiv \frac{G}{\sqrt{2}}\frac{\Lambda^2}{4\pi^2} ;$$

$$\frac{M''}{M'} \lessapprox \frac{|\vec{p}_{\nu_\mu}|}{\Lambda} \ll 1 \qquad (117)$$

Further, use of the boson-current $H_{\text{semi-weak}}$ of Eq. (24a) (see Fig. 9b) does not modify the quadratic divergence of the integral over p_μ in the analogue to M' — this follows since the boson propagators:

$$g^2\,\frac{\delta_{\alpha\kappa} + \dfrac{\left((p_{\nu_\mu}-p_\mu)_\alpha(p_{\nu_\mu}-p_\mu)_\kappa\right)}{m_w^2}}{(p_{\nu_\mu}-p_\mu)^2 + m_w^2}$$

and

$$g^2\,\frac{\delta_{\beta\lambda} + \dfrac{(p_\mu-p'_{\nu_\mu})_\beta(p_\mu-p'_{\nu_\mu})_\lambda}{m_w^2}}{(p_\mu-p'_{\nu_\mu})^2 + m_w^2}$$

become $\frac{G}{\sqrt{2}} \frac{(p_\mu)_\alpha (p_\mu)_\kappa}{p_\mu^2}$

and $\frac{G}{\sqrt{2}} \frac{(p_\mu)_\beta (p_\mu)_\lambda}{p_\mu^2}$

for $p_\mu^2 \gg m_w^2$ (and fixed p_{ν_μ}, p'_{ν_μ})

and so do not introduce any additional convergence.

Eq. (117) yields, with $p \equiv |\vec{p}_{\nu_\mu}| = |-\vec{p}_e| \cong \frac{\sqrt{s}}{2} \cong \frac{\sqrt{2 m_e E_\nu}}{2}$,

and remembering Eqs. (80) and (81),

$$\sigma_{TOTAL}(\nu_\mu + e^- \to \nu_\mu + e^-) \cong \xi^2 \frac{G^2}{\pi} 4p^2 \cong \xi^2 \frac{G^2}{\pi} s \cong \xi^2 \frac{G^2}{\pi}(2m_e E_\nu)$$

$$\cong \xi^2 (1.5 \times 10^{-41}\ cm^2)\ E_\nu^{\text{in GeV}} =$$

$$= \xi^2 \sigma_{TOTAL}(\nu_e + e^- \to \nu_e + e^-) ;$$

$$p < p_{UNITAR.LIM} = \frac{1}{\xi}\left(\sqrt{\frac{\pi}{2}} \frac{1}{\sqrt{G}}\right) = \frac{1}{\xi} 350\ GeV. \quad (118)$$

Thus, since (see Eq. (66b) et seq.)

$$\{\sigma_{TOTAL}(\nu_\mu + e^- \to \nu_\mu + e^-)\}_{EXPERIMENTAL} < 3 \times 10^{-42}\ cm^2 ,\ E_\nu^{\text{in GeV}} \cong 1 \quad (119)$$

we obtain the upper limits:

$$\xi < \tfrac{1}{2} \quad , \quad \Lambda < 1500 \text{ GeV} . \qquad (120)$$

We note that these values of Λ and ξ are much larger than those obtained if one should identify Λ with the $P_{UNITAR.\ LIM.}$ for $\nu_e + e^- \to \nu_e + e^-$ — such an identification gives Eq. (82)

$$\Lambda = \sqrt{\tfrac{\pi}{2}}\,\frac{1}{\sqrt{G}} = 350 \text{ GeV} , \quad \xi = \frac{G}{\sqrt{2}}\,\frac{\left(\sqrt{\frac{\pi}{2}}\frac{1}{\sqrt{G}}\right)^2}{4\pi^2} = \frac{1}{\sqrt{2}\cdot 8\pi} = 3\times 10^{-2}$$

and the corresponding $\dfrac{\sigma_{TOTAL}(\nu_\mu + e^- \to \nu_\mu + e^-)}{\sigma_{TOTAL}(\nu_e + e^- \to \nu_e + e^-)}$ are then of order 10^{-3}. Under these circumstances the observation of $\nu_\mu + e^- \to \nu_\mu + e^-$ would be extremely difficult and would become entirely out of question if ξ^2 were as small as $(10^{-4})^2 = 10^{-8}$ (see Eq.(144) below.)

We next consider the elastic plus inelastic scattering of muon-neutrinos by protons:

$$\nu_\mu + p[q_p q_p q_n] \to \nu_\mu + p[q_p q_p q_n],\ N^{*+}[q_p q_p q_n], \cdots \qquad (122)$$

and, in view of the Theorem in Eq. (107), write

$$\frac{\sigma_{TOTAL}(\nu_\mu + p[q_{\mathfrak{p}} q_{\mathfrak{p}} q_n] \to \nu_\mu + p[q_{\mathfrak{p}} q_{\mathfrak{p}} q_n], N^{*+}[q_{\mathfrak{p}} q_{\mathfrak{p}} q_n], \cdots)}{\sigma_{TOTAL}(\nu_\mu + p[q_{\mathfrak{p}} q_{\mathfrak{p}} q_n] \to \mu^- + N^{*++}[q_{\mathfrak{p}} q_p q_p], \cdots)} \cong$$

$$\cong \frac{\sigma_{TOTAL}(\nu_\mu + q_n \to \nu_\mu + q_n) + 2\,\sigma_{TOTAL}(\nu_\mu + q_{\mathfrak{p}} \to \nu_\mu + q_p)}{\sigma_{TOTAL}(\nu_\mu + q_n \to \mu^- + q_p)} \qquad (123)$$

where (see Figs. 10 and 11)

$$\nu_\mu + q_n \longrightarrow \mu^- + q_p \longrightarrow \nu_\mu' + q_n' \qquad (124)$$

$$\nu_\mu + q_p \longrightarrow \nu_\mu + q_n + \mu^+ + \nu_\mu' \longrightarrow \nu_\mu' + q_p' \qquad (125)$$

and the quarks are considered as bound in $p, N^{*+}, \ldots, N^{*++}, \ldots$

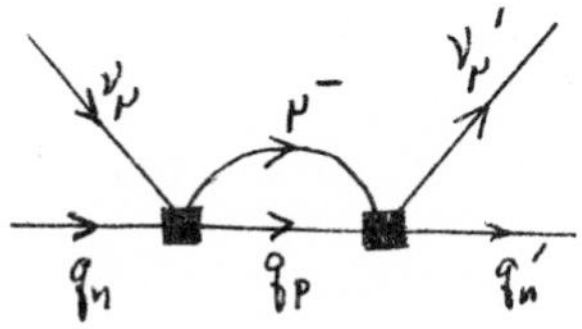

Fig. 10

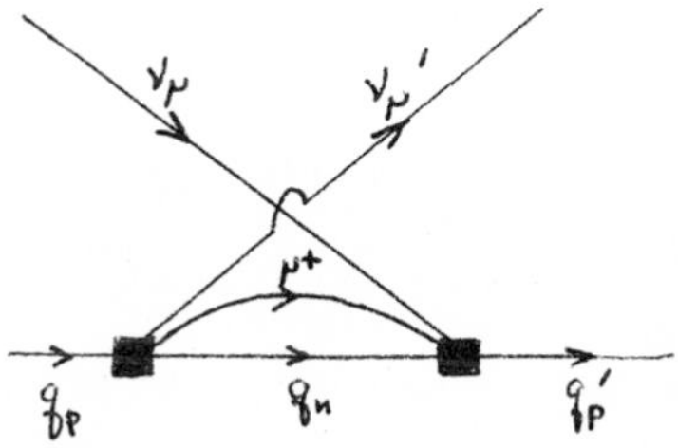

Fig. 11

Then, in analogy with Eqs. (116), (117), (118) (see also Eqs. (102), (108), (110)),

$$\sigma_{TOTAL}(\nu_\mu + q_n \rightarrow \nu_\mu + q_n) \cong \sigma_{TOTAL}(\nu_\mu + q_p \rightarrow \nu_\mu + q_p) \cong$$

$$\cong \xi^2 \, \sigma_{TOTAL}(\nu_\mu + q_n \rightarrow \mu^- + q_p) \cong \xi^2 \left\{ (1.5 \times 10^{-41} \, cm^2) \left(\frac{(m_q)_{EFFECTIVE}}{m_e} \right) \left(E_\nu^{\text{in GeV}} \right) \right\}$$

$$= \xi^2 \left\{ (0.9 \times 10^{-38} \, cm^2)(E_\nu^{\text{in GeV}}) \right\} \qquad (126)$$

so that

$$\left(\frac{\sigma_{TOTAL}(\nu_\mu + p \longrightarrow \nu_\mu + p, N^{*+}, \ldots)}{\sigma_{TOTAL}(\nu_\mu + p \longrightarrow \mu^- + N^{*++}, \ldots)} \right)_{THEORETICAL} \cong 3\xi^2 \qquad (127)$$

and since[27]

$$\left(\frac{\sigma_{\text{TOTAL}} (\nu_\mu + p \to \nu_\mu + p, N^{*+}, \dots)}{\sigma_{\text{TOTAL}} (\nu_\mu + p \to \mu^- + N^{*++}, \dots)} \right)_{\text{EXPERIMENTAL}} < 0.2 \qquad (128)$$

we obtain the upper limits

$$\xi < \frac{1}{4} \ , \quad \Lambda < 1100 \text{ GeV} \ , \qquad (129)$$

which are not very different from those in Eq. (120).

We now treat 2nd-order weak, i.e., 4th-order semiweak, processes such as the semileptonic decay $K_L^0 \to \mu^+ + \mu^-$ or the semileptonic elastic scattering $\nu_\mu + p \to \nu_\mu + p$ where the hadron involved makes a transition from a particular initial state (i.e., $|K_L^0\rangle$) to a particular final state (i. e., $|\text{vacuum}\rangle$). In this case the theorem in Eq. (107) does not apply since no sum is taken over a large number of kinematically accessible final states. However, one can still obtain explicit expressions for the relevant decay rates and cross sections by use of the equal-time commutation relations for the strangeness-nonchanging hadron weak currents. Thus, considering on the basis of the H_{weak} of Eq. (23) (the H_{semiweak} of Eq. (24a) again gives essentially the same results) the decay process

$$A \rightarrow C_n^{(\pm)} + \mu^{\mp} + \bar{\nu}_\mu(\nu_\mu) \longrightarrow B + \mu^- + \mu^+ \qquad (130)$$

(here, e.g., $|A\rangle = |K_L^0\rangle$; $|C_n^{(\pm)}\rangle = |\pi^\pm\rangle, |\rho^\pm\rangle, \ldots, |K^\pm\rangle, \cdots$; $|B\rangle = |\text{vacuum}\rangle$)

we have for the corresponding matrix element (see Figs. 12 and 13)

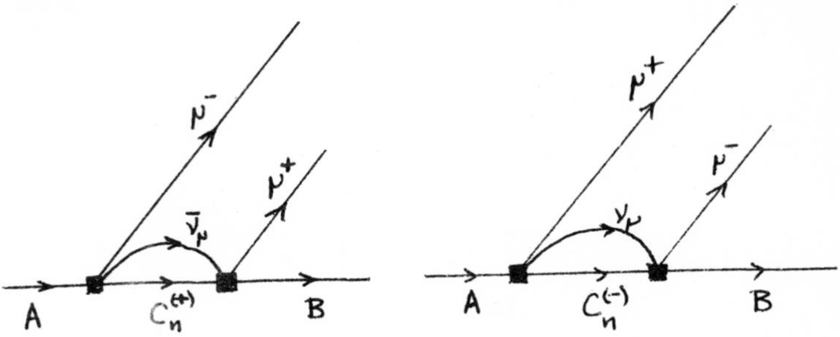

Fig. 12

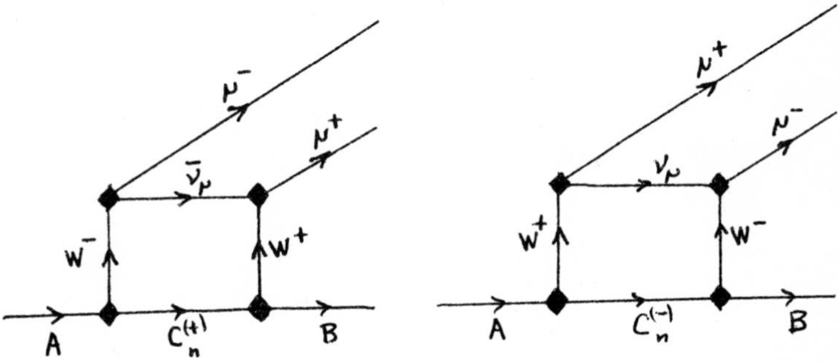

Fig. 13

$$M(A\to B+\mu^-+\mu^+) = \sum_{n;\vec{p}_n,s_n}\sum_{\vec{p}_\nu,s_\nu}\left(\frac{\langle\mu^-\mu^+B|\frac{G}{\sqrt{2}}\int\{l_\beta^{(\mu\nu_\mu)}(\vec{y})\}^+ h_\beta(\vec{y})d^3y|\mu^-\bar{\nu}_\mu C_n^{(+)}\rangle}{(E_{\bar{\nu}}+E_{\mu^-}+E_{C_n^{(+)}})-E_A}\times\right.$$

$$\times\langle\mu^-\bar{\nu}_\mu C_n^{(+)}|\frac{G}{\sqrt{2}}\int l_\alpha^{(\mu\nu_\mu)}(\vec{x})\{h_\alpha(\vec{x})\}^+ d^3x|A\rangle+\langle\mu^-\mu^+B|\frac{G}{\sqrt{2}}\int l_\alpha^{(\mu\nu_\mu)}(\vec{x})\{h_\alpha(x)\}^+d^3x|\mu^+\nu_\mu C_n^{(-)}\rangle\times$$

$$\left.\times\frac{\langle\mu^+\nu_\mu C_n^{(-)}|\frac{G}{\sqrt{2}}\int\{l_\beta^{(\mu\nu_\mu)}(\vec{y})\}^+ h_\beta(\vec{y})d^3y|A\rangle}{(E_\nu+E_{\mu^+}+E_{C_n^{(-)}})-E_A}\right);\quad \langle\mu^-\bar{\nu}_\mu C_n^{(+)}|l_\alpha^{(\mu\nu_\mu)}(\vec{x})\{h_\alpha(\vec{x})\}^+|A\rangle=$$

$$=\langle\mu^-\bar{\nu}_\mu|l_\alpha^{(\mu\nu)}(\vec{x})|\mathrm{vacuum}\rangle\langle C_n^{(+)}|\{h_\alpha(\vec{x})\}^+|A\rangle=$$

$$=\left(e^{-i(\vec{p}_{\mu^-}+\vec{p}_{\bar{\nu}})\cdot\vec{x}}\left(\bar{u}_{\mu^-}\gamma_\alpha(1+\gamma_5)u^*_{\bar{\nu}_\mu}\right)\right)\left(e^{-i(-\vec{p}_n+\vec{p}_A)\cdot\vec{x}}\langle C_n^{(+)}|\{h_\alpha(0)\}^+|A\rangle\right);$$

$$h_\alpha(\vec{x}) = \cos\Theta_C\, h_\alpha^{(q_nq_p)}(\vec{x})+\sin\Theta_C\, h_\alpha^{(q_\lambda q_p)}(\vec{x});$$

$$h_\alpha^{(q_nq_p)}(\vec{x}) = \left\{h_\alpha^{(q_nq_p)}(\vec{x})\right\}_{\text{POLAR PART}}+\left\{h_\alpha^{(q_nq_p)}(\vec{x})\right\}_{\text{AXIAL PART}}\equiv V_\alpha(\vec{x})+A_\alpha(\vec{x})=\mathcal{I}_\alpha^{(-)}(\vec{x})+A_\alpha(\vec{x}) \qquad (131)$$

where we use 2nd-order perturbation theory in a form which is not manifestly covariant in order to attain (hopefully) a somewhat clearer exposition. Eq. (131) gives[14]

$$M(A\to B+\mu^-+\mu^+)=\left(\frac{G}{\sqrt{2}}\right)^2\iint d^3x\,d^3y\,e^{-i(\vec{p}_{\mu^-}\cdot\vec{x}+\vec{p}_{\mu^+}\cdot\vec{y})}\int\frac{d^3\vec{p}_\nu}{(2\pi)^3}\,l_{\alpha\beta}\left(\frac{e^{-i\vec{p}_\nu\cdot(\vec{x}-\vec{y})}\langle B|h_\beta(\vec{y})h_\alpha^+(\vec{x})|A\rangle}{\{D^{(+)}\}_{AV}}+\right.$$

$$\left.-\frac{e^{-i\vec{p}_\nu\cdot(\vec{y}-\vec{x})}\langle B|h_\alpha^+(\vec{x})h_\beta(\vec{y})|A\rangle}{\{D^{(-)}\}_{AV}}\right);\quad l_{\alpha\beta}\equiv\left(\bar{u}_{\mu^-}\gamma_\alpha\left(-i\frac{\vec{\gamma}\cdot\vec{p}_\nu}{E_\nu}\right)\gamma_\beta(1+\gamma_5)u^*_{\mu^+}\right);$$

$$\{D^{(\pm)}\}_{AV} \equiv E_\nu + E_{\mu^\mp} + \{E_{C_n^{(\pm)}}\}_{AV} - E_A =$$

$$= E_\nu + \left[\left(\frac{E_{\mu^-} + E_{\mu^+}}{2}\right) + \left(\frac{\{E_{C_n^{(+)}}\}_{AV} + \{E_{C_n^{(-)}}\}_{AV}}{2}\right) - E_A \right] +$$

$$\pm \left[\left(\frac{E_{\mu^-} - E_{\mu^+}}{2}\right) + \left(\frac{\{E_{C_n^{(+)}}\}_{AV} - \{E_{C_n^{(-)}}\}_{AV}}{2}\right)\right] = E_\nu +$$

$$+ \left[\left(\frac{\{E_{C_n^{(+)}}\}_{AV} + \{E_{C_n^{(-)}}\}_{AV}}{2}\right) - \left(\frac{E_A + E_B}{2}\right)\right] \pm \left[\left(\frac{E_{\mu^-} - E_{\mu^+}}{2}\right) + \left(\frac{\{E_{C_n^{(+)}}\}_{AV} - \{E_{C_n^{(-)}}\}_{AV}}{2}\right)\right] =$$

$$= E_\nu + \epsilon' \pm \epsilon'' \qquad (132)$$

where the relative minus sign between the $|C_n^{(+)}\rangle$ and $|C_n^{(-)}\rangle$ contributions arises from the anticommutation of the creation operators for μ^- and μ^+. Further, remembering the general character of the hadronic form factors in terms of which $\langle C_n^{(+)} | h_\alpha^+(0) | A\rangle$ and $\langle C_n^{(-)} | h_\alpha(0) | A\rangle$ are expressed, we see that the major contributions to the $\sum_{n;\vec{p}_n, s_n} \cdots$ come from $|C_n^{(\pm)}\rangle$ with $m_{C_n^{(\pm)}} \cong m_A$ (to insure "good overlap") and

$$(p_{C_n^{(\pm)}} - p_A)^2 = -m_{C_n^{(\pm)}}^2 - m_A^2 + 2m_A\sqrt{(\vec{p}_n)^2 + m_{C_n^{(\pm)}}^2}$$

small; thus

$$\frac{|\epsilon'|}{E_\nu} \ll 1, \quad \frac{|\epsilon''|}{E_\nu} \ll 1$$

and

$$\frac{1}{\{D^{(\pm)}\}_{AV}} \cong \frac{1}{E_\nu}\left(1 - \frac{\epsilon'}{E_\nu} \mp \frac{\epsilon''}{E_\nu}\right). \qquad (133)$$

Combination of Eq. (132) and Eq. (133) then yields

$$M(A\to B+\mu^-+\mu^+)\cong\left(\frac{G}{\sqrt{2}}\right)^2\iint d^3\vec{x}\,d^3\vec{y}\;e^{-i(\vec{p}_{\mu^-}\cdot\vec{x}+\vec{p}_{\mu^+}\cdot\vec{y})}\times$$

$$\times\int\frac{d\vec{p}_\nu}{(2\pi)^3E_\nu}\Big\{l_{\alpha\beta;4}\cos(\vec{p}_\nu\cdot(\vec{x}-\vec{y}))\Big(\Big(1-\frac{\epsilon'}{E_\nu}\Big)\langle B|[h_\beta(\vec{y}),h^+_\alpha(\vec{x})]_-|A\rangle+$$

$$-\Big(\frac{\epsilon''}{E_\nu}\Big)\langle B|[h_\beta(\vec{y}),h^+_\alpha(\vec{x})]_+|A\rangle\Big)-\vec{l}_{\alpha\beta}\cdot\frac{\vec{p}_\nu}{E_\nu}\sin(\vec{p}_\nu\cdot(\vec{x}-\vec{y}))\times$$

$$\times\Big(\Big(1-\frac{\epsilon'}{E_\nu}\Big)\langle B|[h_\beta(\vec{y}),h^+_\alpha(\vec{x})]_+|A\rangle-\Big(\frac{\epsilon''}{E_\nu}\Big)\langle B|[h_\beta(\vec{y}),h^+_\alpha(\vec{x})]_-|A\rangle\Big)\Big\};$$

$$l_{\alpha\beta;4}\equiv(\bar{u}_{\mu^-}\gamma_\alpha\gamma_4\gamma_\beta(1+\gamma_5)u^*_{\mu^+}),\quad \vec{l}_{\alpha\beta}\equiv(\bar{u}_{\mu^-}\gamma_\alpha\vec{\gamma}\gamma_\beta(1+\gamma_5)u^*_{\mu^+});$$

$$[h_\beta(\vec{y}),h^+_\alpha(\vec{x})]_\mp\equiv h_\beta(\vec{y})h^+_\alpha(\vec{x})\mp h^+_\alpha(\vec{x})h_\beta(\vec{y})\qquad(134)$$

where we can in addition neglect ϵ'/E_ν compared to 1 in $\left(1-\frac{\epsilon'}{E_\nu}\right)$ since even with this neglect we still retain all possible combinations of $\cos(\vec{p}_\nu\cdot(\vec{x}-\vec{y}))$ and $\sin(\vec{p}_\nu\cdot(\vec{x}-\vec{y}))$ with $\langle B|[h_\beta(\vec{y}),h^+_\alpha(\vec{x})]_-|A\rangle$ and $\langle B|[h_\beta(\vec{y}),h^+_\alpha(\vec{x})]_+|A\rangle$.

We next note that the equal-time commutation relations of the weak hadron currents are such that we can write (see in this connection Eq. (114) et seq.)

$$\langle B|[h_\beta(\vec{y}),h^+_\alpha(\vec{x})]_-|A\rangle=\langle B|f^{(-)}_{\beta\alpha}(\vec{x})\,\delta(\vec{x}-\vec{y})|A\rangle=$$

$$=e^{-i(\vec{p}_B-\vec{p}_A)\cdot\vec{x}}\,\delta(\vec{x}-\vec{y})\langle B|f^{(-)}_{\beta\alpha}(0)|A\rangle\qquad(135)$$

while in general

$$\langle B|[h_\beta(\vec{y}), h_\alpha^\dagger(\vec{x})]_+|A\rangle = \langle B|f_{\beta\alpha}^{(+)}(\vec{x})\,\delta(\vec{x}-\vec{y})|A\rangle +$$
$$+ \langle B|g_{\beta\alpha}^{(+)}(\vec{X},\vec{\xi})|A\rangle = e^{-i(\vec{p}_B-\vec{p}_A)\cdot\vec{x}}\,\delta(\vec{x}-\vec{y})\langle B|f_{\beta\alpha}^{(+)}(0)|A\rangle +$$
$$+ e^{-i(\vec{p}_B-\vec{p}_A)\cdot\vec{X}}\langle B|g_{\beta\alpha}^{(+)}(\vec{0},\vec{\xi})|A\rangle\,;$$
$$\vec{X} \equiv \frac{\vec{x}+\vec{y}}{2}\,,\quad \vec{\xi} \equiv \vec{x}-\vec{y} \tag{136}$$

where $f_{\beta\alpha}^{(\mp)}(\vec{x})$, $g_{\beta\alpha}^{(+)}(\vec{X},\vec{\xi})$ are specified (in general, q-number) functions of their arguments, e.g., $[A_4(\vec{y}), A_4^\dagger(\vec{x})]_- =$ $= -2i\,V_4^{(3)}(\vec{x})\,\delta(\vec{x}-\vec{y}) = 2\,I_0^{(3)}(\vec{x})\,\delta(\vec{x}-\vec{y})$ which determines one of the parts of $f_{44}^{(-)}(\vec{x})$.

Insertion of Eqs. (135) and (136) into Eq. (134) gives

$$M(A\to B+\mu^-+\mu^+) \cong \left(\delta_{\vec{p}_A=\vec{p}_B+\vec{p}_{\mu^-}+\vec{p}_{\mu^+}}\right)\left(\frac{G}{\sqrt{2}}\right)^2\int\frac{d\vec{p}_\nu}{(2\pi)^3 E_\nu}\times$$
$$\times\Big\{\,l_{\alpha\beta;4}\Big(\langle B|f_{\beta\alpha}^{(-)}(0)|A\rangle - \frac{\epsilon''}{E_\nu}\langle B|f_{\beta\alpha}^{(+)}(0)|A\rangle - \frac{\epsilon''}{E_\nu}\int d^3\xi\,\cos(\vec{p}_\nu\cdot\vec{\xi})\,\times \tag{137}$$
$$\times e^{-i(\vec{p}_{\mu^-}-\vec{p}_{\mu^+})\cdot\frac{\vec{\xi}}{2}}\langle B|g_{\beta\alpha}^{(+)}(\vec{0},\vec{\xi})|A\rangle\Big) - \Big(\vec{l}_{\alpha\beta}\cdot\frac{\vec{p}_\nu}{E_\nu}\Big)\Big(\int d^3\xi\,\sin(\vec{p}_\nu\cdot\vec{\xi})\,e^{-i(\vec{p}_{\mu^-}-\vec{p}_{\mu^+})\cdot\frac{\vec{\xi}}{2}}\langle B|g_{\beta\alpha}^{(+)}(0,\vec{\xi})|A\rangle\Big)\Big\}$$

so that, introducing the cut-off parameter Λ to render the integral over $\vec{p}_\nu$ finite (compare Eq. (117)) we have

$$M(A \to B + \mu^- + \mu^+) \cong \left(\delta_{\vec{P}_A = \vec{P}_B + \vec{P}_{\mu^-} + \vec{P}_{\mu^+}}\right)\left(\frac{G}{\sqrt{2}}\right)^2 \Bigg\{ l_{\alpha\beta,4}\Bigg(\langle B|f^{(-)}_{\beta\alpha}(0)|A\rangle \frac{\Lambda^2}{4\pi^2} +$$

$$- \langle B|f^{(+)}_{\beta\alpha}(0)|A\rangle \frac{\epsilon''\Lambda}{2\pi^2} - \int d^3\xi \, e^{-i(\vec{p}_{\mu^-} - \vec{p}_{\mu^+})\cdot\frac{\vec{\xi}}{2}} \frac{\langle B|g^{(+)}_{\beta\alpha}(0,\vec{\xi})|A\rangle}{|\vec{\xi}|} \frac{\epsilon''}{4\pi}\Bigg) +$$

$$- \vec{l}_{\alpha\beta} \cdot \left(\int d^3\xi \, e^{-i(\vec{p}_{\mu^-} - \vec{p}_{\mu^+})\cdot\frac{\vec{\xi}}{2}} \frac{\langle B|g^{(+)}_{\beta\alpha}(0,\vec{\xi})|A\rangle}{|\vec{\xi}|^3} \frac{\vec{\xi}}{4\pi}\right)\Bigg\} \equiv$$

$$\equiv M' + M'' ; \tag{138}$$

$$M' \equiv \left(\delta_{\vec{P}_A = \vec{P}_B + \vec{P}_{\mu^-} + \vec{P}_{\mu^+}}\right) \frac{G}{\sqrt{2}} \left(\frac{G}{\sqrt{2}} \frac{\Lambda^2}{4\pi^2}\right) l_{\alpha\beta,4} \langle B|f^{(-)}_{\beta\alpha}(0)|A\rangle \equiv$$

$$\equiv \left(\delta_{\vec{P}_A = \vec{P}_B + \vec{P}_{\mu^-} + \vec{P}_{\mu^+}}\right) \frac{G}{\sqrt{2}} \xi \, l_{\alpha\beta;4} \langle B|f^{(-)}_{\beta\alpha}(0)|A\rangle ;$$

$$\frac{M''}{M'} \approx \left(\frac{\epsilon''}{\Lambda}, \frac{\epsilon''}{\{|\vec{\xi}|\}_{AV}\Lambda^2}, \frac{1}{\{|\vec{\xi}|\}^2_{AV}\Lambda^2}\right) \ll 1, \text{ since } \epsilon'' \approx m_A - m_B, \frac{1}{\{|\vec{\xi}|\}_{AV}} \approx m_A .$$

We therefore see that, because of the presence of the quadratically divergent term M', the semileptonic matrix element $M(A \to B + \mu^- + \mu^+)$ in Eq. (138) is as divergent as the purely leptonic matrix element $M(\nu_\mu + e^- \to \nu_\mu + e^-)$ in Eq. (117). This quadratic divergence in $M(A \to B + \mu^- + \mu^+)$ disappears if we replace $f^{(-)}_{\beta\alpha}(\vec{x})\,\delta(\vec{x} - \vec{y})$ by $g^{(-)}_{\beta\alpha}(\vec{x}, \vec{y}) = g^{(-)}_{\beta\alpha}(\vec{X}, \vec{\xi})$ with $g^{(-)}_{\beta\alpha}(\vec{X}, \vec{0})$ finite so that the restrictions on the values of the

$$\langle C_n^{(+)} | h_\alpha^{\dagger}(0) | A \rangle, \; \langle C_n^{(-)} | h_\alpha(0) | A \rangle$$

arising from the specification of h_α, $h_\alpha^{\dagger}$ via the equal-time commutation relations of Eq. (135) results in the divergence of the sum of products of hadronic form factors divided by energy denominators which constitutes the original expression for $M(A \to B + \mu^- + \mu^+)$ in Eq. (131).

We now apply Eq. (138) to $K_L^0 \to \mu^- + \mu^+$ so that $|A\rangle = |K_L^0\rangle$ and $|B\rangle = |\text{vacuum}\rangle$; this gives, evaluating the various

$$[h_\beta(\vec{y}), \{h_\alpha(\vec{x})\}^{\dagger}]_- = (\cos\theta_C)^2 [h_\beta^{(q_n q_p)}(\vec{y}), \{h_\alpha^{(q_n q_p)}(\vec{x})\}^{\dagger}]_- +$$

$$+ \cos\theta_C \sin\theta_C \left([h_\beta^{(q_n q_p)}(\vec{y}), \{h_\alpha^{(q_\lambda q_p)}(\vec{x})\}^{\dagger}]_- + [h_\beta^{(q_\lambda q_p)}(\vec{y}), \{h_\alpha^{(q_n q_p)}(\vec{x})\}^{\dagger}]_- \right) +$$

$$(\sin\theta_C)^2 [h_\beta^{(q_\lambda q_p)}(\vec{y}), \{h_\alpha^{(q_\lambda q_p)}(\vec{x})\}^{\dagger}]_-$$

in accordance to the anti-commutation rules of the $\psi_{q_p}(\vec{x})$, $\psi_{q_n}(\vec{x})$, $\psi_{q_\lambda}(\vec{x})$ in order to find $f_{\beta\alpha}^{H}(0)$
($h_\beta^{(q_n q_p)}(\vec{y}) = \bar{\psi}_{q_n}(\vec{y})\gamma_\beta(1+\gamma_5)\psi_{q_p}(\vec{y})$, $[\psi_{q_p}(\vec{y}), \{\psi_{q_p}(\vec{x})\}^{\dagger}]_+ = \delta(\vec{x}-\vec{y})$ etc. - see comment after Eq. (114))

$$M(K_L^0 \to \mu^- + \mu^+) \cong \left(\delta_{\vec{P}_{K_L^0} = \vec{P}_{\mu^-} + \vec{P}_{\mu^+}}\right) \frac{G}{\sqrt{2}} \, \xi \, (\bar{u}_{\mu^-} \gamma_\alpha (1+\gamma_5) u^*_{\mu^+}) 4 \cos\theta_C \sin\theta_C \times$$

$$\times \langle \text{vacuum} | h_\alpha^{(q_\lambda q_n)}(0) | K_L^0 \rangle ; \; h_\alpha^{(q_\lambda q_n)} = \bar{\psi}_{q_\lambda} \gamma_\alpha (1+\gamma_5) \psi_{q_n} ; |K_L^0\rangle = \sqrt{\tfrac{1}{2}}|K^0\rangle + \sqrt{\tfrac{1}{2}}|\bar{K}^0\rangle ; \tag{139}$$

whence, using the Wigner-Eckart theorem,

$$\mathcal{M}(K_L^0 \to \mu^- + \mu^+) \cong \left(\delta_{\vec{P}_{K_L^0} = \vec{P}_{\mu^-} + \vec{P}_{\mu^+}}\right) \frac{G}{\sqrt{2}} \xi \left(\bar{u}_{\mu^-} \gamma_\alpha (1+\gamma_5) u^*_{\mu^+}\right) 4\cos\theta_c \sin\theta_c \frac{1}{\sqrt{2}} \langle \text{vacuum} | h_\alpha^{(q_n q_n)}(0) | K^+ \rangle \tag{140}$$

which is to be compared with

$$\mathcal{M}(K^+ \to \nu_\mu + \mu^+) = \left(\delta_{\vec{P}_{K^+} = \vec{P}_{\nu_\mu} + \vec{P}_{\mu^+}}\right) \frac{G}{\sqrt{2}} \sin\theta_c \left(\bar{u}_{\nu_\mu} \gamma_\alpha (1+\gamma_5) u^*_{\mu^+}\right) \langle \text{vacuum} | h_\alpha^{(q_n q_p)}(0) | K^+ \rangle . \tag{141}$$

Eqs. (140) and (141) yield for the calculated ratio of the decay rates

$$\left\{ \frac{\Gamma(K_L^0 \to \mu^- + \mu^+)}{\Gamma(K^+ \to \nu_\mu + \mu^+)} \right\}_{\text{THEORETICAL}} \cong 2 \left(\frac{\xi \cdot 4\cos\theta_c}{\sqrt{2}} \right)^2 \tag{142}$$

while the upper limit on the corresponding measured ratio is[28]

$$\left\{ \frac{\Gamma(K_L^0 \to \mu^- + \mu^+)}{\Gamma(K^+ \to \nu_\mu + \mu^+)} \right\}_{\text{EXPERIMENTAL}} < 7.4 \times 10^{-8} \tag{143}$$

so that

$$\xi < 10^{-4} , \quad \Lambda < 25 \text{ GeV} , \tag{144}$$

a far smaller upper limit on Λ and ξ than those obtained by identifying Λ with $\rho_{\text{UNITAR. LIM.}}$

for $\nu_e + e^- \rightarrow \nu_e + e^-$ $\left(\Lambda = \sqrt{\frac{\pi}{2}} \frac{1}{\sqrt{G}} = 350\,\text{GeV}, \right.$

$$\xi = \frac{G}{\sqrt{2}} \frac{\left(\sqrt{\frac{\pi}{2}} \frac{1}{\sqrt{G}}\right)^2}{4\pi^2} = \frac{1}{\sqrt{2} \cdot 8\pi} = 3 \times 10^{-2}$$

see Eq. (82) and discussion after Eq. (120)). We emphasize that, as already mentioned in the Introduction, a value of ξ as small as 10^{-4} would imply that higher-order weak corrections to 1st-order weak selection rules are quite negligible (as an example of such a 1st-order weak selection rule; we mention the "$\Delta I = 1$" rule in $\nu_\mu + p \rightarrow \mu^- + N^{*++}$, $\nu_\mu + n \rightarrow \mu^- + N^{*+}$ which follows from the isovector character of $h_\alpha^{(q_n q_p)}$ and which predicts

$$\left\{ \frac{\sigma(\nu_\mu + p \rightarrow \mu^- + N^{*++})}{\sigma(\nu_\mu + n \rightarrow \mu^- + N^{*+})} \right\} = 3 . \Bigg)$$

We also remark that if we admit the existence of a neutral intermediate boson W^0 coupled to a neutral strangeness-changing hadron weak current with a semiweak coupling constant

$$g^{(0)} \sin \theta_c^{(0)} = m_{W^0} \sqrt{\frac{G^{(0)}}{\sqrt{2}}} \sin \theta_c^{(0)}$$

and to a neutral lepton weak current with a semiweak coupling constant $g^{(0)} = m_{W^0} \sqrt{\frac{G^{(0)}}{\sqrt{2}}}$ then $K_L^0 \rightarrow \mu^- + \mu^+$ can proceed as a 2nd-order semiweak, i.e. 1st-order weak, process: $K_L^0 \rightarrow \text{vacuum} + W^0 \rightarrow \text{vacuum} + \mu^- + \mu^+$

and we have

$$\left\{\frac{\Gamma(K_L^0 \to W^0 \to \mu^- + \mu^+)}{\Gamma(K^+ \to W^+ \to \nu_\mu + \mu^+)}\right\}_{\text{THEORETICAL}} \cong \left(\frac{G^{(0)}}{G}\right)^2 . \qquad (145)$$

Thus, comparing with Eq. (143),

$$\frac{G^{(0)}}{G} < \sqrt{7\times10^{-8}} \cong 3\times10^{-4} , \qquad (146)$$

which is the best available limit on the possible presence of neutral weak currents. In addition, we note that $K_L^0 \to \mu^- + \mu^+$ can proceed as a 1st-order weak, 4th-order electromagnetic process: $K_L^0 \to \text{vacuum} + \gamma + \gamma \to \text{vacuum} + \mu^- + \mu^+$,

with an estimated rate corresponding to

$$\left\{\frac{\Gamma(K_L^0 \to \gamma + \gamma \to \mu^- + \mu^+)}{\Gamma(K^+ \to \nu_\mu + \mu^+)}\right\}_{\text{THEORETICAL}} \approx \alpha^4 = 0.3\times10^{-8} \qquad (147)$$

which is a factor of about 20 below the presently available measured upper limit.

A treatment entirely similar to the one just described for the 2nd-order weak process $K_L^0 \to \mu^- + \mu^+$ can be given for the 2nd-order weak processes:

$$
\begin{array}{l}
K^+ \to C_n^{(+)} + \bar{\nu}_e + e^- \longrightarrow \pi^+ + \bar{\nu}_e + \bar{\nu}_e \\
\quad \searrow C_n^{(-)} + \nu_e + e^+ \nearrow
\end{array} \tag{148}
$$

$$|C_n^{(-)}\rangle = |\pi^0\rangle, |\rho^0\rangle, \cdots, |K^0\rangle, \cdots; |C_n^{(+)}\rangle = |\pi^+\pi^+\rangle, |\rho^+\pi^+\rangle, \cdots, |K^+\pi^+\rangle, \cdots,$$

and

$$
\begin{array}{l}
\nu_\mu + p \nearrow C_n^{(+)} + \mu^- \searrow \nu_\mu' + p' \\
\quad \searrow \nu_\mu + C_n^{(-)} + \mu^+ + \nu_\mu' \nearrow
\end{array} \tag{149}
$$

$$|C_n^{(+)}\rangle = |N^{*++}\rangle, \cdots; \quad |C_n^{(-)}\rangle = |n\rangle, |N^{*0}\rangle, \cdots.$$

We get

$$\left\{ \frac{\Gamma(K^+ \to \pi^+ + \bar{\nu}_e + \nu_e)}{\Gamma(K^+ \to \pi^0 + e^+ + \nu_e)} \right\}_{\text{THEORETICAL}} \cong \left(\frac{\xi \cdot 4 \cos\theta_c}{\sqrt{2}} \right)^2 \tag{150}$$

$$\left\{ \frac{\sigma(\nu_\mu + p \to \nu_\mu + p)}{\sigma(\nu_\mu + n \to \nu_\mu + n)} \right\}_{\text{THEORETICAL}} \cong \left(\frac{\xi \cdot 4 \cos\theta_c}{2} \right)^2 \tag{151}$$

and since

$$\left\{\frac{\Gamma(K^+ \to \pi^+ + \bar{\nu}_e + \nu_e)}{\Gamma(K^+ \to \pi^0 + e^+ + \nu_e)}\right\}_{\text{EXPERIMENTAL}} < 1.2 \times 10^{-5} \; [29] \tag{152}$$

$$\left\{\frac{\sigma(\nu_\mu + p \to \nu_\mu + p)}{\sigma(\nu_\mu + n \to \nu_\mu + n)}\right\}_{\text{EXPERIMENTAL}} < 0.1 \; [30] \tag{153}$$

obtain

$$\xi < 10^{-3}, \quad \Lambda < 75 \text{ GeV} \tag{154}$$

$$\xi < \frac{1}{6}, \quad \Lambda < 900 \text{ GeV} \tag{155}$$

Finally, we mention that a rather dubious estimate of

$\left\{\frac{m_{K_L^0} - m_{K_S^0}}{m_{K^0}}\right\}$ on the basis of the

$\left\{\frac{G}{\sqrt{2}} \int h_\alpha^\dagger(\vec{x}) h_\alpha(\vec{x}) d^3x \right\}$ part of the H_{weak} of Eq. (23)

$(K^0 \to \pi^+ + \pi^- \to K^0, \text{ etc.})$

yields

$$\left\{\frac{m_{K_L^0} - m_{K_S^0}}{m_{K^0}}\right\}_{\text{THEORETICAL}} \approx \left(\frac{G m_\pi^2}{\sqrt{2}}\right) \xi \left(\cos\theta_C \sin\theta_C\right)^2 \tag{156}$$

and since

$$\left\{\frac{m_{K_L^0} - m_{K_S^0}}{m_{K^0}}\right\}_{\text{EXPERIMENTAL}} \cong 6 \times 10^{-16} \tag{157}$$

we have

$$\xi \approx 10^{-7}, \quad \Lambda \approx 1\ \mathrm{GeV}. \qquad (158)$$

This value of Λ is by far the lowest so far obtained and is in fact so small as to cast doubt on the validity of Eq. (156).

For convenience we summarize our results for $\xi \equiv \frac{G}{\sqrt{2}} \frac{\Lambda^2}{4\pi^2}$ and Λ as follows:

	Process	ξ	Λ in GeV	Reference
purely leptonic	$\left(\frac{m_{\nu_e}}{m_e}\right)$ with $\eta_e = 10^{-3}$	$< 6 \times 10^{-3}$	< 180	Eq. (70)
	Unitarity limit in $\nu_e + e^- \to \nu_e + e^-$	3×10^{-2}	350	Eq. (81) and discussion after Eq. (120)
	$\nu_\mu + e^- \to \nu_\mu + e^-$	$< 1/2$	1500	Eq. (120)
semileptonic	$\nu_\mu + p \to \nu_\mu + p, N^{*+}, \ldots$	$< 1/4$	< 1100	Eq. (129)
	$\nu_\mu + p \to \nu_\mu + p$	$< 1/6$	< 900	Eq. (155)
	$K^+ \to \pi^+ + \bar{\nu}_e + \nu_e$	$< 10^{-3}$	< 75	Eq. (154)
	$K_L^0 \to \mu^- + \mu^+$	$< 10^{-4}$	< 25	Eq. (144)
nonleptonic	$\frac{m_{K_L^0} - m_{K_S^0}}{m_{K^0}}$	10^{-7}	1	Eq. (158)

Finally, we speculate a bit regarding the physical origin of the cut-off parameter Λ. Thus, if the cut-off in all processes treated arises from the higher-order effects of the weak interactions themselves, one would presumably have $\xi = \frac{G}{\sqrt{2}}\frac{\Lambda^2}{4\pi^2}$ in Eqs. (118), (126), (142), (150), (151) and (156) replaced by

$$\xi' \approx \lim_{\Lambda' \to \infty} \left(\frac{\frac{G}{\sqrt{2}}\frac{(\Lambda')^2}{4\pi^2}}{1 + G(\Lambda')^2} \right) = \frac{1}{\sqrt{2}}\frac{1}{4\pi^2} = 2\times 10^{-2}$$

which corresponds to $\Lambda = \frac{1}{\sqrt{G}} = 300$ GeV. (as is in fact required by dimensional reasoning since no reciprocal length other than $\frac{1}{\sqrt{G}}$ then appears – compare in this connection Eq. (82)). This possibility appears however to be excluded by the "low" value of Λ : $\Lambda \lesssim \frac{1}{10}\left(\frac{1}{\sqrt{G}}\right)$ obtained from the upper limit on the measured rate of $\Gamma(K_L^0 \to \mu^- + \mu^+)$, (Eqs. (142) - (144)).

On the other hand, it may be that the low value of Λ in the semileptonic (and nonleptonic) processes is associated with the non-pointlike character of the quarks which would result in the replacement of $\langle B | [h_\beta(\vec{y}), \{h_\alpha(x)\}^\dagger]_- | A\rangle =$

$$= \langle B | f^{(-)}_{\beta\alpha}(\vec{x})\, \delta(\vec{x}-\vec{y}) | A\rangle = e^{-i(\vec{p}_B - \vec{p}_A)\cdot\vec{x}}\, \delta(\vec{x}-\vec{y}) \langle B | f^{(-)}_{\beta\alpha}(0) | A\rangle$$

by

$$\langle B | [h_\beta(\vec{y}), \{h_\alpha(\vec{x})\}^\dagger]_- | A\rangle = \langle B | g^{(-)}_{\beta\alpha}(\vec{X}, \vec{\xi}) | A\rangle =$$

$$= e^{-i(\vec{p}_B - \vec{p}_A)\cdot\vec{X}} \langle B | g^{(-)}_{\beta\alpha}(0,\xi)|A\rangle \approx e^{-i(\vec{p}_B - \vec{p}_A)\cdot\vec{X}} \left(\frac{\Lambda^3 e^{-\Lambda^2 \xi^2}}{\pi^{3/2}} \right) \langle B | g^{(-)}_{\beta\alpha}(0,0)|A\rangle$$

with

$$\Lambda^{-1} \gtrsim 10\sqrt{G} \cong 10^{-15}\ \text{cm}$$

a measure of the effective weak-interaction radius of the quark; under these circumstances, the cut-off parameter in the purely leptonic processes would still be $\approx \frac{1}{\sqrt{G}}$. Alternatively, Λ might be $\lesssim \frac{1}{10}\left(\frac{1}{\sqrt{G}}\right)$ both for the semileptonic and for the purely leptonic processes — if such were the case, this low value of Λ might arise from a relatively strong boson-boson interaction which would render all matrix elements finite where more than one boson was exchanged between two lepton lines or a lepton line and a hadron line (see Figs. 9b and 13) and one would have

$$\Lambda \approx m_W \lesssim \frac{1}{10}\frac{1}{\sqrt{G}} .$$

It is clear that many other hypotheses may be entertained with respect to the physical origin of Λ (see also discussion on pp. 13-14) but, barring a real revelation regarding the fundamental character of the weak interactions, it would appear as if the first order of business is to determine, especially by further experimental study of

$$\nu_\mu + e^- \rightarrow \nu_\mu + e^- ,$$

whether $(\Lambda)_{\text{PURELY LEPTONIC}}$ is or is not appreciably greater than $(\Lambda)_{\text{SEMILEPTONIC}} \lesssim \frac{1}{10}\left(\frac{1}{\sqrt{G}}\right)$.

We conclude this section with a quantitative discussion of the 2nd-order weak process of nuclear double-beta decay (see Eqs. (47) – (52) et seq. and Eqs. (60) - (66)). We first consider the lepton-nonconserving "no-neutrino" nuclear double-beta decay[14]

$$A \begin{matrix} \nearrow & C_n + e_1^- + \bar{\nu}_e & \searrow \\ \searrow & C_n + e_2^- + \bar{\nu}_e & \nearrow \end{matrix} B + e_1^- + e_2^- \qquad (159)$$

where, e.g.

$$|A\rangle = |{}_{52}Te^{130}_{78}\rangle \ , \quad |C_n\rangle = |\{{}_{53}I^{130}_{77}\}_{N^{TH}\,STATE}\rangle, \quad |B\rangle = |{}_{54}Xe^{130}_{76}\rangle \qquad (160)$$

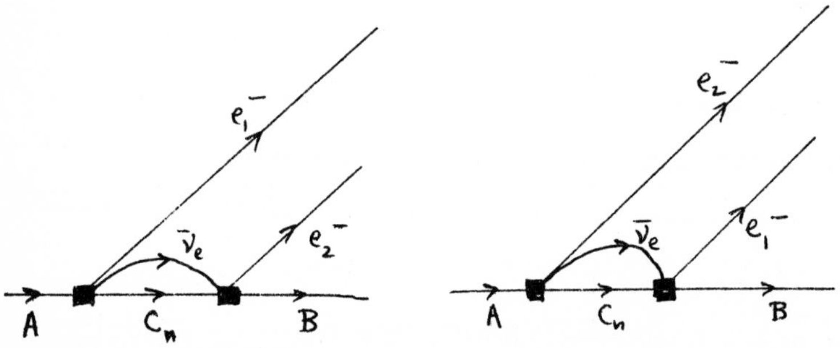

Fig. 14

and work out its matrix element (see Fig. 14) on the basis of the H_{weak} of Eq. (23) but with the $\ell_\alpha^{(e\nu_e)}(\vec{x})$ of Eqs. (63) and (60)

$$\left(\ell_\alpha^{(e\nu_e)}(\vec{x}) = \bar{\psi}_e(\vec{x})\gamma_\alpha\left[(1+\gamma_5)+\eta_e(1-\gamma_5)\right]\psi_{\nu_e};\quad \psi^{\dagger}_{\nu_e}(x) = \psi_{\nu_e}(x).\right)$$

Then a procedure completely analogous to that in Eqs. (131) - (138) yields an equation identical in appearance with Eq. (138), viz:

$$M(A \to B + e_1^- + e_2^-) \cong \left(\delta_{\vec{P}_A = \vec{P}_B + \vec{P}_{e_1^-} + \vec{P}_{e_2^-}}\right)\left(\frac{G}{\sqrt{2}}\right)^2 \Big\{ \ell_{\alpha\beta,4} \times$$

$$\left(\langle B|f^{(-)}_{\beta\alpha}(0)|A\rangle \frac{\Lambda^2}{4\pi^2} - \langle B|f^{(+)}_{\beta\alpha}(0)|A\rangle \frac{\epsilon''\Lambda}{2\pi^2} - \int d^3\xi\, e^{-i(\vec{P}_{e_1^-}-\vec{P}_{e_2^-})\cdot\frac{\vec{\xi}}{2}} \times\right.$$

$$\left. \times \frac{\langle B|g^{(+)}_{\beta\alpha}(0,\vec{\xi})|A\rangle}{|\vec{\xi}|}\frac{\epsilon''}{4\pi}\right) - \vec{\ell}_{\alpha\beta}\cdot\left(\int d^3\xi\, e^{-i(\vec{P}_{e_1^-}-\vec{P}_{e_2^-})\cdot\frac{\vec{\xi}}{2}} \frac{\langle B|g^{(+)}_{\beta\alpha}(0,\vec{\xi})|A\rangle\,\vec{\xi}}{|\vec{\xi}|^3\;4\pi}\right)\Big\}$$

$$(161)$$

but with

$$\ell_{\alpha\beta;4} \equiv \left(\bar{u}_{e_1^-}\gamma_\alpha\gamma_4\gamma_\beta\left(\tfrac{1}{2}\right)\left[(1+\gamma_5)+\eta_e(1-\gamma_5)\right]\left[(1-\gamma_5)+\eta_e(1+\gamma_5)\right]u^*_{e_2^-}\right) =$$

$$= \left(\bar{u}_{e_1^-}\gamma_\alpha\gamma_4\gamma_\beta\, 2\eta_e\, u^*_{e_2^-}\right),$$

$$\vec{\ell}_{\alpha\beta} \equiv \left(\bar{u}_{e_1^-}\gamma_\alpha\vec{\gamma}\gamma_\beta\left(\tfrac{1}{2}\right)\left[(1+\gamma_5)+\eta_e(1-\gamma_5)\right]\left[(1-\gamma_5)+\eta_e(1+\gamma_5)\right]u^*_{e_2^-}\right) = \left(\bar{u}_{e_1^-}\gamma_\alpha\vec{\gamma}\gamma_\beta\, 2\eta_e u^*_{e_2^-}\right);$$

$$\epsilon'' \equiv \tfrac{1}{2}\left[E_{e_1^-} - E_{e_2^-}\right]; \qquad (162)$$

and

$$[h_\beta^\dagger(\vec{y}), h_\alpha^\dagger(\vec{x})]_- = f_{\beta\alpha}^{(-)}(\vec{x})\,\delta(\vec{x}-\vec{y}),$$

$$[h_\beta^\dagger(\vec{y}), h_\alpha^\dagger(\vec{x})]_+ = f_{\beta\alpha}^{(+)}(\vec{x})\,\delta(\vec{x}-\vec{y}) + g_{\beta\alpha}^{(+)}(\vec{X}, \vec{\xi}). \tag{163}$$

Now the equal-time commutation relations of the weak hadron currents are such that $h_\beta^\dagger(\vec{y})$ commutes with $h_\alpha^\dagger(\vec{x})$ so that here $f_{\beta\alpha}^{(-)}(\vec{x}) = 0$. (164)

Further, an impulse approximation expression for $h_\alpha^\dagger(\vec{x})$ is

$$h_\alpha^\dagger(\vec{x}) \cong \{h_\alpha^{(q_n q_p)}(\vec{x})\}^\dagger = \{\bar{\Psi}_{q_n}\gamma_\alpha(1+\gamma_5)\Psi_{q_p}\}^\dagger =$$

$$= \{\bar{\Psi}_{q_N}(\vec{x})\,\tau^{(-)}\gamma_\alpha(1+\gamma_5)\Psi_{q_N}(\vec{x})\}^\dagger = \Psi_{q_N}^\dagger(\vec{x})\,\tau^{(+)}\gamma_4\gamma_\alpha(1+\gamma_5)\Psi_{q_N}(\vec{x}) \cong$$

$$\cong \sum_j \tau_j^{(+)}(\Gamma_\alpha)_j\,\delta(\vec{x}-\vec{r}_j); \quad \Gamma_\alpha \equiv \gamma_4\gamma_\alpha(1+\gamma_5) \tag{165}$$

where the $\sum_j \cdots$ runs over all the presumed basic constituents of the hadrons (nuclei) involved, i.e. over all the nucleons or over all the quarks; as indicated by our notation (see footnote 9) we assume that these basic constituents (p, n) or (q_p, q_n) have only two charge states and hence are characterized by isospin 1/2 so that the isospin step-up operators $\tau_j^{(+)}$ satisfy

$$\tau_j^{(+)} \tau_j^{(+)} = 0 \tag{166}$$

Thus, from Eqs. (163) - (165)

$$[h_\beta^\dagger(\vec{y}), h_\alpha^\dagger(\vec{x})]_+ = 2 h_\alpha^\dagger(\vec{x}) h_\beta^\dagger(\vec{y}) =$$

$$2 \sum_j \tau_j^{(+)} \tau_j^{(+)} (\Gamma_\alpha)_j (\Gamma_\beta)_j \delta(\vec{x}-\vec{r}_j) \delta(\vec{x}-\vec{y}) +$$

$$2 \sum_{j,k} (1-\delta_{jk}) \tau_j^{(+)} \tau_k^{(+)} (\Gamma_\alpha)_j (\Gamma_\beta)_k \delta(\vec{x}-\vec{r}_j) \delta(\vec{y}-\vec{r}_k)$$

$$= 0 + 2 \sum_{j,k} (1-\delta_{jk}) \tau_j^{(+)} \tau_k^{(+)} (\Gamma_\alpha)_j (\Gamma_\beta)_k \delta(\vec{x}-\vec{r}_j) \delta(\vec{y}-\vec{r}_h) \tag{167}$$

whence, comparing with Eq. (163),

$$f_{\beta\alpha}^{(+)}(\vec{x}) = 0 \tag{168}$$

and

$$g_{\beta\alpha}^{(+)}(\vec{X}, \vec{\xi}) = 2 \sum_{j,k} (1-\delta_{jk}) \tau_j^{(+)} \tau_k^{(+)} (\Gamma_\alpha)_j (\Gamma_\beta)_k \delta(\vec{x}-\vec{r}_j) \delta(\vec{y}-\vec{r}_k) \tag{169}$$

Eqs. (161), (164) and (168) show that the terms proportional to Λ^2 and to Λ in $M(A \to B + e_1^- + e_2^-)$ actually vanish so that the 2nd-order weak matrix element of lepton-nonconserving no-neutrino nuclear double-beta decay is finite providing only that the basic hadronic constituents have isospin 1/2 (Eqs. (166) - (168)). Physically, this finiteness arises from the fact that a given constituent (i.e. a neutron or a q_n -type quark) cannot emit and then reabsorb the virtual neutrino while emitting the two electrons since both electrons have the same sign of charge and the constituent has only two charge states; thus the virtual neutrino emitted with the first electron by one constituent is reabsorbed together with emission of the second electron by a different constituent and the reciprocal of the average distance between the constituents plays the role of a natural cut-off in the integration over $\vec{p}_\nu$. The appropriate form of $M(A \to B + e_1^- + e_2^-)$ is then, substituting Eqs. (164), (168), (169) and (162) into Eq. (161) and remembering Eq. (136) and also that

$$\delta_{\vec{p}_A = \vec{p}_B + \vec{p}_{e_1^-} + \vec{p}_{e_2^-}} = \int d^3X \, e^{-i(\vec{p}_B + \vec{p}_{e_1^-} + \vec{p}_{e_2^-} - \vec{p}_A)\cdot\vec{X}} ,$$

$$M(A\to B+e_1^-+e_2^-) \cong -\left(\frac{G}{\sqrt{2}}\right)^2 \frac{1}{4\pi}\iint d^3x\,d^3y\, e^{-i(\vec{p}_{e_1^-}\cdot\vec{x}+\vec{p}_{e_2^-}\cdot\vec{y})} \times$$

$$\times\left\langle \Psi_B(\cdots\vec{r}_j,(\tau_3)_j,(\sigma_3)_j,\cdots)\right| 2\sum_{j,k}\frac{(1-\delta_{jk})\tau_j^{(+)}\tau_k^{(+)}(\Gamma_\alpha)_j(\Gamma_\beta)_k\,\delta(\vec{x}-\vec{r}_j)\times}{|\vec{x}-\vec{y}|}$$

$$\times\,\delta(\vec{y}-\vec{r}_k)\left|\Psi_A(\cdots\vec{r}_j,(\tau_3)_j,(\sigma_3)_j,\cdots)\right\rangle\left(l_{\alpha\beta;4}\,\epsilon''+\vec{l}_{\alpha\beta}\cdot\frac{(\vec{x}-\vec{y})}{|\vec{x}-\vec{y}|^2}\right)=$$

$$=-\left(\frac{G}{\sqrt{2}}\right)^2\left(\frac{n_e}{2\pi}\right)\left\langle\Psi_B(\cdots\vec{r}_j,(\tau_3)_j,(\sigma_3)_j,\cdots)\right|\sum_{j,k}\frac{(1-\delta_{jk})\tau_j^{(+)}\tau_k^{(+)}(\Gamma_\alpha)_j(\Gamma_\beta)_k\,e^{-i(\vec{p}_{e_1^-}\cdot\vec{r}_j+\vec{p}_{e_2^-}\cdot\vec{r}_k)}}{|\vec{r}_j-\vec{r}_k|}\times$$

$$\left((\bar{u}_{e_1^-}\gamma_\alpha\gamma_4\gamma_\beta u^*_{e_2^-})(E_{e_1^-}-E_{e_2^-})+2(\bar{u}_{e_1^-}\gamma_\alpha\vec{\gamma}\gamma_\beta u^*_{e_2^-})\cdot\frac{|\vec{r}_j-\vec{r}_k|}{|\vec{r}_j-\vec{r}_k|^2}\right)\left|\Psi_A(\cdots\vec{r}_j,(\tau_3)_j,(\sigma_3)_j\cdots)\right\rangle \tag{170}$$

and, since in all cases of practical interest Ψ_B and Ψ_A have the same parity and are also such that

$$\langle\Psi_B|(\vec{p}_{e_1^-}\cdot\vec{r}_j)\vec{r}_j|\Psi_A\rangle \cong \frac{1}{3}\vec{p}_{e_1^-}\langle\Psi_B||\vec{r}_j|^2|\Psi_A\rangle \ll \hat{p}_{e_1^-}\langle\Psi_B||\vec{r}_j||\Psi_A\rangle$$

etc.,

$$M(A\to B+e_1^-+e_2^-)\cong-\frac{G}{\sqrt{2}}\left(\frac{n_e}{2\pi}\frac{G}{\sqrt{2}}\right)\left\langle\Psi_B(\cdots\vec{r}_j,(\tau_3)_j,(\sigma_3)_j\cdots)\right|\sum_{j,k}\frac{(1-\delta_{jk})\tau_j^{(+)}\tau_k^{(+)}(\Gamma_\alpha)_j}{|\vec{r}_j-\vec{r}_k|}\times$$

$$\times(\Gamma_\beta)_k\left|\Psi_A(\cdots\vec{r}_j,(\tau_3)_j,(\sigma_3)_j,\cdots)\right\rangle\left(\bar{u}_{e_1^-}\gamma_\alpha\left[\gamma_4(E_{e_1^-}-E_{e_2^-})-\frac{2i\vec{\gamma}}{3}\cdot(\vec{p}_{e_1^-}-\vec{p}_{e_2^-})\right]\gamma_\beta u^*_{e_2^-}\right). \tag{171}$$

Eq. (171) for the matrix element of the 2nd-order weak process of lepton-nonconserving "no-neutrino" nuclear double-beta decay is to be compared to the matrix element for the 1st-order weak process of nuclear single-beta decay:

$C \to D + e^- + \bar{\nu}_e$ which, for an "allowed" transition, is

$$M(C \to D + e^- + \bar{\nu}_e) = \frac{G}{\sqrt{2}} \langle \Psi_D(\ldots \vec{r}_j, (\tau_3)_j, (\sigma_3)_j, \ldots) | \sum_j \tau_j^{(+)} \times$$

$$\times (\Gamma_\alpha)_j | \Psi_C(\ldots \vec{r}_j, (\tau_3)_j, (\sigma_3)_j, \ldots) \rangle \left(\bar{u}_{e^-} \gamma_\alpha (1+\gamma_5) u^*_{\bar{\nu}_e} \right) \qquad (172)$$

Eqs. (171) and (172) yield for the ratio of the decay rates

$$\frac{\Gamma(A \to B + e_1^- + e_2^-)}{\Gamma(C \to D + e_1^- + \bar{\nu}_e)} \cong \eta_e^2 \left(\left(\frac{G}{\sqrt{2}}\right)^2 \frac{\{|\vec{r}_j - \vec{r}_k|^{-1}\}_{Av} (m_A - m_B)}{2\pi} \right)^2 \times$$

$$\times \frac{|\langle \Psi_B | \sum_{j,k} (1-\delta_{jk}) \tau_j^{(+)} \tau_k^{(+)} \vec{\sigma}_j \cdot \vec{\sigma}_k | \Psi_A \rangle|^2}{|\langle \Psi_C | \sum_j \tau_j^{(+)} \vec{\sigma}_j | \Psi_C \rangle|^2} \cong$$

$$\eta_e^2 \left(\frac{G}{\sqrt{2}} \left(\frac{m_\pi}{A^{1/3}}\right) \frac{(m_A - m_B)}{2\pi} \right)^2 \frac{|\langle \Psi_B | \sum_j (1-\delta_{jk}) \tau_j^{(+)} \tau_k^{(+)} \vec{\sigma}_j \cdot \vec{\sigma}_k | \Psi_A \rangle|^2}{|\langle \Psi_C | \sum_j \tau_j^{(+)} \vec{\sigma}_j | \Psi_C \rangle|^2} \qquad (173)$$

with $(m_C - m_D)$ assumed $\cong (m_A - m_B)$ and with the second (approximate) equality appropriate to the basic nuclear constituents being taken as nucleons.[31] With $A \equiv$ nuclear mass number = 130, $(m_A - m_B) = (m_{Te^{130}} - m_{Xe^{130}}) = 3.5\,\text{MeV} = 2\times10^{-2}\, m_\pi$, and[14]

$$\frac{\left|\langle \Psi_B \,|\, \sum_{j,k} (1-\delta_{jk})\, \tau_j^{(+)} \tau_k^{(+)}\, \vec{\sigma}_j\cdot\vec{\sigma}_k \,|\, \Psi_A\rangle\right|^2}{\left|\langle \Psi_D \,|\, \sum_j \tau_j^{(+)} \vec{\sigma}_j \,|\, \Psi_C\rangle\right|^2} \approx 10^{-2}$$

we have

$$\frac{\Gamma(A \to B + e_1^- + e_2^-)}{\Gamma(C \to D + e_1^- + \bar{\nu}_e)} \approx \eta_e^2\, 10^{-18} \qquad (174)$$

and, since $\Gamma(C \to D + e_1^- + \bar{\nu}_e) \approx 10^{-4}$ sec. for an allowed nuclear single-beta decay with $m_C - m_D = 3.5$ MeV and $A \approx 150$,

$$\left(\frac{1}{\Gamma(A \to B + e_1^- + e_2^-)}\right) \approx \frac{10^{22}\,\text{sec}}{\eta_e^2} . \qquad (175)$$

Eq. (175) has been used to set the upper limit on η_e given in Eq. (44). (See Eqs. (44) - (48) et seq.)

Finally, we remark that the matrix element for the 2nd-order weak process of lepton-conserving "two-neutrino" nuclear double-beta decay (see Fig. 15)

$$A \nearrow C_n + e_1^- + (\bar{\nu}_e)_1 \searrow B + e_1^- + e_2^- + (\bar{\nu}_e)_1 + (\bar{\nu}_e)_2$$
$$A \searrow C_n + e_2^- + (\bar{\nu}_e)_2 \nearrow \tag{176}$$

is finite since no integration over the momentum of a virtual lepton is present in this case (compare Fig. 15 with Figs. 14 and 12) and since

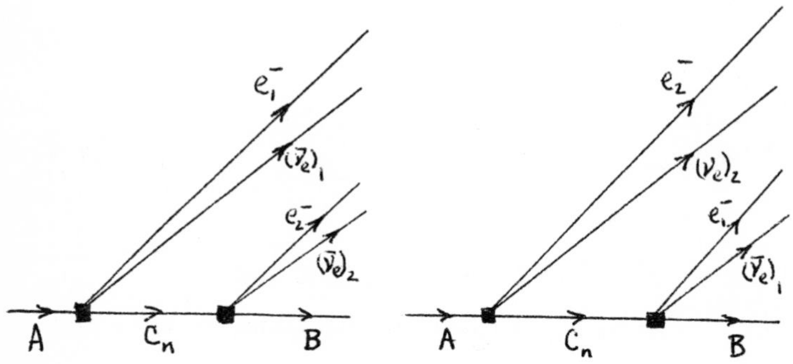

Fig. 15

the sum over the contributions of the various $|C_n\rangle$ can be carried out by closure. The expression for the ratio of the "two-neutrino" double-beta decay rate to a corresponding allowed single-beta decay rate can be crudely approximated by

$$\frac{\Gamma\,(A \to B + e_1^- + e_2^- + (\bar{\nu}_e)_1 + (\bar{\nu}_e)_2)}{\Gamma\,(C \to D + e^- + \bar{\nu}_e)} \approx$$

$$\approx \left(\frac{\frac{G}{\sqrt{2}} \, (m_A - m_B)^3 \, \frac{4\pi}{(2\pi)^3}}{(m_A - m_B)} \right)^2 \frac{|\langle \Psi_B | \sum_{j,k} (1-\delta_{jk}) \tau_j^{(+)} \tau_k^{(+)} \vec{\sigma}_j \cdot \vec{\sigma}_k | \Psi_A \rangle|^2}{|\langle \Psi_D | \sum_j \tau_j^{(+)} \vec{\sigma}_j | \Psi_C \rangle|^2} \approx$$

$$\approx 10^{-25} \quad \text{for } (m_A - m_B) = 3.5 \text{ MeV}, \qquad (177)$$

which is in rough agreement with experiment (see Eqs. (47)-(49) et seq.).

4. "Abnormalities" in the Weak Currents and How to Discover Them

Up to now we have used an H_{WEAK} (Eq. (23)) or an $H_{SEMIWEAK}$ (Eq. (24a)) which involves the conventional weak current (Eq. (23)):

$$J_\alpha(x) = \ell_\alpha(x) + h_\alpha(x) = \left(\ell_\alpha^{(e\nu_e)}(x) + \ell_\alpha^{(\mu\nu_\mu)}(x)\right) + \left(\cos\theta_C\, h_\alpha^{(g_n g_p)}(x) + \sin\theta_C\, h_\alpha^{(g_\Lambda g_p)}(x)\right)$$

$$\ell_\alpha^{(e\nu_e)}(x) = \bar{\Psi}_e(x)\, \gamma_\alpha (1+\gamma_5) \Psi_{\nu_e}(x)$$

$$\ell_\alpha^{(\mu\nu_\mu)}(x) = \bar{\Psi}_\mu(x)\, \gamma_\alpha (1+\gamma_5) \Psi_{\nu_\mu}(x)$$

$$h_\alpha^{(g_n g_p)}(x) = \left\{ h_\alpha^{(g_n g_p)}(x) \right\}_{\text{POLAR PART}} + \left\{ h_\alpha^{(g_n g_p)}(x) \right\}_{\text{AXIAL PART}} \equiv V_\alpha(x) + A_\alpha(x)$$

$$V_\alpha(x) = \bar{\Psi}_{q_n}(x)\gamma_\alpha\psi_{q_p}(x) = \bar{\Psi}_{q_N}(x)\tau^{(-)}\gamma_\alpha\psi_{q_N}(x) = I_\alpha^{(-)}(x)$$
$$A_\alpha(x) = \bar{\Psi}_{q_n}(x)\gamma_\alpha\gamma_5\psi_{q_p}(x) = \bar{\Psi}_{q_N}(x)\tau^{(-)}\gamma_\alpha\gamma_5\psi_{q_N}(x) = A_\alpha^{(n)(r)}(x)$$
$$h_\alpha^{(q_n q_p)}(x) = \{h_\alpha^{(q_n q_p)}(x)\}_{\text{POLAR PART}} + \{h_\alpha^{(q_n q_p)}(x)\}_{\text{AXIAL PART}} \equiv \mathcal{V}_\alpha(x) + \mathcal{A}_\alpha(x)$$
$$\mathcal{V}_\alpha(x) = \bar{\Psi}_{q_n}(x)\gamma_\alpha\psi_{q_p}(x)\ ;\ \mathcal{A}_\alpha(x) = \bar{\Psi}_{q_n}(x)\gamma_\alpha\gamma_5\psi_{q_p}(x) = \mathcal{A}_\alpha^{(n)}(x)$$

(178)

where, as shown in Eqs. (26) - (30), $l_\alpha^{(e\nu_e)}(x)$, $l_\alpha^{(\mu\nu_\mu)}(x)$, $I_\alpha^{(-)}(x)$, $A_\alpha^{(n)(r)}(x)$, $\mathcal{V}_\alpha(x)$ and $\mathcal{A}_\alpha(x)$ are "normal" (n), i.e., odd under time reversal (T) while $I_\alpha^{(-)}(x)$ and $A_\alpha^{(n)(r)}(x)$ are isovectors and are "regular" (r), i.e., odd under charge symmetry ($e^{i\pi I^{(2)}}$). This specification of the space-time transformation behavior of the various weak currents, e.g. of $A_\alpha^{(n)(r)}(x)$, is complete since all weak currents are odd under CPT and since $A_\alpha^{(n)(r)}(x)$ is even under P and a space-time vector (under rotations) — thus the "normality" (i.e. oddness) of $A_\alpha^{(n)(r)}(x)$ under T implies its "normality" (i.e., evenness) under C ; the specification of the isospace transformation behavior of $A_\alpha^{(n)(r)}(x)$ is also complete since it is an isovector and "regular."

As discussed in connection with Eqs. (26) - (30) the comprehensive experimental evidence in favor of CVC, i.e.

in favor of the identification of $V_\alpha(x)$ with $I_\alpha^{(-)}(x)$ (and of the generalization of this to the appropriate identification of $\mathcal{V}_\alpha(x)$), discourages attempts to augment the above expression for $V_\alpha(x)$ by $\mathcal{T}$-abnormal or $e^{i\pi I^{(2)}}$-irregular terms though augmentation by an isotensor term is not immediately excluded (see below). Also, no modification of the above expressions for $\ell_\alpha^{(e\nu_e)}(x)$, $\ell_\alpha^{(\mu\nu_\mu)}(x)$ is suggested by available experiments apart from the already discussed augmentation of the expression for $\ell_\alpha^{(e\nu_e)}(x)$ by a term which violates lepton conservation (and $\mathcal{T}$-invariance if η_e is complex) viz.: $\ell_\alpha^{(e\nu_e)}(x) = \bar{\psi}_e(x)\gamma_\alpha(1+\gamma_5)\psi_{\nu_e}(x)$

$$\rightarrow \ell_\alpha^{(e\nu_e)}(x) = \bar{\psi}_e(x)\gamma_\alpha\left[(1+\gamma_5)+\eta_e(1-\gamma_5)\right]\psi_{\nu_e}(x) \;;\; \psi_{\nu_e}^{\dagger}(x) = \psi_{\nu_e}(x)$$

(see Eqs. (60)-(64) et seq. and Eqs. (159)-(175)). On the other hand, and as emphasized in the introduction, the augmentation of $A_\alpha^{(n)(r)}(x)$ by terms that are $\mathcal{T}$-abnormal and/or $e^{i\pi I^{(2)}}$ -irregular, i.e. by terms that we have called $A_\alpha^{(a)(r)}(x)$, $A_\alpha^{(n)(i)}(x)$, $A_\alpha^{(a)(i)}(x)$ should be seriously considered since on the basis of available experiments only the term $A_\alpha^{(a)(i)}(x)$ can be excluded (see below). We therefore write (see Eqs. (28)-(30) and footnote 11) so that if $J_\alpha^{(a)}(x) \neq 0$, H_{WEAK} is not invariant under $\mathcal{T}$ and terms odd under $\mathcal{T}$ appear in the expressions for the various transition probabilities of 1st-order weak processes (even in the limit of vanishing final state interaction). As in the case of $V_\alpha(x)$, it is also possible to augment $A_\alpha(x)$ by an isotensor term (see below). Finally, and as indicated in Eq. (179), the $\mathcal{T}$ -normal strangeness-changing axial hadron weak current, $a_\alpha^{(n)}(x)$, can be augmented by a $\mathcal{T}$ -abnormal term, $a_\alpha^{(a)}(x)$, though for the sake of simplicity, we shall confine our explicit discussion to the strangeness-nonchanging case.

We begin our detailed discussion by setting down the neutron → proton matrix elements of $\{V_\alpha(x)\}^\dagger$ and $\{A_\alpha(x)\}^\dagger$ which enter linearly into the expressions for the matrix elements $\langle e^- \bar{\nu}_e p | H_{WEAK} | n \rangle$ (neutron beta decay) and $\langle \mu^- p | H_{WEAK} | \nu_\mu n \rangle$ (neutrino-neutron "quasi-elastic" collision). On the basis of Lorentz covariance we have:

$$\langle p | \{V_\alpha(0)\}^\dagger | n \rangle = \langle p | I_\alpha^{(+)}(0) | n \rangle =$$

$$= \left(\bar{u}_p \left[\gamma_\alpha F_V(q^2; n \to p) - \frac{\sigma_{\alpha\beta} q_\beta}{2m_p} F_M(q^2; n \to p) \right] u_n \right) ;$$

$$A_\alpha(x) = \left\{A_\alpha^{(n)(r)}(x) + A_\alpha^{(n)(i)}(x)\right\} + \left\{A_\alpha^{(a)(r)}(x) + A_\alpha^{(a)(i)}(x)\right\} \equiv A_\alpha^{(n)}(x) + A_\alpha^{(a)}(x)$$

$$a_\alpha(x) = a_\alpha^{(n)}(x) + a_\alpha^{(a)}(x)$$

$$J_\alpha(x) = \left\{ \ell_\alpha(x) + \cos\theta_c\left(I_\alpha^{(-)}(x) + A_\alpha^{(n)}(x)\right) + \sin\theta_c\left(\mathcal{V}_\alpha(x) + a_\alpha^{(n)}(x)\right)\right\} +$$

$$+ \left\{ \cos\theta_c\left(A_\alpha^{(a)}(x)\right) + \sin\theta_c\left(a_\alpha^{(a)}(x)\right)\right\} \equiv$$

$$\equiv J_\alpha^{(n)}(x) + J_\alpha^{(a)}(x)$$

$$H_{WEAK} = \frac{G}{\sqrt{2}} \int \left\{J_\alpha^{(n)}(x) + J_\alpha^{(a)}(x)\right\}^\dagger \left\{J_\alpha^{(n)}(x) + J_\alpha^{(a)}(x)\right\} d^3x =$$

$$= \frac{G}{\sqrt{2}} \int \left(\left\{J_\alpha^{(n)}(x)\right\}^\dagger J_\alpha^{(n)}(x) + \left\{J_\alpha^{(a)}(x)\right\}^\dagger J_\alpha^{(a)}(x)\right) d^3x +$$

$$+ \frac{G}{\sqrt{2}} \int \left(\left\{J_\alpha^{(n)}(x)\right\}^\dagger J_\alpha^{(a)}(x) + \left\{J_\alpha^{(a)}(x)\right\}^\dagger J_\alpha^{(n)}(x)\right) d^3x \equiv$$

$$\equiv \left\{H_{WEAK}\right\}_{T\text{-}EVEN} + \left\{H_{WEAK}\right\}_{T\text{-}ODD}$$

(179)

$$q \equiv p_p - p_n = -(p_{e^-} + p_{\bar{\nu}_e}) \; ; \; \sigma_{\alpha\beta} \equiv \frac{i^{-1}}{2}(\gamma_\alpha \gamma_\beta - \gamma_\beta \gamma_\alpha) \qquad (180)$$

$$\langle p | \{A_\alpha(0)\}^\dagger | n \rangle = \langle p | \{A_\alpha^{(n)}(0)\}^\dagger | n \rangle + \langle p | \{A_\alpha^{(a)}(0)\}^\dagger | n \rangle =$$

$$= \sum_{y=r,i} \langle p | \{A_\alpha^{(n)(y)}(0)\}^\dagger | n \rangle + \sum_{y=r,i} \langle p | \{A_\alpha^{(a)(y)}(0)\}^\dagger | n \rangle =$$

$$= \sum_{y=r,i} \Big(\bar{u}_p \Big[\gamma_\alpha \gamma_5 F_A^{(n)(y)}(q^2; n \to p) - \Big(\frac{\sigma_{\alpha\beta} q_\beta \gamma_5}{2m_p} \Big) F_T^{(n)(y)}(q^2; n \to p) +$$

$$+ \Big(\frac{i q_\alpha \gamma_5}{m_\pi} \Big) \Big(\frac{m_p + m_n}{m_\pi} \Big) F_P^{(n)(y)}(q^2; n \to p) \Big] u_n \Big) + \sum_{y=r,i} \Big(\bar{u}_p \Big[\gamma_\alpha \gamma_5 F_A^{(a)(y)}(q^2; n \to p) +$$

$$- \Big(\frac{\sigma_{\alpha\beta} q_\beta \gamma_5}{2m_p} \Big) F_T^{(a)(y)}(q^2; n \to p) + \Big(\frac{i q_\alpha \gamma_5}{m_\pi} \Big) \Big(\frac{m_p + m_n}{m_\pi} \Big) F_P^{(a)(y)}(q^2; n \to p) \Big] u_n \Big) \qquad (181)$$

$$\langle e^- \bar{\nu}_e p | H_{WEAK} | n \rangle = \frac{G}{\sqrt{2}} \big(\bar{u}_{e^-} \gamma_\alpha (1+\gamma_5) u^*_{\bar{\nu}_e} \big) \big(\langle p | \{V_\alpha(0)\}^\dagger | n \rangle + \langle p | \{A_\alpha(0)\}^\dagger | n \rangle \big)$$

$$\langle \mu^- p | H_{WEAK} | \nu_\mu n \rangle = \frac{G}{\sqrt{2}} \big(\bar{u}_{\mu^-} \gamma_\alpha (1+\gamma_5) u_{\nu_\mu} \big) \big(\langle p | \{V_\alpha(0)\}^\dagger | n \rangle + \langle p | \{A_\alpha(0)\}^\dagger | n \rangle \big) \qquad (182)$$

where $F_V(q^2; n \to p)$, $F_M(q^2; n \to p)$, $\sum_{\substack{x=n,a \\ y=r,i}} F_A^{(x)(y)}(q^2; n \to p)$, $\sum_{\substack{x=n,a \\ y=r,i}} F_T^{(x)(y)}(q^2; n \to p)$,

and $\sum_{\substack{x=n,a \\ y=r,i}} F_P^{(x)(y)}(q^2; n \to p)$

are polar-vector, weak-magnetism, axial-vector, pseudo-tensor and pseudoscalar $n \to p$ form factors. Since

$$\left(\bar{u}_{e^-}\gamma_\alpha(1+\gamma_5)u^*_{\bar{\nu}_e}\right)\frac{q_\alpha}{m_\pi} = -\left(\bar{u}_{e^-}\gamma_\alpha(1+\gamma_5)u^*_{\bar{\nu}_e}\right)\frac{(P_{e^-}+P_{\bar{\nu}_e})_\alpha}{m_\pi} =$$

$$= \frac{-i m_e}{m_\pi}\left(\bar{u}_{e^-}(1+\gamma_5)u^*_{\bar{\nu}_e}\right) \ll 1$$

the term proportional to $\sum_{\substack{x=n,a\\ y=r,i}} F_P^{(x)(y)}(q^2; n\to p)$ can be dropped in $n \to p + e^- + \bar{\nu}_e$ and in most configurations of $\nu_\mu + n \to \mu^- + p$. Since $V_\alpha(x) = I_\alpha^{(-)}(x)$ and $A_\alpha^{(n)(r)}(x)$, $A_\alpha^{(n)(i)}(x)$ are $\mathcal{T}$-normal, $F_V(q^2; n\to p)$, $F_M(q^2; n\to p)$, $F_A^{(n)(r)}(q^2; n\to p)$, $F_T^{(n)(r)}(q^2; n\to p)$, $F_A^{(n)(i)}(q^2; n\to p)$, $F_T^{(n)(i)}(q^2; n\to p)$ must all be real while the $\mathcal{T}$-abnormality of $A_\alpha^{(a)(r)}(x)$ and $A_\alpha^{(a)(i)}(x)$ implies that $F_A^{(a)(r)}(q^2; n\to p)$, $F_T^{(a)(r)}(q^2; n\to p)$, $F_A^{(a)(i)}(q^2; n\to p)$, $F_T^{(a)(i)}(q^2; n\to p)$ are all imaginary, i. e.

$$F_A^{(x)(y)}(q^2; n\to p) = a_{xy}\left(F_A^{(x)(y)}(q^2; n\to p)\right)^*,$$
$$F_T^{(x)(y)}(q^2; n\to p) = a_{xy}\left(F_T^{(x)(y)}(q^2; n\to p)\right)^*, \qquad (183)$$
$$a_{nr} = a_{ni} = -a_{ar} = -a_{ai} = 1 .$$

Further, the $e^{i\pi I^{(2)}}$-regularity of $A_\alpha^{(n)(r)}(x)$ and $A_\alpha^{(a)(r)}(x)$ and the $e^{i\pi I^{(2)}}$-irregularity of $A_\alpha^{(n)(i)}(x)$ and $A_\alpha^{(a)(i)}(x)$ yield[32]

$$F_A^{(x)(y)}(q^2; n \to p) = b_{xy}\left(F_A^{(x)(y)}(q^2; p' \to n')\right)^* =$$
$$= b_{xy} F_A(q^2; n' \to p'),$$
$$F_T^{(x)(y)}(q^2; n \to p) = -b_{xy}\left(F_T^{(x)(y)}(q^2; p' \to n')\right)^* =$$
$$= b_{xy} F_T^{(x)(y)}(q^2; n' \to p'),$$
$$b_{nr} = b_{ar} = -b_{ni} = -b_{ai} = 1, \tag{184}$$

where the second equality on the right is a consequence of $\langle D | J_\alpha | C\rangle = \langle C | \{J_\alpha\}^\dagger | D\rangle^*$ and where p' and n' are charge-symmetric to p and n. Thus, since n and p are to a very high degree of approximation the two members of the nucleon ($A = 1$) isodoublet, $p' = n$ and $n' = p$, and Eq. (184) becomes

$$F_A^{(x)(y)}(q^2; n \to p) = b_{xy}\left(F_A^{(x)(y)}(q^2; n \to p)\right)^* = b_{xy} F_A^{(x)(y)}(q^2; p \to n),$$
$$F_T^{(x)(y)}(q^2; n \to p) = -b_{xy}\left(F_T^{(x)(y)}(q^2; n \to p)\right)^* = b_{xy} F_T^{(x)(y)}(q^2; p \to n);$$
$$b_{nr} = b_{ar} = -b_{ni} = -b_{ai} = 1. \tag{185}$$

Combination of Eqs. (183) and (185) gives

$$F_A^{(n)(r)}(q^2; n\to p) = \left(F_A^{(n)(r)}(q^2; n\to p)\right)^*, \quad F_T^{(n)(r)}(q^2; n\to p) = 0$$

$$F_A^{(a)(r)}(q^2; n\to p) = 0, \quad F_T^{(a)(r)}(q^2; n\to p) = -\left(F_T^{(a)(r)}(q^2; n\to p)\right)^*$$

$$F_A^{(n)(i)}(q^2; n\to p) = 0, \quad F_T^{(n)(i)}(q^2; n\to p) = \left(F_T^{(n)(i)}(q^2; n\to p)\right)^*$$

$$F_A^{(a)(i)}(q^2; n\to p) = -\left(F_A^{(a)(i)}(q^2; n\to p)\right)^*, \quad F_T^{(a)(i)}(q^2; n\to p) = 0 \tag{186}$$

Eq. (186) (and Eqs. (181), (182)) show that the presence of (the second class) currents $A_\alpha^{(a)(r)}(x)$ and $A_\alpha^{(n)(i)}(x)$ is undetectible in transitions between the two members of the nucleon isodoublet at $q = 0$.

To continue, we consider the beta decay processes:

$$\Sigma^+ \to \Lambda + e^+ + \nu_e, \quad \Sigma^- \to \Lambda + e^- + \bar{\nu}_e$$

and obtain the precise analogues of Eqs. (183) and (184), viz:

$$F_A^{(x)(y)}(q^2; \Sigma^\pm \to \Lambda) = a_{xy}\left(F_A^{(x)(y)}(q^2; \Sigma^\pm \to \Lambda)\right)^*,$$

$$F_T^{(x)(y)}(q^2; \Sigma^\pm \to \Lambda) = a_{xy}\left(F_T^{(x)(y)}(q^2; \Sigma^\pm \to \Lambda)\right)^*,$$

$$a_{nr} = a_{ni} = -a_{ar} = -a_{ai} = 1, \tag{187}$$

and

$$F_A^{(x)(y)}(q^2;\Sigma^{\pm}\to\Lambda) = b_{xy}\left(F_A^{(x)(y)}(q^2;\Lambda'\to(\Sigma^{\pm})')\right)^* =$$
$$= b_{xy} F_A^{(x)(y)}(q^2;(\Sigma^{\pm})'\to\Lambda'),$$
$$F_T^{(x)(y)}(q^2;\Sigma^{\pm}\to\Lambda) = -b_{xy}\left(F_T^{(x)(y)}(q^2;\Lambda'\to(\Sigma^{\pm})')\right)^* = b_{xy} F_T^{(x)(y)}(q^2;(\Sigma^{\pm})'\to\Lambda'),$$
$$b_{nr} = b_{ar} = -b_{ni} = -b_{ai} = 1\,, \qquad (188)$$

where $(\Sigma^{\pm})'$, Λ' are charge-symmetric to $\Sigma^{\pm}$, Λ so that $(\Sigma^{\pm})' = \Sigma^{\mp}$, $\Lambda' = \Lambda$ ($\Sigma^+, \Sigma^0, \Sigma^-$ are the three members of the Σ isotriplet and Λ is an isosinglet); thus

$$F_A^{(x)(y)}(q^2;\Sigma^{\pm}\to\Lambda) = b_{xy}\left(F_A^{(x)(y)}(q^2;\Lambda\to\Sigma^{\mp})\right)^* = b_{xy} F_A^{(x)(y)}(q^2;\Sigma^{\mp}\to\Lambda),$$
$$F_T^{(x)(y)}(q^2;\Sigma^{\pm}\to\Lambda) = -b_{xy}\left(F_T^{(x)(y)}(q^2;\Lambda\to\Sigma^{\mp})\right)^* = b_{xy} F_T^{(x)(y)}(q^2;\Sigma^{\mp}\to\Lambda),$$
$$b_{nr} = b_{ar} = -b_{ni} = -b_{ai} = 1\,, \qquad (189)$$

so that in the case of a transition between two hadrons belonging to different isomultiplets the $e^{i\pi I^{(2)}}$-transformation behavior of $A_\alpha^{(x)(y)}$ does not provide a "reality" condition as in the isodoublet case (compare Eq. (189) with Eq. (185)). Thus, combination of Eqs. (189) and (187) does not, as in the case of the combination of Eqs. (185) and (183), result in the juxtaposition of two "clashing" reality conditions with the consequent vanishing of some of the $\Sigma^{\pm} \to \Lambda$ form factors (in contrast to Eq. (186) which predicts the vanishing of some of the $n \to p$ form factors).

As a third example, we consider the neutrino-nucleon inelastic collisions:

$$\nu_\mu + p \to \mu^- + N^{*++} \quad , \quad \bar{\nu}_\mu + n \to \mu^+ + N^{*-}$$

Here, as in the $\Sigma^{\pm} \leftrightarrow \Lambda$ case, the initial and final hadrons are in different isomultiplets and, in addition, have different spins (here, as before, $N^* \equiv N^*(I = 3/2,\ J = 3/2)$). In analogy with Eqs. (187), (189), and considering explicitly only these form factors which are not multiplied by a kinematic term that $\to 0$ as $q \to 0$, we have

$$F_A^{(x)(y)}(q^2; p \to N^{*++}) = a_{xy}\left(F_A^{(x)(y)}(q^2; p \to N^{*++})\right)^* \qquad (190)$$

$$F_A^{(x)(y)}(q^2; p \to N^{*++}) = b_{xy}\left(F_A^{(x)(y)}(q^2; N^{*-} \to n)\right)^* = b_{xy} F_A(q^2; n \to N^{*-}) \qquad (191)$$

so that (as in the $\Sigma^{\pm} \to \Lambda$ case) combination of Eqs. (190) and (191) will not result in the vanishing of some of the $p \to N^{*++}$ form factors.

To complete this general discussion of the relations among the {initial hadron} $\rightarrow$ {final hadron} form factors implied by the $\mathcal{T}$ - and $e^{i\pi I^{(2)}}$ - transformation behavior of $A_\alpha^{(x)(y)}(x)$ we treat the case of nuclear beta decay: $N_i \rightarrow N_f + e^- + \bar{\nu}_e$ (or $N_i \rightarrow N_f + e^+ + \nu_e$). In this case we have [33, 10]

$$F_A^{(x)(y)}(q^2; N_i \rightarrow N_f) = F_A^{(x)(y)}(q^2; n \rightarrow p) \langle N_f | \tau \vec{\sigma} | N_i \rangle + f_A^{(x)(y)}(q^2; N_i \rightarrow N_f)$$

$$F_T^{(x)(y)}(q^2; N_i \rightarrow N_f) = F_T^{(x)(y)}(q^2; n \rightarrow p) \langle N_f | \tau \vec{\sigma} | N_i \rangle + f_T^{(x)(y)}(q^2; N_i \rightarrow N_f)$$

$$\langle N_f | \tau \vec{\sigma} | N_i \rangle \equiv \pm \left\{ \left(\frac{J(N_i)}{J(N_i)+1} \right) \sum_{M(N_f)=-J(N_f)}^{+J(N_f)} \left| \left\langle \Psi_{N_f; M(N_f)} (\ldots \vec{r}_j, (\sigma_3)_j, (\tau_3)_j, \ldots) \right| \sum_{j=1}^{A} \tau_j^{(+)} \times \right. \right.$$

$$\left. \left. \times \vec{\sigma}_j \, e^{i \vec{q} \cdot \vec{r}_j} \left| \Psi_{N_i; M(N_i)} (\ldots \vec{r}_j, (\sigma_3)_j, (\tau_3)_j, \ldots) \right\rangle \right|^2 \right\}^{1/2} . \qquad (192)$$

Here, the first term on the right side arises from the impulse approximation to $\left\{ A_\alpha^{(x)(y)}(x) \right\}^\dagger$ i.e.

$$\left\{ A_\alpha^{(x)(y)}(x) \right\}^\dagger = \sum_{j=1}^{A} \tau_j^{(+)} \, i(\gamma_4 \gamma_\alpha \gamma_5)_j \, \delta(\vec{x} - \vec{r}_j) \qquad (193)$$

(see Eq. (165)). The basic nuclear constituents being taken as nucleons (treated nonrelativistically) the second term on the right side represents the corrections to this nucleon-type impulse approximation. The corrections in question arise from the beta decay of (virtual) mesons exchanged between the nucleons and from the beta decay of any N^* present (albeit with a small probability) in the nuclei [14] so that one of the terms in $f_A^{(x)(y)}(q^2; N_i \to N_f)$ is proportional to $F_A^{(x)(y)}(q^2; \rho^- \to \pi^0)$ and another to $F_A^{(x)(y)}(q^2; N^{*-} \to n)$. The general order of magnitude of $\left|\sum_{\substack{x=n,a\\y=r,i}} f_A^{(x)(y)}(q^2; N_i \to N_f)\right|$ is:

$$\left|\sum_{\substack{x=n,a\\y=r,i}} f_A^{(x)(y)}(q^2; N_i \to N_f)\right| = 0.01 - 0.1 \quad , \qquad (194)$$

as deduced from the comparison of $\{\Gamma(H^3 \to He^3 + e^- + \bar{\nu}_e)\}_{\text{THEORETICAL}}$ with $\{\Gamma(H^3 \to He^3 + e^- + \bar{\nu}_e)\}_{\text{EXPERIMENTAL}}$ using Eq. (192) for the computation of $\{\Gamma\}_{\text{THEORETICAL}}$ [33].

Combination of Eqs. (182), (186) and (192) yields

$$F_A^{(n)(r)}(q^2; N_i \to N_f) = \left(F_A^{(n)(r)}(q^2; N_i \to N_f)\right)^* , \quad F_T^{(n)(r)}(q^2; N_i \to N_f) = f_T^{(n)(r)}(q^2; N_i \to N_f) = \left(f_T^{(n)(r)}(q^2; N_i \to N_f)\right)^* ,$$

$$F_A^{(a)(r)}(q^2; N_i \to N_f) = f_A^{(a)(r)}(q^2; N_i \to N_f) = -\left(f_A^{(a)(r)}(q^2; N_i \to N_f)\right)^* , \quad F_T^{(a)(r)}(q^2; N_i \to N_f) = -\left(F_T^{(a)(r)}(q^2; N_i \to N_f)\right)^* ,$$

$$F_A^{(n)(i)}(q^2; N_i \to N_f) = f_A^{(n)(i)}(q^2; N_i \to N_f) = \left(f_A^{(n)(i)}(q^2; N_i \to N_f)\right)^* , \quad F_T^{(n)(i)}(q^2; N_i \to N_f) = \left(F_T^{(n)(i)}(q^2; N_i \to N_f)\right)^* ,$$

$$F_A^{(a)(i)}(q^2; N_i \to N_f) = -\left(F_A^{(a)(i)}(q^2; N_i \to N_f)\right)^* , \quad F_T^{(a)(i)}(q^2; N_i \to N_f) = f_T^{(a)(i)}(q^2; N_i \to N_f) = -\left(f_T^{(a)(i)}(q^2; N_i \to N_f)\right)^* , \qquad (195)$$

so that, on the basis of Eqs. (195) and (194), the presence of $A_\alpha^{(a)(r)}(x)$ and $A_\alpha^{(n)(i)}(x)$ in nuclear beta decay $(q \cong 0)$ would be difficult to detect even if

$$A_\alpha^{(a)(r)}(x) \approx A_\alpha^{(n)(i)}(x) \approx A_\alpha^{(n)(r)}(x)$$

and even if the two nuclei are in different isomultiplets. Also, if the nuclei N_i and N_f are mirror nuclei and so, to a sufficient approximation, are the two members of the $A(N_i)$ isodoublet and, in addition, have spin $\frac{1}{2}$, e.g., ${}_{10}Ne_9^{19}$ and ${}_9F_{10}^{19}$, Eq. (195) must agree exactly with Eq. (186), whence:

$$f_A^{(a)(r)}(q^2; N_i \to N_f) = 0,\ f_A^{(n)(i)}(q^2; N_i \to N_f) = 0,\ f_T^{(n)(r)}(q^2; N_i \to N_f) = 0,$$
$$f_T^{(a)(i)}(q^2; N_i \to N_f) = 0;\ I(N_i) = I(N_f) = \tfrac{1}{2},\ J(N_i) = J(N_f) = \tfrac{1}{2}. \qquad (196)$$

We proceed to discuss actual and possible tests of time-reversal invariance in the 1st-order weak semileptonic strangeness-nonchanging decay and collision processes mentioned, viz: $n \to p + e^- + \bar{\nu}_e$, $\nu_\mu + n \to p + \mu^-$, $\Sigma^\pm \to \Lambda + e^\pm + \nu_e(\bar{\nu}_e)$, $\nu_\mu + p \to \mu^- + N^{*++}$,

and $$N_i \to N_f + e^- + \bar{\nu}_e \quad (N_i \to N_f + e^+ + \nu_e).$$

In the most straightforward of such tests one examines the T-odd terms in the decay transition probability, e.g. the term

$$D\,\frac{\langle \vec{J}_{INI}\rangle}{J_{INI}}\cdot(\hat{p}_e\times\hat{p}_\nu);$$

$$D \cong \frac{2\left(F_V - F_M\left(\frac{m_{INI}-m_{FIN}}{2m_p}\right)\right) Im \sum_{\substack{x=n,a\\y=r,i}} \left(F_A^{(x)(y)} + F_T^{(x)(y)}\left(\frac{m_{INI}-m_{FIN}}{2m_p}\right)\right)}{\left(F_V - F_M\left(\frac{m_{INI}-m_{FIN}}{2m_p}\right)\right)^2 + \left(\frac{J_{INI}+1}{J_{INI}}\right)\left|\sum_{\substack{x=n,a\\y=r,i}} \left(F_A^{(x)(y)} + F_T^{(x)(y)}\left(\frac{m_{INI}-m_{FIN}}{2m_p}\right)\right)\right|^2} \tag{197}$$

Here, F_V, F_M, $F_A^{(x)(y)}$, $F_T^{(x)(y)}$ are defined, when $J_{INI} = J_{FIN} = \frac{1}{2}$ as e.g. in the case ${}_{10}Ne_9^{19} \rightarrow {}_9F_{10}^{19} + e^+ + \nu_e$, by equations whose form is the precise analogue of Eqs. (180) and (181), viz:

$$\langle F^{19}|V_\alpha(0)|Ne^{19}\rangle = \Big(\bar{u}_{F^{19}}\big[\gamma_\alpha F_V(q^2; Ne^{19}\rightarrow F^{19}) + \\ -\left(\frac{\sigma_{\alpha\beta}q_\beta}{2m_p}\right)F_M(q^2; Ne^{19}\rightarrow F^{19})\big]\,u_{Ne^{19}}\Big) \tag{198}$$

$$\langle F^{19}|A_\alpha(0)|Ne^{19}\rangle = \sum_{\substack{x=n,a\\y=r,i}} \Big(\bar{u}_{F^{19}}\big[\gamma_\alpha\gamma_5 F_A^{(x)(y)}(q^2; Ne^{19}\rightarrow F^{19}) + \\ -\left(\frac{\sigma_{\alpha\beta}q_\beta}{2m_p}\right)F_T^{(x)(y)}(q^2; Ne^{19}\rightarrow F^{19}) + \left(\frac{iq_\alpha\gamma_5}{m_\pi}\right)\left(\frac{m_{F^{19}}+m_{Ne^{19}}}{m_\pi}\right)F_P^{(x)(y)}(q^2; Ne^{19}\rightarrow F^{19})\big]u_{Ne^{19}}\Big) \tag{199}$$

and, when $J_{INI} = 1$, $J_{FIN} = 0$, as e.g., in the case ${}_{15}P_{17}^{32} \rightarrow {}_{16}S_{16}^{32} + e^- + \bar{\nu}_e$, by equations of the form:

$$\langle S^{32}|\{V_\alpha(0)\}^\dagger|P^{32}\rangle = i\,\epsilon_{\alpha\xi\eta\beta}\left(\frac{Q_\xi}{m_{P^{32}}+m_{S^{32}}}\right)\left(\frac{q_\eta}{2m_p}\right)S_\beta\,F_M(q^2;P^{32}\to S^{32});$$

$$F_V(q^2;P^{32}\to S^{32})=0;\quad Q\equiv P_{S^{32}}+P_{P^{32}},\quad q\equiv P_{S^{32}}-P_{P^{32}}=-(P_{e^-}+P_{\bar\nu_e});$$

$$S\cdot P_{P^{32}}=0,\quad S\cdot S=J_{INI}(J_{INI}+1)=2 \text{ in } P^{32}\text{-rest frame}, \quad (200)$$

$$\langle S^{32}|\{A_\alpha(0)\}^\dagger|P^{32}\rangle=\sum_{\substack{x=n,a\\ y=v,i}}\Bigg(S_\alpha F_A^{(x)(y)}(q^2;P^{32}\to S^{32})+$$
$$-\left(\frac{Q_\alpha}{m_{P^{32}}+m_{S^{32}}}\right)\left(\frac{q}{2m_p}\right)\cdot S\,F_T^{(x)(y)}(q^2;P^{32}\to S^{32})+$$
$$+\,i\,\frac{q_\alpha}{m_\pi}\left(\frac{q}{m_\pi}\right)\cdot S\,F_P^{(x)(y)}(q^2;P^{32}\to S^{32})\Bigg)\quad . \quad (201)$$

We first treat $n\to p+e^-+\bar\nu_e$. Here, using Eqs. (197) and (186) and with $q^2\cong 0,\ \left(\frac{m_n-m_p}{2m_p}\right)=7\times10^{-4}$;

$$F_V(0;n\to p)=1 \text{ (from CVC)},\quad F_M(0;n\to p)=(\mu_p-1)-\mu_n=$$
$$=(2.79-1)-(-1.91)=3.70 \text{ (from CVC)};$$

$$|F_A^{(n)(v)}(0;n\to p)+F_A^{(a)(i)}(0;n\to p)|=1.23 \left(\text{from } \Gamma(n\to p+e^-+\bar\nu_e)_{EXPTL}\right),$$

we have

$$\left|\left\{D(n\to p+e^-+\bar{\nu}_e)\right\}\right|_{\text{THEORETICAL}} \cong \frac{2\left(\left|F_A^{(a)(i)}(0;n\to p)\right| \pm \left|F_T^{(a)(r)}(0;n\to p)\right| \left(\frac{m_n-m_p}{2m_p}\right)\right)}{1+3\times(1.23)^2} \quad (202)$$

while[34]

$$\left|\left\{D(n\to p+e^-+\bar{\nu}_e)\right\}_{\text{EXPERIMENTAL}}\right| < 0.01\,, \quad (203)$$

so that $\left|F_A^{(a)(i)}(0;n\to p)\right| < 0.03$

and we see that the (first-class) current $A_\alpha^{(a)(i)}(x)$ is absent. Then, even with $A_\alpha^{(a)(r)}(x) \approx A_\alpha^{(n)(r)}(x)$ so that

$$\left|F_T^{(a)(r)}(0;n\to p)\right| \cong \left|F_A^{(n)(r)}(0;n\to p)\right| \cong 1\,, \quad (204)$$

we get from Eqs. (202), (204)

$$\left|\left\{D(n\to p+e^-+\bar{\nu}_e)\right\}_{\text{THEORETICAL}}\right| \cong \frac{2(7\times10^{-4})}{5.5} = 3\times10^{-4}\,, \quad (205)$$

which is much smaller than the upper limit on $\left|\left\{D(n\to p+e^-+\bar{\nu}_e)\right\}_{\text{EXPERIMENTAL}}\right|$ in Eq. (203). Thus, the upper limit on $\left|\left\{D(n\to p+e^-+\bar{\nu}_e)\right\}_{\text{EXPERIMENTAL}}\right|$ does not exclude a $\mathcal{T}$-abnormal (second-class) current $A_\alpha^{(a)(r)}(x) \approx$ $\approx A_\alpha^{(n)(r)}(x)$ and so does not exclude a maximal violation of $\mathcal{T}$-invariance in a suitably chosen 1st-order semileptonic strangeness-nonchanging process.

Essentially the same conclusions as we have just obtained from the discussion of $\mathcal{D}(n \to p + e^- + \bar{\nu}_e)$ can also be obtained from the discussion of

$$\mathcal{D}(Ne^{19} \to F^{19} + e^+ + \nu_e) ,$$

the decay $Ne^{19} \to F^{19} + e^+ + \nu_e$ being the only other 1st-order weak semileptonic strangeness-nonchanging process where an experimental search has been made for T-odd terms in the decay or collision transition probability. In this case, supposing that $A_\alpha^{(a)(i)}(x) = 0$ and $A_\alpha^{(a)(r)}(x) \approx A_\alpha^{(n)(r)}(x)$, using Eqs. (197), (198), (199), (195), (192), (194), and (204) and with $\left(\frac{m_i - m_f}{2m_p}\right) = 1 \times 10^{-3}$;

$$F_V(0; Ne^{19} \to F^{19}) = 1 \text{ (from CVC)}; \; \Big| F_A^{(n)(r)}(0; n \to p) \langle F^{19} | \tau \vec{\sigma} | Ne^{19} \rangle +$$

$$+ f_A^{(n)(r)}(0; Ne^{19} \to F^{19}) + f_A^{(a)(r)}(0; Ne^{19} \to F^{19}) + f_A^{(n)(i)}(0; Ne^{19} \to F^{19}) \Big| \cong$$

$$\Big| F_A^{(n)(r)}(0; n \to p) \langle F^{19} | \tau \vec{\sigma} | Ne^{19} \rangle \Big| = 0.9 \left(\text{from } \Gamma(Ne^{19} \to F^{19} + e^+ + \nu_e)_{\text{EXPTL}}\right),$$

we have

$$\Big| \{ \mathcal{D}(Ne^{19} \to F^{19} + e^+ + \nu_e) \}_{\text{THEORETICAL}} \Big| \cong \frac{2\left(|f_A^{(a)(r)}(0; Ne^{19} \to F^{19})| \pm |F_A^{(n)(r)}(0; n \to p) \langle F^{19} | \tau \vec{\sigma} | Ne^{19} \rangle| \left(\frac{m_i - m_f}{2m_p}\right)\right)}{1 + 3 \times (0.9)^2} \tag{206}$$

where

$$\Big| f_A^{(a)(r)}(0; Ne^{19} \to F^{19}) \Big| \cong \Big| \begin{matrix}\text{Amplitude}\\ \text{Admixture}\end{matrix} \text{ of } I = 3/2 \text{ in } |Ne^{19}\rangle, |F^{19}\rangle \Big| \, \Big| f_A^{(n)(r)}(0; Ne^{19} \to F^{19}) \Big| \approx$$

$$\approx (0.02) \times (0.05) = 1 \times 10^{-3} . \tag{207}$$

Here it is to be noted that, in agreement with Eq. (196), $f_A^{(a)(r)}(o; Ne^{19} \to F^{19})$ would be zero if $I(Ne^{19})$ and $I(F^{19})$ were each exactly $1/2$, and that, apart from this isodoublet restriction, and in view of Eq. (194), one expects

$$\left| f_A^{(a)(r)}(o; N_i \to N_f) \right| \cong \left| f_A^{(n)(r)}(o; N_i \to N_f) \right| \cong 0.05 \tag{208}$$

since

$$\left| F_A^{(a)(r)}(o; \rho^- \to \pi^0) \right| \cong \left| F_A^{(n)(r)}(o; \rho^- \to \pi^0) \right|, \left| F_A^{(a)(r)}(o; N^{*-} \to n) \right| \cong \left| F_A^{(n)(r)}(o; N^{*-} \to n) \right| \tag{209}$$

if $A_\alpha^{(a)(r)} \approx A_\alpha^{(n)(r)}$. Eqs. (206) and (207) yield

$$\left| \left\{ D(Ne^{19} \to F^{19} + e^+ + \nu_e) \right\}_{\text{THEORETICAL}} \right| \cong 1 \times 10^{-3}, \tag{210}$$

again very much smaller than the upper limit on $\left| \left\{ D(Ne^{19} \to F^{19} + e^+ + \nu_e) \right\}_{\text{EXPERIMENTAL}} \right|$ viz:[35]

$$\left| \left\{ D(Ne^{19} \to F^{19} + e^+ + \nu_e) \right\}_{\text{EXPERIMENTAL}} \right| < 0.01, \tag{211}$$

even though we have once more assumed $A_{(\alpha)}^{(a)(r)} \approx A_{(\alpha)}^{(n)(r)}$.

We proceed to discuss the $\Sigma^{\pm} \to \Lambda + e^{\pm} + \nu_e(\bar{\nu}_e)$ beta decay. Here, using Eq. (197) and (187), and with $\left(\frac{m_\Sigma - m_\Lambda}{2m_p}\right) = 4\times 10^{-2}$ and

$$F_V(q^2; \Sigma^{\pm} \to \Lambda) \cong F_V(0; \Sigma^{\pm} \to \Lambda) = 0$$

from CVC, we have

$$\left|\left\{D(\Sigma^{\pm} \to \Lambda + e^{\pm} + \nu_e(\bar{\nu}_e)\right\}_{\text{THEORETICAL}}\right| \cong \frac{2F_M(0;\Sigma^{\pm}\to\Lambda)\left(\frac{m_\Sigma - m_\Lambda}{2m_p}\right)\left|F_A^{(a)(r)}(0;\Sigma^{\pm}\to\Lambda)\right|}{3\left|F_A^{(n)(r)}(0,\Sigma^{\pm}\to\Lambda) + F_A^{(a)(r)}(0;\Sigma^{\pm}\to\Lambda) + F_A^{(n)(i)}(0,\Sigma^{\pm}\to\Lambda)\right|^2} \tag{212}$$

We now assume as before that $A_\alpha^{(a)(r)}(x) \approx A_\alpha^{(n)(r)}(x)$ and, suppose in addition that an SU3 analogue to the SU2 theorem in Eqs. (184)-(186) exists;[36] then

$$F_A^{(a)(r)}(0;\Sigma^{\pm}\to\Lambda) \equiv \beta F_A^{(n)(r)}(0;\Sigma^{\pm}\to\Lambda) \cong \beta\left(F_A^{(n)(r)}(0;\Sigma^{\pm}\to\Lambda) + F_A^{(a)(r)}(0;\Sigma^{\pm}\to\Lambda) + F_A^{(n)(i)}(0;\Sigma^{\pm}\to\Lambda)\right) \tag{213}$$

where $\beta = -\beta^*$ is a parameter which essentially measures the amplitude of "wrong" SU3-state admixture in $|\Sigma^{\pm}\rangle$ and $|\Lambda\rangle$. Thus

$$\left|\left\{D(\Sigma^{\pm} \to \Lambda + e^{\pm} + \nu_e(\bar{\nu}_e))\right\}_{\text{THEORETICAL}}\right| \cong \frac{2F_M(0;\Sigma^{\pm}\to\Lambda)\left(\frac{m_\Sigma - m_\Lambda}{2m_p}\right)|\beta|}{3\left|F_A^{(n)(r)}(0;\Sigma^{\pm}\to\Lambda) + F_A^{(a)(r)}(0;\Sigma^{\pm}\to\Lambda) + F_A^{(n)(i)}(0;\Sigma^{\pm}\to\Lambda)\right|} \tag{214}$$

so that, with $F_M(0; \Sigma^\pm \to \Lambda) = \sqrt{2}\left(\frac{1}{2}\sqrt{3}\,|\mu_n|\right) = 2.3$

(from CVC and SU3) and

$$\left| F_A^{(n)(r)}(0; \Sigma^\pm \to \Lambda) + F_A^{(a)(r)}(0; \Sigma^\pm \to \Lambda) + F_A^{(n)(i)}(0; \Sigma^\pm \to \Lambda) \right| = 0.6$$

(from $\{\Gamma(\Sigma^\pm \to \Lambda + e^\pm + \nu_e(\bar{\nu}_e))\}_{\text{EXPERIMENTAL}}$) Eq. (214) yields

$$\left| \{D(\Sigma^\pm \to \Lambda + e^\pm + \nu_e(\bar{\nu}_e))\}_{\text{THEORETICAL}} \right| \cong 0.12\,|\beta| \,, \qquad (215)$$

which does not encourage any immediate experimental attempts since $|\beta|$ can hardly be taken greater than $\frac{1}{4}$.[36] One interesting feature of the expression for $\left|\{D(\Sigma^\pm \to \Lambda + e^\pm + \nu_e(\bar{\nu}_e))\}_{\text{THEORETICAL}}\right|$ is however to be noted —— from Eqs. (214)-(212) we see that for a given amount of SU3 symmetry breaking (e.g. $|\beta| = \frac{1}{4}$ and $\left(\frac{m_\Sigma - m_\Lambda}{2m_p}\right) = 4 \times 10^{-2}$), $\left|\{D(\Sigma^\pm \to \Lambda + e^\pm + \nu_e(\bar{\nu}_e))\}_{\text{THEORETICAL}}\right|$ is inversely proportional to

$$\left| F_A^{(n)(r)}(0; \Sigma^\pm \to \Lambda) + F_A^{(a)(r)}(0; \Sigma^\pm \to \Lambda) + F_A^{(n)(i)}(0; \Sigma^\pm \to \Lambda) \right| .$$

We next discuss a beta decay process where a rather sizeable value of the T-odd asymmetry coefficient D is expected if $A_\alpha^{(n)(r)}(x) \approx A_\alpha^{(a)(r)}(x)$, namely ${}_{15}P_{17}^{32} \to {}_{16}S_{16}^{32} + e^- + \bar{\nu}_e$, where

$$I(P^{32}) = 1 \,, \quad I(S^{32}) = 0 \,, \quad J(P^{32}) = 1 \,, \quad J(S^{32}) = 0 .$$

Here, using Eqs. (197), (200), (201), (195), (192), and (194) and with $\left(\frac{m_i - m_f}{2m_p}\right) = 1 \times 10^{-3}$, we have

$$\left|\left\{D(P^{32}\rightarrow S^{32}+e^{-}+\bar{\nu}_e)\right\}_{\text{THEORETICAL}}\right| \cong \frac{2\,F_M(0;P^{32}\rightarrow S^{32})\left(\frac{m_i-m_f}{2m_p}\right)\left|f_A^{(a)(r)}(0;P^{32}\rightarrow S^{32})\right|}{2\left|F_A^{(n)(v)}(0;n\rightarrow p)\langle S^{32}|\tau\vec{\sigma}|P^{32}\rangle+f_A^{(n)(r)}(0,P^{32}\rightarrow S^{32})+f_A^{(a)(r)}(0;P^{32}\rightarrow S^{32})+f_A^{(n)(i)}(0;P^{32}\rightarrow S^{32})\right|^2} \tag{216}$$

and since $\left|F_A^{(n)(v)}(0;n\rightarrow p)\langle S^{32}|\tau\vec{\sigma}|P^{32}\rangle+f_A^{(n)(r)}(0;P^{32}\rightarrow S^{32})+f_A^{(a)(r)}(0;P^{32}\rightarrow S^{32})+f_A^{(n)(i)}(0;P^{32}\rightarrow S^{32})\right| =$

$$= \left|1.2\langle S^{32}|\tau\vec{\sigma}|P^{32}\rangle+f_A^{(n)(r)}(0;P^{32}\rightarrow S^{32})+f_A^{(a)(r)}(0;P^{32}\rightarrow S^{32})+f_A^{(n)(i)}(0;P^{32}\rightarrow S^{32})\right| = 4\times 10^{-3}$$

(from $\left\{\Gamma(P^{32}\rightarrow S^{32}+e^{-}+\bar{\nu}_e)\right\}_{\text{EXPERIMENTAL}}$), it is reasonable to suppose that

$$\left|\langle S^{32}|\tau\vec{\sigma}|P^{32}\rangle\right| \lessapprox \left|f_A^{(n)(r)}(0;P^{32}\rightarrow S^{32})+f_A^{(a)(r)}(0;P^{32}\rightarrow S^{32})+f_A^{(n)(i)}(0;P^{32}\rightarrow S^{32})\right|. \tag{217}$$

Then, since consistent with Eq. (208) we can set

$$\left|f_A^{(a)(r)}(0,P^{32}\rightarrow S^{32})\right| \approx \left|f_A^{(n)(r)}(0;P^{32}\rightarrow S^{32})\right|, \tag{218}$$

we obtain

$$\left|f_A^{(a)(r)}(0;P^{32}\rightarrow S^{32})\right| \approx \frac{1}{\sqrt{2}}\left|F_A^{(n)(v)}(0;n\rightarrow p)\langle S^{32}|\tau\vec{\sigma}|P^{32}\rangle+f_A^{(n)(r)}(0;P^{32}\rightarrow S^{32})+f_A^{(a)(r)}(0;P^{32}\rightarrow S^{32})+f_A^{(n)(i)}(0;P^{32}\rightarrow S^{32})\right| \tag{219}$$

and substituting Eq. (219) into Eq. (216)

$$\left|\left\{D(P^{32}\to S^{32}+e^-+\bar{\nu}_e)\right\}_{\text{THEORETICAL}}\right| \approx$$

$$\approx \frac{2F_M(0;P^{32}\to S^{32})\left(\frac{m_i-m_f}{2m_p}\right)\frac{1}{\sqrt{2}}}{2\left|F_A^{(n)(r)}(0;n\to p)\langle S^{32}|\tau\vec{\sigma}|P^{32}\rangle + f_A^{(n)(r)}(0,P^{32}\to S^{32}) + f_A^{(a)(r)}(0;P^{32}\to S^{32}) + f_A^{(W)(i)}(0;P^{32}\to S^{32})\right|} =$$

$$= \frac{1}{4}\left(\frac{F_M(0;P^{32}\to S^{32})}{\sqrt{2}}\right) = \frac{1}{4}\,\mu((S^{32})^*\to S^{32}), \qquad (220)$$

where $\mu((S^{32})^*\to S^{32})$ is the transition magnetic moment from that excited state of S^{32} which is in the same isomultiplet as the ground state of P^{32} to the ground state of S^{32}. No experimental determination of $\mu((S^{32})^*\to S^{32})$ has as yet been made (e.g. via

$$\Gamma((S^{32})^*\to S^{32}+\gamma) = \frac{1}{3}\left[\mu((S^{32})^*\to S^{32})\right]^2 \alpha\,\frac{E_\gamma^3}{m_p^2}\,)$$

but the estimate (remember Eqns (217) and (194))

$$\mu((S^{32})^*\to S^{32}) \cong \left|\left\langle \Psi_{S^{32}}(\ldots\vec{r}_j,(\sigma_3)_j,(\tau_3)_j\ldots)\right|\sum_{j=1}^{A}\frac{\tau_j^{(3)}}{2}\left((\mu_p-\mu_n)\vec{\sigma}_j+\right.\right.$$

$$\left.\left.+\,i^{-1}\,\vec{r}_j\times\text{grad}_{\vec{r}_j}\right)e^{i\vec{q}\cdot\vec{r}_j}\left|\Psi_{(S^{32})^*}(\ldots\vec{r}_j,(\sigma_3)_j,(\tau_3)_j\ldots)\right\rangle\right| \cong$$

$$\cong \left|\left\langle\Psi_{S^{32}}\right|\frac{1}{2}\left(\vec{L}_{\text{PROT.}}-\vec{L}_{\text{NEUT.}}\right)\left|\Psi_{(S^{32})^*}\right\rangle\right| \approx 1 \qquad (221)$$

appears not unreasonable. Eqs. (221) and (220) yield

$$\left|\left\{D(P^{32}\to S^{32}+e^-+\bar{\nu}_e)\right\}_{\text{THEORETICAL}}\right| \approx \frac{1}{4} \qquad (222)$$

which is comfortably large and which, hopefully, will spark an experimental search for a nonvanishing $\mathcal{T}$-odd asymmetry in $P^{32} \to S^{32} + e^- + \bar{\nu}_e$.

We note that $\left| \{ D(P^{32} \to S^{32} + e^- + \bar{\nu}_e) \}_{\text{THEORETICAL}} \right|$ is inversely proportional to

$$\left| F_A^{(n)(r)}(0; n \to p) \langle S^{32} | \tau \vec{\sigma} | P^{32} \rangle + f_A^{(n)(r)}(0; P^{32} \to S^{32}) + f_A^{(a)(r)}(0; P^{32} \to S^{32}) + f_A^{(n)(i)}(0; P^{32} \to S^{32}) \right|$$

(compare Eqs. (214), (215) et seq.). This result will be valid for other allowed beta decays as long as an equation similar to Eq. (217) holds, i.e. as long as the allowed beta decay matrix element in impulse approximation is anomalously small and the corresponding allowed beta decay strongly "unfavored." Thus, everything else being equal, the larger the "(ft)" of the allowed beta decay the more accurately Eq. (217) may be expected to hold and, because $|D| \sim \{ \text{"}(ft)\text{"} \}^{1/2}$, the larger the $\mathcal{T}$-odd asymmetry anticipated. This fact is obviously important in the planning of experimental studies.

We now very briefly discuss neutrino-nucleon quasi-elastic and inelastic processes: $\nu_\mu + n \to \mu^- + p$ and $\nu_\mu + p \to \mu^- + N^{*++}$ at large $E_\nu (\gtrsim m_p)$ and large $q^2 (\equiv (p_p - p_n)^2 = (p_{\nu_\mu} - p_{\mu^-})^2 \cong 2E_\nu E_{\mu^-}(1-\cos\theta) \cong 2E_\nu^2(1-\cos\theta) \approx m_p^2)$.

In this case, and considering first the quasi-elastic process, the kinematic term $\left(\frac{\bar{u}_p \sigma_{\alpha\beta} q_\beta \gamma_5 u_n}{2m_p}\right)$ multiplying $\sum_{\substack{x=n,a\\y=r,i}} F_T^{(x)(y)}(q^2; n\to p)$ (Eq. (181)) is of the same order of magnitude as the kinematic term $(\bar{u}_p \gamma_\alpha \gamma_5 u_n)$ multiplying $\sum_{\substack{x=n,a\\y=r,i}} F_A^{(x)(y)}(q^2; n\to p)$ or the kinematic term $(\bar{u}_p \gamma_\alpha u_n)$ multiplying $F_V(q^2; n\to p)$. Thus, even though $F_A^{(a)(r)}(q^2; n\to p) = 0$ (Eq. (186)), the $\mathcal{T}$-odd term:

$$D'(\nu_\mu + n \to \mu^- + p)\ \frac{\vec{S}_{\mu^-}}{S_{\mu^-}} \cdot (\hat{p}_\mu \times \hat{p}_\nu)$$

should be appreciable if $A_\alpha^{(a)(r)}(x) \approx A_\alpha^{(n)(r)}(x)$, since $D'(\nu_\mu + n \to \mu^- + p)$ will contain, e.g., a term of the order (compare Eqs. (197), (186), (202) - (204))

$$\left|D'(\nu_\mu + n \to \mu^- + p)\right| \approx \frac{2\left|F_A^{(n)(r)}(q^2; n\to p)\right| \left|\operatorname{Im} F_T^{(a)(r)}(q^2; n\to p)\right| \left(\frac{\sqrt{q^2}}{m_p}\right)}{\left(F_V(q^2; n\to p)\right)^2 + \left(F_A^{(n)(r)}(q^2; n\to p)\right)^2} =$$

$$= \frac{2\left|F_A^{(n)(r)}(q^2; n\to p)\right| \left|F_T^{(a)(r)}(q^2; n\to p)\right| \left(\frac{\sqrt{q^2}}{m_p}\right)}{\left(F_V(q^2; n\to p)\right)^2 + \left(F_A^{(n)(r)}(q^2; n\to p)\right)^2} =$$

$$\approx \frac{2\left|F_A^{(n)(r)}(q^2;n\to p)\right|\left|F_A^{(n)(r)}(q^2;n\to p)\right|\left(\frac{\sqrt{q^2}}{m_p}\right)}{\left(F_V(q^2;n\to p)\right)^2+\left(F_A^{(n)(r)}(q^2;n\to p)\right)^2} \approx \left(\frac{\sqrt{q^2}}{m_p}\right) \approx 1 \qquad (223)$$

Similarly, in the inelastic process, the $\mathcal{T}$-odd term

$$D'(\nu_\mu + p \to \mu^- + N^{*++})\,\frac{\vec{S}_{\mu^-}}{S_{\mu^-}}\cdot(\vec{p}_\mu \times \vec{p}_\nu)$$

will also be appreciable if $A_\alpha^{(a)(r)}(x) \approx A_\alpha^{(n)(r)}(x)$ since $D'(\nu_\mu + p \to \mu^- + N^{*++})$ will contain, e.g., a term of the order (compare Eqs. (197), (190), (209))

$$\left|D'(\nu_\mu + p \to \mu^- + N^{*++})\right| \approx$$

$$\approx \frac{2\left|F_V(q^2;p\to N^{*++})\right|\left|\operatorname{Im} F_A^{(a)(r)}(q^2;p\to N^{*++})\right|\left(\frac{\sqrt{q^2}}{2E_\nu}\right)}{\left(F_V(q^2;p\to N^{*++})\right)^2+\left|F_A^{(n)(r)}(q^2;p\to N^{*++})+F_A^{(a)(r)}(q^2;p\to N^{*++})+F_A^{(n)(i)}(q^2;p\to N^{*++})\right|^2} =$$

$$= \frac{2\left|F_V(q^2;p\to N^{*++})\right|\left|F_A^{(a)(r)}(q^2;p\to N^{*++})\right|\left(\frac{\sqrt{q^2}}{2E_\nu}\right)}{\left(F_V(q^2;p\to N^{*++})\right)^2+\left|F_A^{(n)(r)}(q^2;p\to N^{*++})+F_A^{(a)(r)}(q^2;p\to N^{*++})+F_A^{(n)(i)}(q^2;p\to N^{*++})\right|^2}$$

$$\approx \frac{1}{\sqrt{2}}\left(\frac{\sqrt{q^2}}{2E_\nu}\right) \approx 1 \qquad (224)$$

Therefore, it is clear that the study of neutrino-nucleon quasi-elastic and inelastic processes at large E_ν and q^2 is bound to be fruitful in the investigation of $\mathcal{T}$-invariance violation by H_{WEAK}.

In concluding this discussion of the possible presence of $\mathcal{T}$-abnormal hadron weak currents $A_\alpha^{(a)(r)}(x)$ and $a_\alpha^{(a)}(x)$

we make the following comments:

(I) In a theory where $A_\alpha^{(a)(r)}(x)$, $A_\alpha^{(n)(i)}(x)$ and $a_\alpha^{(a)}(x)$ are present the conventional equal-time commutation relations and PCAC relations satisfied by the hadron weak currents[2] apply only to $A_\alpha^{(n)(r)}(x)$ and $a_\alpha^{(n)}(x)$ and not to $\left(A_\alpha^{(n)(r)}(x) + A_\alpha^{(a)(r)}(x) + A_\alpha^{(n)(i)}(x)\right)$ and $\left(a_\alpha^{(n)}(x) + a_\alpha^{(a)}(x)\right)$ (i.e., $\left[A_4^{(n)(r)}(\vec{y}), \{A_4^{(n)(r)}(\vec{x})\}^\dagger\right]_- = 2 I_0^{(3)}(\vec{x})\,\delta(\vec{x}-\vec{y})$, etc.) Thus the Adler-Weisberger sum-rule and the Goldberger-Treiman relation will refer to $\left|F_A^{(n)(r)}(0; N_i \to N_f)\right|$ rather than to $\left|F_A^{(n)(r)}(0; N_i \to N_f) + F_A^{(a)(r)}(0; N_i \to N_f) + F_A^{(n)(i)}(0; N_i \to N_f)\right|$, etc. However, in the only case where a quantitative prediction has proved possible, viz., $N_i = n$, $N_f = p$, we have

$$\left|F_A^{(n)(r)}(0; N_i \to N_f) + F_A^{(a)(r)}(0; N_i \to N_f) + F_A^{(n)(i)}(0; N_i \to N_f)\right| = \left|F_A^{(n)(r)}(0; N_i \to N_f)\right|$$

(Eq. (186)) so that no change is anticipated. It is clear that various assumptions can be made regarding the equal-time commutation relations and the PCAC relations for $A_\alpha^{(a)(r)}(x)$ (and $A_\alpha^{(n)(i)}(x)$) without contradicting presently available experimental data.

(II) The possible presence of the T-abnormal strangeness-changing axial hadron weak current $a_\alpha^{(a)}(x)$ may be tested by an experimental search for the T-odd terms in the transition probability for $\Lambda \to p + e^- + \bar{\nu}_e$ and for $K^{\pm} \to \pi^+ + \pi^- + e^{\pm} + \nu_e (\bar{\nu}_e)$. Analogously to Eq. (197), we have the T-odd term: $D(\Lambda \to p + e^- + \bar{\nu}_e) \left(\frac{\langle \vec{J}_\Lambda \rangle}{J_\Lambda} \right) \cdot (\hat{p}_e \times \hat{p}_\nu)$ with $\{ D(\Lambda \to p + e^- + \bar{\nu}_e) \}_{\text{THEORETICAL}} \cong$

$$\cong \frac{2 F_V(0; \Lambda \to p) \, | F_A^{(a)}(0; \Lambda \to p) |}{(F_V(0, \Lambda \to p))^2 + 3 \, | F_A^{(n)}(0; \Lambda \to p) + F_A^{(a)}(0; \Lambda \to p) |^2} \qquad (225)$$

and if $a_\alpha^{(a)}(x) \approx a_\alpha^{(n)}(x)$ and, in addition an SU3 analogue to the SU2 theorem in Eqs. (184)-(186) exists,[36] then (compare Eq. (213))

$$F_A^{(a)}(0; \Lambda \to p) = \beta F_A^{(n)}(0; \Lambda \to p) \cong \beta \left(F_A^{(n)}(0, \Lambda \to p) + F_A^{(a)}(0; \Lambda \to p) \right);$$
$$\beta = -\beta^* \qquad (226)$$

so that $\left| \{ D(\Lambda \to p + e^- + \bar{\nu}_e) \}_{\text{THEORETICAL}} \right| \cong$

$$\cong \frac{2 F_V(0, \Lambda \to p) \, F_A^{(n)}(0; \Lambda \to p) \, |\beta|}{(F_V(0; \Lambda \to p))^2 + 3 (F_A^{(n)}(0; \Lambda \to p))^2} \qquad (227)$$

Then, since $\left\{ \frac{F_A^{(n)}(0; \Lambda \to p)}{F_V(0; \Lambda \to p)} \right\} \cong 0.75$ [2], we obtain

$$\left| \left\{ D(\Lambda \to p + e^- + \bar{\nu}_e) \right\}_{\text{THEORETICAL}} \right| \cong 0.6 |\beta| \lessapprox 0.15^{[36]} , \qquad (228)$$

(subject to corrections $\sim \left(\frac{m_\Lambda - m_p}{m_p}\right), \left(\frac{m_\Lambda - m_p}{m_p}\right)^2, \ldots$). Such a relatively large $\mathcal{T}$-odd term should be detectible in an experiment on the 10^3 event level.

Similarly, if $a_\alpha^{(a)}(x) \approx a_\alpha^{(n)}(x)$, the $\mathcal{T}$-odd term:

$$D(K^\pm \to \pi^+ + \pi^- + e^\pm + \nu_e(\bar{\nu}_e)) \left(\frac{(\vec{p}_{\pi^+} - \vec{p}_{\pi^-})}{|\vec{p}_{\pi^+} - \vec{p}_{\pi^-}|} \right) \cdot (\hat{p}_e \times \hat{p}_\nu)$$

is characterized by a relatively large $D(K^\pm \to \pi^+ + \pi^- + e^\pm + \nu_e(\bar{\nu}_e))$ corresponding to $F_A^{(a)}(q^2; K^\pm \to \pi^+ + \pi^-) \approx F_A^{(n)}(q^2; K^\pm \to \pi^+ + \pi^-)$ since $K^\pm$ and $\pi^+ + \pi^-$ lie in different SU3 multiplets so that no SU3 analogue to the SU2 theorem of Eqs. (184)-(186) exists in this case. We therefore see that the $K^\pm \to \pi^+ + \pi^- + e^\pm + \nu_e(\bar{\nu}_e)$ process is especially favorable for an unambiguous test of the possible presence of relatively large $\mathcal{T}$-abnormal hadron weak currents.

(III) The absence of any $\mathcal{T}$-odd term of the form $D(K^+ \to \pi^0 + \mu^+ + \nu_\mu)\left(\frac{\vec{S}_{\mu^+}}{S_{\mu^+}}\right)\cdot(\hat{p}_\mu \times \hat{p}_\nu)$ in $K^+ \to \pi^0 + \mu^+ + \nu_\mu$ is expected in our formulation since this decay is mediated purely by the strangeness-changing polar hadron weak current which we assume to be $\mathcal{T}$-normal (see above).

(IV) The whole above discussion supposes that a rather large $\mathcal{T}$-invariance violation may be present in suitable 1st-order weak semileptonic decay and collision processes while the observed $\mathcal{T}$-invariance violation ($\leftrightarrow$ CP-invariance violation) in the 1st-order weak nonleptonic K_L^0 processes ($K_L^0 \to \pi^+ + \pi^-, \pi^0 + \pi^0$) is small.[2] We believe that this supposition is not necessarily incorrect since the relationship between the 1st-order weak semileptonic and the 1st-order weak nonleptonic processes is still so obscure (largely because of the difficulty of calculating matrix elements such as $\langle \pi\pi | \{H_{WEAK}\}_{T\text{-ODD}} | K^0\rangle =$

$$= \langle \pi\pi | \frac{G}{\sqrt{2}} \int d^3x \left\{ \cos\theta_C \left(I_\alpha^{(-)}(x) + A_\alpha^{(n)}(x) \right) + \sin\theta_C \left(\mathcal{V}_\alpha(x) + a_\alpha^{(n)}(x) \right) \right\}^\dagger \times$$

$$\times \left\{ \cos\theta_C \left(A_\alpha^{(a)}(x) \right) + \sin\theta_C \left(a_\alpha^{(a)}(x) \right) \right\} + \text{herm. conj.} \, | K^0\rangle, \text{etc.})$$

that it is rather hard to establish a quantitative connection between the two.

We proceed to treat the question of the possible presence of the $e^{i\pi I^{(2)}}$ -irregular (second class) current $A_\alpha^{(n)(i)}(x)$ and consider first the relation between the charge-symmetric $\Sigma^+ \to \Lambda + e^+ + \nu_e$ and $\Sigma^- \to \Lambda + e^- + \bar{\nu}_e$ decays. If $\Gamma_{CORR}(\Sigma^\pm \to \Lambda + e^\pm + \nu_e(\bar{\nu}_e))$ are the indicated decay rates corrected for phase-space (i. e.,
$\Gamma_{CORR}(\Sigma^\pm \to \Lambda + e^\pm + \nu_e(\bar{\nu}_e)) \cong \left\{ \frac{\Gamma(\Sigma^\pm \to \Lambda + e^\pm + \nu_e(\bar{\nu}_e))}{(m_{\Sigma^\pm} - m_\Lambda)^5} \right\}$)
we have

$$\left\{ \frac{\Gamma_{CORR}(\Sigma^+ \to \Lambda + e^+ + \nu_e)}{\Gamma_{CORR}(\Sigma^- \to \Lambda + e^- + \bar{\nu}_e)} \right\}_{THEORETICAL} \cong$$

$$\cong \frac{\left| \mathcal{F}^{(n)(r)}(\Sigma^+) + \mathcal{F}^{(a)(r)}(\Sigma^+) + \mathcal{F}^{(n)(i)}(\Sigma^+) \right|^2}{\left| \mathcal{F}^{(n)(r)}(\Sigma^-) + \mathcal{F}^{(a)(r)}(\Sigma^-) + \mathcal{F}^{(n)(i)}(\Sigma^-) \right|^2} \qquad (229)$$

where

$$\mathcal{F}^{(x)(y)}(\Sigma^\pm) \equiv F_A^{(x)(y)}(0; \Sigma^\pm \to \Lambda) + \left(\frac{m_{\Sigma^\pm} - m_\Lambda}{2m_p} \right) F_T^{(x)(y)}(0; \Sigma^\pm \to \Lambda) \qquad (230)$$

and where, it will be recalled, $F_V(0; \Sigma^\pm \to \Lambda) = 0$ (from CVC) and $\left(\frac{m_{\Sigma^\pm} - m_\Lambda}{2m_p} \right) = 4 \times 10^{-2}$. Eqs. (229), (230) and (187)-(189) yield

$$\left\{ \frac{\Gamma_{CORR}(\Sigma^+ \to \Lambda + e^+ + \nu_e)}{\Gamma_{CORR}(\Sigma^- \to \Lambda + e^- + \bar{\nu}_e)} \right\}_{THEORETICAL} \cong \frac{\left(\mathcal{F}^{(n)(r)}(\Sigma^+) + \mathcal{F}^{(n)(i)}(\Sigma^+) \right)^2 + \left| \mathcal{F}^{(a)(r)}(\Sigma^+) \right|^2}{\left(\mathcal{F}^{(n)(r)}(\Sigma^+) - \mathcal{F}^{(n)(i)}(\Sigma^+) \right)^2 + \left| \mathcal{F}^{(a)(r)}(\Sigma^+) \right|^2}, \qquad (231)$$

which is basic in what follows.

To continue, we suppose that $A_\alpha^{(n)(l)}(x) \approx A_\alpha^{(n)(r)}(x)$ as well as $A_\alpha^{(a)(r)}(x) \approx A_\alpha^{(n)(r)}(x)$ and that the SU3 analogue to the SU2 theorem of Eqs. (184)-(186) exists.[36] Then (compare Eq. (232) with Eq. (213) and Eq. (233) with Eq. (204))

$$F_A^{(a)(r)}(0;\Sigma^+\to\Lambda) \equiv \beta\, F_A^{(n)(r)}(0;\Sigma^+\to\Lambda)$$

$$F_A^{(n)(l)}(0;\Sigma^+\to\Lambda) \equiv \beta'\, F_A^{(n)(r)}(0;\Sigma^+\to\Lambda)$$

$$F_T^{(n)(r)}(0;\Sigma^+\to\Lambda) \equiv \beta''\, F_A^{(n)(r)}(0;\Sigma^+\to\Lambda)$$

$$|\beta| \cong |\beta'| \cong |\beta''| \lesssim \frac{1}{4} \quad [36] \tag{232}$$

where $\beta = -\beta^*$, $\beta' = (\beta')^*$, $\beta'' = (\beta'')^*$ are parameters which essentially measure the amplitude of "wrong" SU3-state admixture into $|\Sigma^+\rangle$ and $|\Lambda\rangle$, and,

$$F_T^{(a)(r)}(0;\Sigma^+\to\Lambda) \equiv \gamma\, F_A^{(n)(r)}(0;\Sigma^+\to\Lambda)$$

$$F_T^{(n)(l)}(0;\Sigma^+\to\Lambda) \equiv \gamma'\, F_A^{(n)(r)}(0;\Sigma^+\to\Lambda)$$

$$|\gamma| \cong |\gamma'| \cong 1 \tag{233}$$

where $\gamma = -\gamma^*$, $\gamma' = (\gamma')^*$. Eqs. (230), (232), (233) yield

$$\mathcal{F}^{(n)(r)}(\Sigma^+) = F_A^{(n)(r)}(0;\Sigma^+\to\Lambda)\left(1+\left(\frac{m_\Sigma - m_\Lambda}{2m_p}\right)\beta''\right) \cong F_A^{(n)(r)}(0;\Sigma^+\to\Lambda)$$

$$\mathcal{F}^{(n)(i)}(\Sigma^+) = F_A^{(n)(r)}(0;\Sigma^+\to\Lambda)\left(\beta' + \left(\frac{m_\Sigma - m_\Lambda}{2m_p}\right)\gamma'\right)$$

$$\mathcal{F}^{(a)(r)}(\Sigma^+) = F^{(n)(r)}(0;\Sigma^+\to\Lambda)\left(\beta + \left(\frac{m_\Sigma - m_\Lambda}{2m_p}\right)\gamma\right) \tag{234}$$

so that, substituting Eq. (234) into Eq. (231),

$$\left\{\frac{\Gamma_{CORR}(\Sigma^+\to\Lambda+e^++\nu_e)}{\Gamma_{CORR}(\Sigma^-\to\Lambda+e^-+\bar\nu_e)}\right\}_{THEORETICAL} \cong$$

$$\cong \frac{\left(1+\left[\beta'+\left(\frac{m_\Sigma-m_\Lambda}{2m_p}\right)\gamma'\right]\right)^2 + \left|\beta+\left(\frac{m_\Sigma-m_\Lambda}{2m_p}\right)\gamma\right|^2}{\left(1-\left[\beta'+\left(\frac{m_\Sigma-m_\Lambda}{2m_p}\right)\gamma'\right]\right)^2 + \left|\beta+\left(\frac{m_\Sigma-m_\Lambda}{2m_p}\right)\gamma\right|^2}$$

$$\cong 1 + 4\left[\beta' + \left(\frac{m_\Sigma - m_\Lambda}{2m_p}\right)\gamma'\right]$$

$$= 1 + 4\left[\left(\frac{F_A^{(n)(i)}(0;\Sigma^+\to\Lambda)}{F_A^{(n)(r)}(0;\Sigma^+\to\Lambda)}\right) + \left(\frac{m_\Sigma - m_\Lambda}{2m_p}\right)\left(\frac{F_T^{(n)(i)}(0;\Sigma^+\to\Lambda)}{F_A^{(n)(r)}(0;\Sigma^+\to\Lambda)}\right)\right] \tag{235}$$

which is to be compared with[37]

$$\left\{\frac{\Gamma_{CORR}(\Sigma^+\to\Lambda+e^++\nu_e)}{\Gamma_{CORR}(\Sigma^-\to\Lambda+e^-+\bar\nu_e)}\right\}_{EXPERIMENTAL} = 1.16 \pm 0.25 \tag{236}$$

Unfortunately, the experimental error is here too large to draw any definite conclusions. However it may be worth mentioning that, if the "wrong" SU3 - admixture coefficient $|\beta'|$ is $\ll \frac{1}{4}$ rather than $\cong \frac{1}{4}$, and if one ignores the experimental error, the theoretical and experimental values of $\left\{\frac{\Gamma_{CORR}(\Sigma^+ \to \Lambda + e^+ + \nu_e)}{\Gamma_{CORR}(\Sigma^- \to \Lambda + e^- + \bar{\nu}_e)}\right\}$ agree for

$$\gamma' \equiv \frac{F_T^{(n)(i)}(0; \Sigma^+ \to \Lambda)}{F_A^{(n)(r)}(0; \Sigma^+ \to \Lambda)} = 1 .$$

A little more light can be thrown on the question of the possible presence of the $e^{i\pi I^{(2)}}$ -irregular current $A_\alpha^{(n)(i)}(x)$ by consideration of the relations between various charge symmetric nuclear beta decays, e.g. ${}_3Li_5^8 \to {}_4Be_4^8 + e^- + \bar{\nu}_e$ and ${}_5B_3^8 \to {}_4Be_4^8 + e^+ + \nu_e$, ${}_3Li_6^9 \to {}_4Be_5^9 + e^- + \bar{\nu}_e$ and ${}_6C_3^9 \to {}_5B_4^9 + e^+ + \nu_e$,

${}_5B_7^{12} \to {}_6C_6^{12} + e^- + \bar{\nu}_e$

and ${}_7N_5^{12} \to {}_6C_6^{12} + e^+ + \nu_e$, etc. Then, in a development entirely similar to that leading from Eq. (229) to Eq. (235), we obtain the analogue of Eq. (235) which, e.g. in the $A = 12$ case, takes the form

$$\left\{\frac{\Gamma_{CORR}(B^{12} \to C^{12} + e^- + \bar{\nu}_e)}{\Gamma_{CORR}(N^{12} \to C^{12} + e^+ + \nu_e)}\right\}_{\text{THEORETICAL}} \cong 1 + 4\left[\left(\frac{F_A^{(n)(i)}(0; B^{12} \to C^{12})}{F_A^{(n)(r)}(0, B^{12} \to C^{12})}\right) + \left(\frac{\left(\frac{m_{B^{12}} + m_{N^{12}}}{2}\right) - m_{C^{12}}}{2m_p}\right)\left(\frac{F_T^{(n)(i)}(0; B^{12} \to C^{12})}{F_A^{(n)(r)}(0, B^{12} \to C^{12})}\right)\right]. \tag{237}$$

Eq. (237), together with Eqs. (192) and (186), yields

$$\left\{\frac{\Gamma_{CORR}(B^{12}\to C^{12}+e^-+\bar{\nu}_e)}{\Gamma_{CORR}(N^{12}\to C^{12}+e^++\nu_e)}\right\}_{THEORETICAL} \cong$$

$$\cong 1+4\left[\left(\frac{f_A^{(n)(i)}(0;B^{12}\to C^{12})}{F_A^{(n)(r)}(0;n\to p)\langle C^{12}|\tau\vec{\sigma}|B^{12}\rangle}\right)+\left(\frac{\left(\frac{m_{B^{12}}+m_{N^{12}}}{2}\right)-m_{C^{12}}}{2m_p}\right)\left(\frac{F_T^{(n)(i)}(0;n\to p)}{F_A^{(n)(r)}(0;n\to p)}\right)\right] \tag{238}$$

with $F_A^{(n)(r)}(0,n\to p)=1.23$, $F_A^{(n)(r)}(0;B^{12}\to C^{12})\cong F_A^{(n)(r)}(0;n\to p)\langle C^{12}|\tau\vec{\sigma}|B^{12}\rangle=$

$$=0.55, \quad \left(\frac{\left(\frac{m_{B^{12}}+m_{N^{12}}}{2}\right)-m_{C^{12}}}{2m_p}\right)=8\times10^{-3}, \tag{239}$$

which is to be compared with

$$\left\{\frac{\Gamma_{CORR}(B^{12}\to C^{12}+e^-+\bar{\nu}_e)}{\Gamma_{CORR}(N^{12}\to C^{12}+e^++\nu_e)}\right\}_{EXPERIMENTAL}=1.10\ . \tag{240}$$

In conducting this comparison we note first of all that if we assume that $A_\alpha^{(n)(i)}(x)=0$, so that $f_A^{(n)(i)}(0;B^{12}\to C^{12})=0$ and $F_T^{(n)(i)}(0;n\to p)=0$, we have to explain the observed 10% deviation of

$$\left\{\frac{\Gamma_{CORR}(B^{12}\to C^{12}+e^-+\bar{\nu}_e)}{\Gamma_{CORR}(N^{12}\to C^{12}+e^++\nu_e)}\right\}$$ from 1 on the basis

of an argument that $|N^{12}\rangle$ deviates from

$e^{i\pi I^{(2)}}|B^{12}\rangle$ by about 5% because of Coulomb-

induced higher-than-1-isospin admixtures; detailed consideration of this possibility appears to show that these isospin admixtures (and other electromagnetic effects) are at best sufficient to explain only half of the observed 10% deviation.[38] We therefore conclude from Eqs. (238), (239) that

$$4\left[\frac{f_A^{(n)(i)}(0;B^{12}\to C^{12})}{F_A^{(n)(r)}(0;n\to p)\langle C^{12}|\tau\vec{\sigma}|B^{12}\rangle}+\left(\frac{\left(\frac{m_{B^{12}}+m_{N^{12}}}{2}\right)-m_{C^{12}}}{2m_p}\right)\left(\frac{F_T^{(n)(i)}(0;n\to p)}{F_A^{(n)(r)}(0;n\to p)}\right)\right]=$$

$$=4\left[\frac{f_A^{(n)(i)}(0;B^{12}\to C^{12})}{0.55}+8\times10^{-3}\left(\frac{F_T^{(n)(i)}(0;n\to p)}{F_A^{(n)(r)}(0;n\to p)}\right)\right]\cong 0.05, \tag{241}$$

which is consistent, e.g., with

$$f_A^{(n)(i)}(0; B'^{12} \to C'^{12}) = \beta' f_A^{(n)(r)}(0; B^{12} \to C^{12}) = \beta'(0.05) << 0.05$$

and $$\left(\frac{F_T^{(n)(i)}(0; n \to p)}{F_A^{(n)(r)}(0; n \to p)} \right) = 1.6$$

(compare with the discussion after Eq. (236) and with Eq. (208)). In a similar way, analysis of several other charge-symmetric decays appears to show that

$$\left(\frac{\Gamma_{CORR}(0; N_i \to N_f + e^- + \bar{\nu}_e)}{\Gamma_{CORR}(0; N_i' \to N_f' + e^+ + \nu_e)} - 1 \right)$$ is roughly proportional to

$\left(\frac{m_i - m_f}{2m_p} \right)$ which would indicate that the term proportional to $f_A^{(n)(i)}(0; N_i \to N_f)$ is less important than the term proportional to $F_T^{(n)(i)}(0; n \to p)$. The best fit of the

$$\left\{ \frac{\Gamma_{CORR}(N_i \to N_f + e^- + \bar{\nu}_e)}{\Gamma_{CORR}(N_i' \to N_f' + e^+ + \nu_e)} \right\}_{\text{EXPERIMENTAL}}$$

to a single value of

$$\left(\frac{F_T^{(n)(i)}(0; n \to p)}{F_A^{(n)(r)}(0; n \to p)} \right)$$ corresponds to $$\left(\frac{F_T^{(n)(i)}(0; n \to p)}{F_A^{(n)(r)}(0; n \to p)} \right) \cong 2. \qquad [39]$$

All in all, one cannot escape the impression that a certain amount of evidence exists for the presence of $A_\alpha^{(n)(i)}(x)$.

In concluding this section we very briefly discuss the possible augmentation of the strangeness-nonchanging hadron weak current, $\left(V_\alpha(x) + A_\alpha(x)\right)$, which is an isovector (Eq. (178)), by an isotensor strangeness-nonchanging hadron weak current, $\left(V_\alpha^{\text{ISOTENSOR}}(x) + A_\alpha^{\text{ISOTENSOR}}(x)\right)$. Since $|p\rangle$ and $|n\rangle$ are the two members of the nucleon isodoublet the matrix element

$$\langle p| \left\{ V_\alpha^{\text{ISOTENSOR}}(x) + A_\alpha^{\text{ISOTENSOR}}(x) \right\}^\dagger |n\rangle = 0$$

while the matrix element $\langle N_f| \left(V_\alpha^{\text{ISOTENSOR}}(x) + A_\alpha^{\text{ISOTENSOR}}(x) \right) |N_i\rangle$ ($|N_i\rangle$ and $|N_f\rangle$ are the initial and final nuclei) will differ from zero only by an amount corresponding to the contribution of corrections arising from the $\langle \pi^0| \left\{ V_\alpha^{\text{ISOTENSOR}}(x) + A_\alpha^{\text{ISOTENSOR}}(x) \right\}^\dagger |\rho^-\rangle$ and $\langle n| \left\{ V_\alpha^{\text{ISOTENSOR}}(x) + A_\alpha^{\text{ISOTENSOR}}(x) \right\}^\dagger |N^{*-}\rangle$ matrix elements associated with the beta decay (muon capture, etc.) of virtual mesons exchanged between the nucleons within the nuclei and the beta decay (muon capture, etc.) of any N^* present within the nuclei; as we have seen above, this amount is not likely to be more than about 5% (compare Eqs. (192)-(196), (208), (209)). It thus appears that the contributions of any $\left(V_\alpha^{\text{ISOTENSOR}}(x) + A_\alpha^{\text{ISOTENSOR}}(x) \right)$ currents should be sought in a detailed study of reactions such as

$$\nu_\mu + p \rightarrow \mu^- + N^{*++}, \quad \nu_\mu + n \rightarrow \mu^- + N^{*+} \qquad (242)$$

where $\left| \left(\frac{\sigma(\nu_\mu + p \rightarrow \mu^- + N^{*++})}{\sigma(\nu_\mu + n \rightarrow \mu^- + N^{*+})} \right) - 3 \right|$

will be significantly different from zero if

$\left(V_\alpha^{\text{ISOTENSOR}}(x) + A_\alpha^{\text{ISOTENSOR}}(x) \right)$ is of the same order as

$\left(V_\alpha(x) + A_\alpha(x) \right)$ (compare discussion after Eq. (144)).

5. Conclusion

We hope that in these lectures we have successfully conveyed the impression that the theory of weak interactions is still very much in flux. Thus we have tried to emphasize that not only is the question of the appropriate theoretical description of the higher-order weak processes quite open but that also in the far better explored realm of first-order weak processes various definitive experimental studies remain to be undertaken and various corresponding important theoretical questions (lepton conservation, time-reversal invariance, isospace-transformation behavior) remain to be answered.

References and Footnotes

1. A fifth interaction, e.g. the so-called "superweak" interaction, may be necessary to explain the CP and T noninvariance observed in the K_L^0 decays.

2. See the recent book: Theory of Weak Interactions in Particle Physics by R. E. Marshak, Riazuddin, and C. P. Ryan (Wiley-Interscience, New York, 1969) which presents a lucid, comprehensive, and detailed treatment.

3. Note that measurements on the elastic scattering process in Eq. (3) require observation of collisions between ν_μ in a neutrino beam and μ^- in a muon beam!!!

4. In the case that the μ^+ and e^+ are decay products of the W^+ (Eqs. (15a), (15b)) their mean energy will be comparable with the mean energy of the μ^-; on the other hand, the μ^+ and e^+ produced directly will have mean energies quite different from the mean energy of the μ^-.

5. In both Fig. 1 and Fig. 2, as well as in all figures below, an electromagnetic vertex is represented by • and a weak vertex by ■.

6. Since we work in the Schrödinger picture t is fixed and may be set equal to zero.

7. We take the γ_α to be hermitian with γ_1, γ_2, γ_3 real and γ_4 imaginary. $\gamma_5 \equiv \gamma_1\gamma_2\gamma_3\gamma_4$ is then also imaginary.

8. In Fig. 3, as well as in all figures below, the semiweak vertex is represented by ◆.

9. $g_{\mathcal{N}}$ is related to g_p and g_n in the same way as a nucleon ($\mathcal{N}$) is related to a proton or a neutron. Also $\tau^{(-)} = \frac{1}{2}(\tau^{(1)} - i\,\tau^{(2)})$ is the isospin step-down op-

erator appropriate to the (q_p , q_n) isodoublet (or the (p, n) isodoublet) i.e. $\tau^{(-)}|q_p\rangle = |q_n\rangle, \ \tau^{(-)}|q_n\rangle = 0$.

10. C. W. Kim and H. Primakoff, Phys. Rev. 180, 1502 (1969).

11. As possible examples of such explicit expressions for $A_\alpha^{(a)(r)}(x)$, $A_\alpha^{(n)(i)}(x)$ and $A_\alpha^{(a)(i)}(x)$ in terms of $\psi_{q r}(x)$ we may mention (compare Eq. (181) below)

$$A_\alpha^{(a)(r)}(x) = C^{(a)(r)} i^{-1} \frac{\partial}{\partial x_\beta}\left(\bar{\psi}_{qr}(x)\,\tau^{(-)}\sigma_{\alpha\beta}\gamma_5\,\psi_{qr}(x)\right); \quad C^{(a)(r)}_{\text{an imaginary constant}}$$

$$A_\alpha^{(n)(i)}(x) = C^{(n)(i)} i^{-1} \frac{\partial}{\partial x_\beta}\left(\bar{\psi}_{qr}(x)\,\tau^{(-)}\sigma_{\alpha\beta}\gamma_5\,\psi_{qr}(x)\right); \quad C^{(n)(i)}_{\text{a real constant}}$$

$$A_\alpha^{(a)(i)}(x) = C^{(a)(i)}\left(\bar{\psi}_{qr}(x)\,\tau^{(-)}\gamma_\alpha\gamma_5\,\psi_{qr}(x)\right); \quad C^{(a)(i)}_{\text{an imaginary constant}}$$

12. See reference 2 on CVC.

13. Since every $A_\alpha^{(x)(y)}(x)$ is odd under CPT and even under P

$$G A_\alpha^{(x)(y)}(x) G^{-1} = \left(Ce^{i\pi I^{(2)}}\right) A_\alpha^{(x)(y)}(x) \left(Ce^{i\pi I^{(2)}}\right)^{-1} =$$

$$= -T\left(e^{i\pi I^{(2)}} A_\alpha^{(x)(y)}(x)\, e^{-i\pi I^{(2)}}\right) T^{-1} = -a_{xy} b_{xy} A_\alpha^{(x)(y)}(x)$$

14. H. Primakoff and S. P. Rosen, Phys. Rev. 184, 1925 (1969).

15. T. Kirsten, O. E. Schaeffer, E. Norton, and R. W. Stoenner, Phys. Rev. Letters 20, 1300 (1968); T. Kirsten, W. Gentner, and O. A. Schaeffer, Z. Physik 202, 273 (1967).

16. N. Takaoka and K. Ogata, Z. Naturforsch. 21a, 84 (1966).

17. H. Primakoff and D. H. Sharp, Phys. Rev. Letters 23, 501 (1969).

18. J. Bernstein, M. Ruderman, and G. Feinberg, Phys. Rev. 132, 1227 (1963); T. D. Lee and J. Bernstein, Phys. Rev. Letters 11, 512 (1963); P. Meyer and D. Schiff, Phys. Letters 8, 217 (1964); W. K. Cheng and S. Bludman, Phys. Rev. 136B, 1787 (1964).

19. F. Reines and H. Gurr, Phys. Rev. Letters 24, 1448 (1970).

20. R. B. Stothers, Phys. Rev. Letters 24, 538 (1970).

21. M. Gell-Mann, M. L. Goldberger, N. M. Kroll, and F. E. Low, Phys. Rev. 179, 1518 (1969).

22. R. W. Brown, A. K. Mann, and J. Smith, Phys. Rev. Letters 25, 257 (1970) have shown that, contrary to general expectations,

$$\left(\frac{\sigma(\nu_\mu + (Z,A) \rightarrow W^+ + \mu^- + \text{all final } (Z,A) \text{ states})}{\sigma(\mu^- + (Z,A) \rightarrow W^- + \nu_\mu + \text{all final } (Z,A) \text{ states})}\right) \gg 1$$

for experimentally relevant E_ν and theoretically "reasonable" m_W.

23. See e.g. R. H. Dalitz, High Energy Physics, edited by C. DeWitt and M. Jacob (Gordon and Breach, New York, 1965).

24. I. Budagov, D. C. Cundy, C. Franzinetti, W. B. Fretter, H. W. K. Hopkins, C. Manfredotti, G. Myatt, F. A. Nezrick, M. Nikolic, T. B. Novey, R. B. Palmer, J. B. M. Pattison, D. H. Perkins, C. A. Ramm, B. Roe, R. Stump, W. Venus, H. W. Wachsmuth, and H. Yoshiki, Phys. Letters 30B, 364 (1969).

25. D. H. Perkins, Topical Conference on Weak Interactions at CERN, Jan. 1969, p. 1, CERN 69-7.

26. S. L. Adler, Phys. Rev. 143, 1144 (1966).

27. H. Steiner, Phys. Rev. Letters 24, 746 (1970); D. C. Cundy, G. Myatt, F. A. Nezrick, G. B. M. Pattison, D. H. Perkins, C. A. Ramm, W. Venus and H. W. Wachsmuth, Phys. Letters 31B, 478 (1970).

28. H. Foeth, M. Holder, E. Radermacher, A. Stande, P. Darriulat, J. Deutsch, K. Kleinknecht, C. Rubbia, K. Tittel, M. I. Ferrero, and C. Grosso, Phys. Letters 30B, 282 (1969).

29. J. H. Klems, R. H. Hildebrand, and R. Stiening, Phys. Rev. Letters 24, 1086 (1970).

30. See the second reference in footnote 27.

31. A treatment where it is supposed that the basic nuclear constituents are quarks and where it is assumed that N^* may be present (albeit with a small probability) in the nuclei is given in reference 14. In this case the $\bar{\nu}_e$ emitted by one quark can be reabsorbed by a second quark within the

same nucleon ($n[q_n q_n q_p] \to p[q_p q_n q_r] + e^- + \bar{\nu}_e \to N^{*++}[q_p q_r q_r] + e^- + e^-$)

and $\{|\vec{r}_j - \vec{r}_k|\}_{AV}$ in Eq. (173) becomes the average interquark distance in a nucleon (or an N^*) $\cong 3\left(\frac{1}{m_p}\right)$; this is to be contrasted with

$\{|\vec{r}_j - \vec{r}_k|\}_{AV} \cong A^{1/3}\left(\frac{1}{m_\pi}\right)$ appropriate to the situation where no N^* are present (second equality on right side in Eq. (173)). The upper limit on η_e in the N^*-model works out to be some ten times smaller than the upper limit on η_e (Eq. (47)) appropriate to the model where only nucleons are present.

32. See reference 10. An idea of the argument involved is given by considering, e.g.

$$\langle p|\{A_3^{(h)(r)}(0)\}^\dagger|n\rangle = \langle (e^{i\pi I^{(2)}}p)|\, e^{i\pi I^{(2)}}\{A_3^{(h)(r)}(0)\}^\dagger e^{-i\pi I^{(2)}}\,|(e^{i\pi I^{(2)}}n)\rangle$$

$$= \langle (e^{i\pi I^{(2)}}n)|\,\{e^{i\pi I^{(2)}}\{A_3^{(h)(r)}(0)\}^\dagger e^{-i\pi I^{(2)}}\}^\dagger\,|(e^{i\pi I^{(2)}}p)\rangle^* =$$

$$= \langle -n'|\; - \{A_3^{(h)(r)}(0)\}^\dagger\; |p'\rangle^*$$

so that

$$\left(F_A^{(h)(r)}(q^2; n\to p)\,(\bar{u}_p\gamma_3\gamma_5 u_n) - F_T^{(h)(r)}(q^2; n\to p)\left(\bar{u}_p\left(\frac{\sigma_{3\beta}q_\beta\gamma_5}{2m_p}\right)u_n\right)\right) =$$

$$= \left(F_A^{(h)(r)}(q^2; p'\to n')\,(\bar{u}_{n'}\gamma_3\gamma_5 u_{p'}) - F_T^{(h)(r)}(q^2; p'\to n')\left(\bar{u}_p\left(\frac{\sigma_{3\beta}q_\beta\gamma_5}{2m_p}\right)u_n\right)\right)^*$$

whence, considering the reality properties of $(\bar{u}_p \gamma_3 \gamma_5 u_n)$ etc., one obtains:

$$F_A^{(n)(r)}(q^2; n \to p) = \left(F_A^{(n)(r)}(q^2; p' \to n')\right)^*,$$

$$F_T^{(n)(r)}(q^2; n \to p) = -\left(F_T^{(n)(r)}(q^2; p' \to n')\right)^*$$

33. H. Primakoff, Springer Tracts in Modern Physics (Ergebnisse der exacten Naturwissenschaften) 53, 6 (1970).

34. B. G. Erozolimsky, L. N. Bondarenko, Yu. A. Mostovoy, B. A. Obinyakov, V. P. Zacharova, and V. A. Titov, Phys. Letters 27B, 577 (1968); M. T. Burgy, V. E. Krohn, T. B. Novey, G. R. Ringo, and V. L. Telegdi, Phys. Rev. 120, 1289 (1960); M. A. Clark, J. M. Robson, and R. Nathans, Phys. Rev. Letters 1, 100, (1958).

35. F. M. Calaprice, E. D. Commins, H. M. Gibbs, and G. L. Wick, Phys. Rev. 184, 1117 (1969).

36. We recall that $\Sigma^{\pm}$ and Λ are members of the same SU3 octet and that SU3 breaking is typically $\lesssim$ 25%. For a general discussion of such an SU3 analogue to the SU2 theorem in Eqs. (184)-(186) see N. Cabibbo, Phys. Letters 12, 137 (1964), N. Cabibbo, Phys. Rev. Letters 14, 965 (1965); L. Maiani, Phys. Letters 26B, 538 (1968).

37. H. Filthuth, Topical Conference on Weak Interactions at CERN, Jan. 1969, p. 131, CERN 69-7.

38. R. J. Blin-Stoyle and M. Rosina, Nucl. Phys. 70, 321 (1965).

39. D. H. Wilkinson, Phys. Letters 31B, 447 (1970). See also P. Hertel, Z. Physik 202, 383 (1967).

The GNS Construction—A Pedagogical Example

Michael C. Reed
Department of Mathematics
Princeton University
Princeton, New Jersey

CONTENTS

The purpose of this lecture is to show why one must "change Hilbert space" in Quantum Field Theory and how this is accomplished using the GNS construction. The example has the nice feature that both before and after the construction the relevant representations of the anti-commutation relations can be written down explicitly making the construction itself seem less abstract. The example has no physical importance (there is no significant interaction) but was chosen so that the difficult mathematical problems (usually involving taking limits) which arise in Wightman-Jaffe-Glimm constructive Quantum Field Theory are trivial; one can therefore understand how the GNS construction is used without having to master the difficult functional analysis which is necessary to construct "real" models. On the way we will discuss infinite tensor products of Hilbert spaces and a few (uncountably many) of the "myriad" representations of the anti-commutation relations to which Professor Haag has referred in his lectures.

I. Infinite Tensor Products of Hilbert Spaces

Suppose that H_K, $K = 1, 2, \ldots$ are Hilbert spaces (separable). We would like to construct an infinite tensor product of these spaces. This infinite tensor product should contain vectors like $\varphi_1 \otimes \varphi_2 \otimes \cdots$ and $\psi_1 \otimes \psi_2 \otimes \cdots$ where $\varphi_K, \psi_K \in H_K$, with the inner product $(\varphi, \psi) = \prod_{K=1}^{\infty} (\varphi_K, \psi_K)$. The trouble is that this infinite product of complex numbers may not converge so that the inner product will not in general make sense. A way around this difficulty was discovered by John von Neumann in 1939; what we outline here is his work [1].

Let $\chi_K \in H_K$, $\|\chi_K\| = 1$. If $\varphi = \varphi_1 \otimes \varphi_2 \otimes \cdots$ and $\psi = \psi_1 \otimes \psi_2 \otimes \cdots$ have the property that $\psi_K = \varphi_K = \chi_K$ except for a finite number of K then we can define the inner product between $\varphi = \bigotimes_{K=1}^{\infty} \varphi_K$ and $\psi = \bigotimes_{K=1}^{\infty} \psi_K$ by $(\varphi, \psi) = \prod_{K=1}^{\infty} (\varphi_K, \psi_K)$ since only a finite number of terms are different from one. This inner product can then be extended by linearity to $\mathcal{D}(\chi)$, the set of finite linear combinations of such vectors as φ and ψ. On $\mathcal{D}(\chi)$, $(\cdot, \cdot)$ is positive definite so we can complete $\mathcal{D}(\chi)$ under the inner product $(\cdot, \cdot)$ and obtain a Hilbert space $\mathcal{H}(\chi)$.

A vector such as χ is called a C_0 - vector. Two C_0 - vectors, $\chi = \bigotimes_{K=1}^{\infty} \chi_K$ and $\chi' = \bigotimes_{K=1}^{\infty} \chi'_K$ are called equivalent (written $\chi \sim \chi'$) if $\sum_{K=1}^{\infty} |1 - (\chi_K, \chi'_K)| < \infty$

One can show that $\sim$ is an equivalence relation on the set of C_0- vectors and furthermore that for each C_0- vector, χ, the space $\mathcal{H}(\chi)$ contains exactly those C_0- vectors which lie in the same equivalence class as χ. Another way of saying this is that the Hilbert spaces, $\mathcal{H}(\chi)$ and $\mathcal{H}(\chi')$ constructed by the above procedure will be the "same" if and only if χ and χ' are equivalent. What is important to remember is that all the C_0-vectors in $\mathcal{H}(\chi)$ have approximately the same tail (as $K \to \infty$) as the vector $\chi = \chi_1 \otimes \chi_2 \otimes \cdots$ which generates $\mathcal{H}(\chi)$.

The complete infinite tensor product space H is defined as the direct sum $H = \sum \oplus \mathcal{H}(\chi)$ of the spaces $\mathcal{H}(\chi)$, in which only one χ from each equivalence class is allowed (so that each $\mathcal{H}(\chi)$ appears only once). Each $\mathcal{H}(\chi)$ is a separable subspace of H and subspaces $\mathcal{H}(\chi)$ and $\mathcal{H}(\chi')$ corresponding to different (inequivalent) C_0-vectors are orthogonal. Essentially, what has been done is to define the product $(\varphi, \psi) = \prod_{K=1}^{\infty} (\varphi_K, \psi_K)$ to be zero whenever it does not make sense; this is what makes the subspaces of H corresponding to different χ and χ' orthogonal. Since there are uncountably many equivalence classes, the space H is inseparable. We have the following picture.

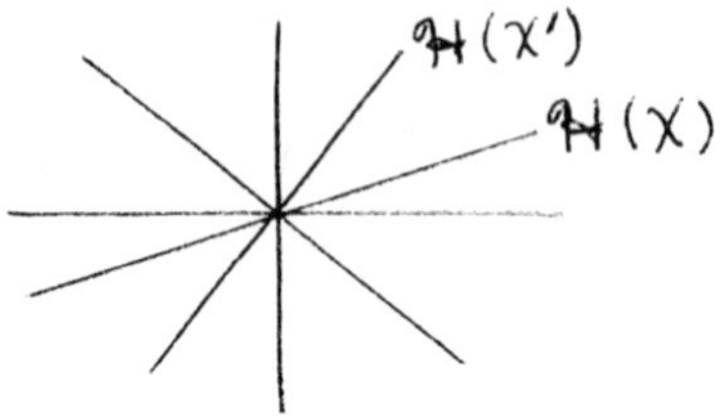

Each line represents a separable space $\mathcal{H}(\chi)$ generated by an equivalence class of C_0- vectors; the vectors in $\mathcal{H}(\chi)$ have tails which look asymptotically like the tail of $\chi = \bigotimes_{k=1}^{\infty} \chi_k$. The whole space H is obtained by adding up orthogonally (which is hard to draw in two dimensions) all these separable subspaces. The spaces $\mathcal{H}(\chi)$ are sometimes called "incomplete" infinite tensor product spaces. This is a little confusing since they are complete Hilbert spaces. The word "incomplete" refers to the fact that $\mathcal{H}(\chi)$ is only a part (that is, a subspace) of H.

Remarks: The two basic references for infinite tensor product spaces are the paper by von Neumann [1] and the paper by Guichardet [2]. Neither is easy to read, the von Neumann paper because its terminology is a little old-fashioned, the Guichardet paper because of the mathematical sophistication expected of the reader. I suggest beginning with the very readable lectures by Wehrl in [3] where proofs are given. As a last resort, a somewhat more detailed sketch than is given in this lecture is available in the appendix to [4]. We mention that a theory of continuous tensor products has arisen; see for example the work of Streater, Guichardet, and Wulfson.

[1] von Neumann, J., Compos. Math. 6 (1939), p. 1.

[2] Guichardet, A., Ann. Sci. Ecole Norm. Sup. (3) 83 (1966), pp. 1 - 52.

[3] Guenin, M., Wehrl, A. and W. Thirring; "Introduction to Algebraic Techniques," CERN lecture note series (1969).

[4] Reed, M., Journ. Func. Anal. 5 (1970), p. 94.

II. The Canonical Anti-Commutation Relations

Let $\mathbb{C}^2_K$, $K = 1, 2, \cdots$ be two dimensional complex vector spaces and H the complete tensor product of the $\mathbb{C}^2_K$. We focus our attention on one of the subspaces

$$\mathcal{H}(v), \quad v = \begin{pmatrix} \alpha_1 \\ \beta_1 \end{pmatrix} \otimes \begin{pmatrix} \alpha_2 \\ \beta_2 \end{pmatrix} \otimes \cdots, \quad |\alpha_i|^2 + |\beta_i|^2 = 1.$$

Let

$$a_K = \begin{pmatrix} -1 & 0 \\ 0 & 1 \end{pmatrix} \otimes \begin{pmatrix} -1 & 0 \\ 0 & 1 \end{pmatrix} \otimes \cdots \otimes \underset{\substack{\uparrow \\ K^{th}\text{ place}}}{\begin{pmatrix} 0 & 0 \\ 1 & 0 \end{pmatrix}} \otimes \begin{pmatrix} 1 & 0 \\ 0 & 1 \end{pmatrix} \otimes \cdots,$$

$$a_K^* = \begin{pmatrix} -1 & 0 \\ 0 & 1 \end{pmatrix} \otimes \begin{pmatrix} -1 & 0 \\ 0 & 1 \end{pmatrix} \otimes \cdots \otimes \begin{pmatrix} 0 & 1 \\ 0 & 0 \end{pmatrix} \otimes \begin{pmatrix} 1 & 0 \\ 0 & 1 \end{pmatrix} \otimes \cdots .$$

$\uparrow$ K^{th} place

a_K and a_K^* are well-defined operators from $\mathcal{H}(v)$ to itself. This is because they affect only the first K components, that is, they do not affect the tails of vectors so the result of applying a_K or a_K^* to a vector in $\mathcal{H}(v)$ is again a vector in $\mathcal{H}(v)$. It is easy to see that a_K^* is the adjoint of a_K and that the family of operators $\{a_K, a_K^*\}_{K=1}^{\infty}$ satisfies the anti-commutation relations:

$$a_K^* a_\ell + a_\ell a_K^* = \delta_{K\ell} \mathbf{1} ,$$

$$a_K a_\ell + a_\ell a_K = 0 = a_K^* a_\ell^* + a_\ell^* a_K^* .$$

So, on each of the separable subspaces, $\mathcal{H}(v)$, we have a representation of the canonical anti-commutation relations. It is a natural question to ask when two such representations are unitarily equivalent; i.e., if $\{a_K', a_K^{*\prime}\}$ is the representation on $\mathcal{H}(v')$, when does there exist a unitary operator $V : \mathcal{H}(v) \to \mathcal{H}(v')$ so that

$$a_K' = V a_K V^{-1} , \quad a_K^{*\prime} = V a_K^* V^{-1} \ ?$$

The theorem of Klauder, McKenna, and Woods [5] states that the representations on $\mathcal{H}(v)$, $v = \bigotimes_{K=1}^{\infty} v_K$, and $\mathcal{H}(v')$, $v' = \bigotimes_{K=1}^{\infty} v_K'$, are unitarily equivalent if and only if

$$\sum_{K=1}^{\infty} \Big| 1 - |(v_K, v_K')| \Big| < \infty .$$

Two $\mathcal{C}_0$- vectors which satisfy this relation are said to be weakly equivalent (written $\overset{w}{\sim}$). It is not hard to show that $\overset{w}{\sim}$ is an equivalence relation on the set of $\mathcal{C}_0$- vectors. Since $\overset{w}{\sim}$ is a weaker relation than $\sim$, each weak equivalence class will contain many inequivalent $\mathcal{C}_0$-vectors, but the representations on the spaces generated by these $\mathcal{C}_0$-vectors will be unitarily equivalent.

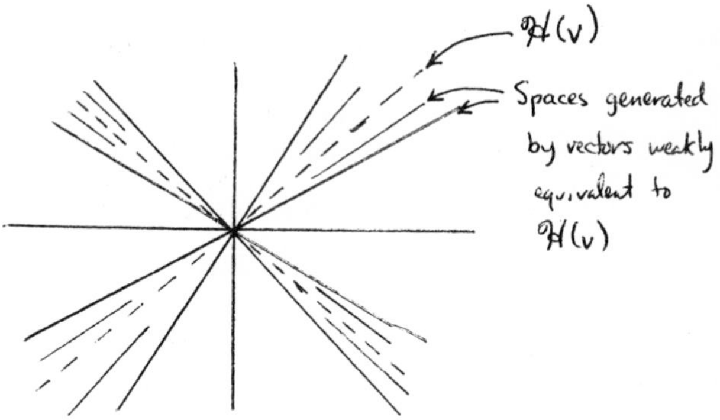

The reader can easily convince himself that there are uncountably many weak equivalence classes so we have exhibited uncountably many unitarily inequivalent representations of the canonical anti-commutation relations. These representations are all irreducible (they are known as the irreducible product representations). They by no means exhaust all possible representations, but they will be sufficient for the example which we present in sections III and IV.

Let us consider again one of the spaces $\mathcal{H}(v)$. Let $\mathcal{B}_N(v)$ be the algebra of all operators on $\mathcal{H}(v)$ which are finite linear combinations of finite products of a_K and a_K^* for $K = 1, 2, \cdots, N$. By an honorable, classical result of Jordan and Wigner [6] this algebra is the algebra of all operators which affect only the first N components (the algebra of 2^N by 2^N matrices). The set theoretic union $\bigcup_{N=1}^{\infty} \mathcal{B}_N(v)$ is an algebra of operators on $\mathcal{H}(v)$. We define $\mathfrak{A}_v$ to be the closure of $\bigcup_{N=1}^{\infty} \mathcal{B}_N(v)$ in the uniform operator topology (in the operator norm). $\mathfrak{A}_v$ is thus a C^*-algebra.

Suppose that $\mathcal{H}(v')$ is another space with the corresponding algebra $\mathfrak{A}_{v'}$. Let τ be the map from $\bigcup_{N=1}^{\infty} \mathcal{B}_N(v')$ to $\bigcup_{N=1}^{\infty} \mathcal{B}_N(v)$ which assigns to each $A \in \bigcup_{N=1}^{\infty} \mathcal{B}_N(v')$ the operator $\tau(A)$ in $\bigcup_{N=1}^{\infty} \mathcal{B}_N(v)$ which acts in exactly the same way on $\mathcal{H}(v)$ as A does on $\mathcal{H}(v')$ (this makes sense since operators in $\bigcup_{N=1}^{\infty} \mathcal{B}_N(v')$ act on only a finite number of components). τ is a linear map which is also an algebraic isomorphism, i.e.,

$\tau(AB) = \tau(A)\tau(B)$. Furthermore, $\|\tau(A)\| = \|A\|$. This means that τ can be extended by continuity to an isometric isomorphism of $\mathcal{A}_{V'}$ onto $\mathcal{A}_V$. That is, $\mathcal{A}_{V'}$ and $\mathcal{A}_V$ are just different representations of the same C^*- algebra, $\mathcal{A}$. This algebra is called the CAR algebra or sometimes the Clifford algebra. Corresponding to each representation of the anti-commutation relations we get a representation of $\mathcal{A}$ and vice-versa. It follows from the Klauder, McKenna, Woods theorem that the representations $\mathcal{A}_V$ on $\mathcal{H}(V)$ and $\mathcal{A}_{V'}$ on $\mathcal{H}(V')$ are unitarily equivalent if and only if $V \overset{w}{\sim} V'$.

Remarks: The classical sources for information about the canonical anti-commutation relations are the papers by Jordan and Wigner [5] and Gårding and Wightman [7]. The construction of the CAR algebra and many nice theorems are contained in R. Power's Princeton thesis (1967) [8]. The paper by Shale and Stinespring [9] is excellent though more difficult to read. The paper by Streit [10] has an elegant proof of the Klauder, McKenna, Woods theorem. After these, the best thing to do is to leaf through back issues of the Communications in Mathematical Physics. There one can find many papers on the CAR algebra and many references to papers in other journals.

[5] Jordan, P. and E. Wigner, Z. Phys. 47 (1928), p. 631.

[6] Klauder, J., McKenna, J. and E. Woods, Jour. Math. Phys. 7, (1967) p. 822.

[7] Gårding, L. and A. S. Wightman, Proc. Nat. Acad. Sci. 40 (1954), p. 617.

[8] Powers, R., Princeton Univ. Thesis (1967).

[9] Shale, D. and W. Stinespring, Ann. Math. 80 (1964), p. 365.

[10] Streit, L., Comm. Math. Physics, 4 (1967), p. 22.

III. The Example

Let $\psi_0 = \binom{0}{1} \otimes \binom{0}{1} \otimes \cdots$ The representation of the anti-commutation relations on the subspace $\mathcal{H}(\psi_0)$ of $\bigotimes_{K=1}^{\infty} \mathbb{C}_K^2$ is called the Fock representation (in a "real" model the index K would run over a lattice in 3-space; for simplicity our index set is the positive integers). The operators $a_K^* a_K$ are well-defined on each space $\mathcal{H}(v)$; they act like $\begin{pmatrix} 1 & 0 \\ 0 & 0 \end{pmatrix}$ on the Kth component and like the identity on the other components. Let $\omega_K = \sqrt{K^2+m^2}$. Then it is well known that the free Hamiltonian

$$h_0 = \sum_{K=1}^{\infty} \omega_K a_K^* a_K$$

makes sense on the space $\mathcal{H}(\psi_0)$. Let us examine why this is so. Recall that $\mathcal{D}(\psi_0)$ is the set of finite linear combinations of product vectors $\psi_1 \otimes \psi_2 \otimes \cdots$ all of whose components, except for a finite number, are equal to $\binom{0}{1}$. For any vector $\psi \in \mathcal{D}(\psi_0)$, $a_K^* a_K \psi = 0$ for all but a finite number of K since a_K will annihilate ψ if all the Kth components in the finite sum equal $\binom{0}{1}$. Thus, on each vector $\psi \in \mathcal{D}(\psi_0)$ the sum $\sum_{K=1}^{\infty} \omega_K a_K^* a_K$ is actually finite. Now, $\mathcal{D}(\psi_0)$ is dense in $\mathcal{H}(\psi_0)$, (by the construction of $\mathcal{H}(\psi_0)$, see Section I) so we have a well-defined operator $h_0 = \sum_{K=1}^{\infty} \omega_K a_K^* a_K$ on a dense domain in $\mathcal{H}(\psi_0)$. It turns out that h_0 is essentially self-adjoint on $\mathcal{D}(\psi_0)$ but we will not worry about such technical things here.

Let us suppose that we are given an "interaction" term $h_I = \sum_{K=1}^{\infty} E_K$, where E_K acts like a symmetric matrix $\begin{pmatrix} e_{11}^K & e_{12}^K \\ e_{21}^K & e_{22}^K \end{pmatrix}$ on the Kth component and like the identity on the other components. The problem is to make sense out of the formal expression

$$h_0 + h_I = \sum_{K=1}^{\infty} \omega_K a_K^* a_K + \sum_{K=1}^{\infty} E_K .$$

In this section we show how to do this by cleverly choosing the right representation (non-Fock) of the anti-commutation relations for a_K and a_K^*, $K = 1, 2, \cdots$. First, we write formally $h_0 + h_I = \sum_{K=1}^{\infty} D_K$, where D_K is the operator which acts like $\begin{pmatrix} e_{11}^K + \omega_K & e_{12}^K \\ e_{21}^K & e_{22}^K \end{pmatrix}$ on the Kth component and like the identity on the other components. Since

$$\| D_K \psi_0 \| = \sqrt{|e_{12}^K|^2 + |e_{22}^K|^2}$$

it is clear that in general (we have made no assumptions on the e_{ij}^K except that the E_K are symmetric; the e_{ij}^K could grow exponentially in K) the sum $\sum_{K=1}^{N} D_K \psi_0$ will diverge as $N \to \infty$. As a matter of fact, in general it will diverge on every vector in Fock space, $\mathcal{H}(\psi_0)$.

So, what we do is this. Since $\begin{pmatrix} e_{11}^K + \omega_K & e_{12}^K \\ e_{21}^K & e_{22}^K \end{pmatrix}$ is symmetric, it has two orthogonal eigenfunctions in $\mathbb{C}_K^2$. Denote the normalized eigenfunction with the lowest corresponding eigenvalue by u_K and the eigenvalue by λ_K. Consider the C_0- vector $u = u_1 \otimes u_2 \otimes \cdots$ and the representation of the anti-commutation relations on $\mathcal{H}(u)$. For each K, u is an eigenfunction of D_K with eigenvalue λ_K, so u is annihilated by $D_K - \lambda_K$. Therefore, the operator $\sum_{K=1}^{\infty} (D_K - \lambda_K)$ is well-defined on $\mathcal{D}(\mu)$ since for each $\psi \in \mathcal{D}(\mu)$ the sum $\sum_{K=1}^{\infty} (D_K - \lambda_K)\psi$ is actually finite. That is, if we choose the representation of a_K, a_K^* on $\mathcal{H}(\mu)$ (rather than the Fock representation) then we can make sense out of $h_c + h_I$ if we are willing to make the numerical subtractions λ_K. It is not difficult to show that $\sum_{K=1}^{\infty} (D_K - \lambda_K)$ is essentially self-adjoint on $\mathcal{D}(\mu)$ so that the "dynamics" $\exp\left(it \sum_{K=1}^{\infty} (D_K - \lambda_K)\right)$ makes sense on $\mathcal{H}(\mu)$.

Remarks: In physics texts the expression $\sum_{K=1}^{\infty} (D_K - \lambda_K)$ might be written $\sum_{K=1}^{\infty} D_K - \sum_{K=1}^{\infty} \lambda_K$. The sum $\sum_{K=1}^{\infty} \lambda_K$, which is in general divergent, is analogous to the vacuum renormalization. The reader may wonder if making different subtractions μ_K the sum $\sum_{K=1}^{\infty} (D_K - \mu_K)$ could make sense on Fock space. Only in special cases will this be true. In general by choosing $\{\mu_K\}_{K=1}^{\infty}$ appropriately we can make sense of $\sum_{K=1}^{\infty} (D_K - \mu_K)$ in any representation whose generating vector is a product of the eigenfunctions of the D_K. We chose $\mathcal{H}(\mu)$ because $\sum_{K=1}^{\infty} (D_K - \lambda_K)$ is a positive operator on $\mathcal{H}(\mu)$. Whether or not this is the only product representation where $\sum_{K=1}^{\infty} (D_K - \mu_K)$ is well-defined and positive depends on the spread between the eigenvalues of D_K as $K \to \infty$. For a theorem about this see [4]. An example similar to the one in this section but using the canonical commutation relations and allowing small off-diagonal coefficients is worked out in [11].

[11] Reed, M., Commun. Math. Phys. 11, p. 346.

IV. The Example - via the GNS Construction

The trouble with the method of solution of the example in section III is that for "real" models the formal Hamiltonian is too complicated for one to be able to cleverly guess the correct representation. This is just where the GNS construction comes in; the correct representation can be obtained by using the GNS construction with the physical vacuum ω which one hopes to obtain as the limit of vector states in the Fock representation. We will show how this works with our example.

So, we begin again with the Fock representation $\mathcal{A}_{\psi_0}$ of the algebra $\mathcal{A}$. We would like to define $\sum_{K=1}^{\infty} D_K$ but the sum does not make sense, so we mimic the current lore of constructive Quantum Field Theory. First we cut off the Hamiltonian obtaining the operators $h^M = \sum_{K=1}^{\infty} \omega_K a_K^* a_K + \sum_{K=1}^{M} E_K$, which are well-defined on $\mathcal{H}(\psi_0)$. Let ψ^M be the ground state of h^M. As $M \to \infty$, ψ^M

does not converge as a vector in $\mathcal{H}(\psi_c)$, but it does converge as a state on $\mathcal{Q}$. We state this as a proposition.

Proposition: Let ω^M be the state on $\mathcal{Q}$ given by the vector ψ^M in the representation $\mathcal{Q}_{\psi_0}$. As $M \to \infty$, ω^M converges (weak *) to a state ω. Furthermore, the representation of $\mathcal{Q}$ obtained through ω via the GNS construction is the representation $\mathcal{Q}_\mu$ on $\mathcal{H}(\mu)$. The cyclic vector corresponding to ω is $\mu = \mu_1 \otimes \mu_2 \otimes \cdots$.

Proof: Suppose $A \in \mathcal{B}_n(\psi_0)$ for some n (see Section II for notation). The ground state for h^M on $\mathcal{H}(\psi_0)$ is

$$\mu_1 \otimes \cdots \otimes \mu_m \otimes \binom{0}{1} \otimes \binom{0}{1} \otimes \cdots$$

Then $\omega^M(A) = (A\psi^M, \psi^M)$ converges as $M \to \infty$ because it is constant for $M > n$ since A only acts on the first n components. Now suppose A is a general element of $\mathcal{Q}_{\psi_c}$ and $A_n \in \mathcal{B}_n(\psi_0)$.

Then

$$\begin{aligned} |\omega^M(A) - \omega^{M'}(A)| \le & |\omega^M(A) - \omega^M(A_n)| + \\ & + |\omega^M(A_n) - \omega^{M'}(A_n)| + \\ & + |\omega^{M'}(A_n) - \omega^{M'}(A)| \\ \le & 2\|A - A_n\| + |\omega^M(A_n) - \omega^{M'}(A_n)|. \end{aligned}$$

Given $\epsilon > 0$, we first choose A_n in some $\mathcal{B}_n(\psi_0)$ so that $\|A - A_n\| \leq \frac{1}{2}\epsilon$.
For $M > n$, $M' > n$, $|\omega^M(A_n) - \omega^{M'}(A_n)| = 0$,
so $|\omega^M(A) - \omega^{M'}(A)| \leq \epsilon$, which proves that $\omega^M(A)$ is a Cauchy sequence.

We define $\omega(A) = \lim_{M \to \infty} \omega^M(A)$.
This $\omega(\cdot)$ is a state on $\mathcal{A}_{\psi_0}$ and therefore on $\mathcal{A}$. Now, $\mu = \mu_1 \otimes \mu_2 \otimes \cdots$ is a vector state in the representation $\mathcal{A}_\mu$ of $\mathcal{A}$. It is easy to check that the state on $\mathcal{A}$ corresponding to μ is the same as ω on $\bigcup_{n=1}^{\infty} \mathcal{B}_n$. Since $\bigcup_{n=1}^{\infty} \mathcal{B}_n$ is dense in $\mathcal{A}$ these states are the same on all of $\mathcal{A}$. But μ is cyclic for $\mathcal{A}_\mu$ (the representation is irreducible so all vectors in $\mathcal{H}(\mu)$ are cyclic) so by the uniqueness of the GNS construction the representation of $\mathcal{A}$ generated by ω is $\mathcal{A}_\mu$ on $\mathcal{H}(\mu)$ and ω corresponds to μ.

The operators h^M are well-defined and self-adjoint on $\mathcal{H}(\psi_0)$ and $\sigma_t^M(A) = e^{ith^M} A e^{-ith^M}$ is a one parameter group of automorphisms of $\mathcal{A}_{\psi_0}$ (and therefore of $\mathcal{A}$). Though the Hamiltonians, h^M, do not converge as $M \to \infty$ the automorphisms σ_t^M do.

<u>Proposition:</u> As $M \to \infty$, the automorphisms σ_t^M converge to a (strongly) continuous one-parameter group of auto-

morphisms σ_t of $\mathcal{A}$. Furthermore, ω is invariant under σ_t and the unitary group $U(t)$ which implements σ_t in the representation $\mathcal{A}_\mu$ is

$$\exp\left(it\sum_{K=1}^{\infty}(D_K-\lambda_K)\right).$$

Proof: The proof that the σ_t^M converge is so similar to the proof that the ω^M converge that we omit it. The proof uses the fact that automorphisms have norm equal to one as operators from $\mathcal{A}$ to itself. The simple proof that the limiting automorphism is continuous uses this fact also as well as the fact that every operator in $\mathcal{A}$ can be approximated arbitrarily closely by an $A_n \in \mathcal{B}_n$ for some n.

To prove that ω is invariant we observe

$$\begin{aligned}|\omega(\sigma_t(A)) - \omega(A)| \leq & |\omega(\sigma_t(A)) - \omega^M(\sigma_t(A))| + \\ & + |\omega^M(\sigma_t(A)) - \omega^M(\sigma_t^{M'}(A))| + \\ & + |\omega^M(\sigma_t^{M'}(A)) - \omega(A)|.\end{aligned}$$

By choosing M and M' large enough the right hand side becomes arbitrarily small. Since the left hand side does not depend on M or M' we have $\omega(\sigma_t(A)) = \omega(A)$. Thus ω is invariant and σ_t is unitarily implementable in the representation of $\mathcal{A}$ generated by ω (the representation

$\mathfrak{A}_\mu$ on $\mathcal{H}(\mu)$ according to the first proposition).
Now, according to the GNS construction the formula for the implementing unitary operator $U(t)$ is

$$U(t)(A\mu) = \sigma_t(A)\mu, \quad A \in \mathfrak{A}_\mu .$$

Suppose $A \in \mathcal{B}_n(\mu)$. Then $\sigma_t(A) = \sigma_t^n(A)$ and

$$U(t)(A\mu) = \sigma_t(A)\mu = \sigma_t^n(A)\mu =$$

$$= e^{ith^n} A e^{-ith^n} \mu =$$

$$= e^{it\sum_{K=1}^{n}(D_K - \lambda_K)} A\mu .$$

So, on the dense set $\{A\mu, A \in \mathcal{B}_n(\mu)$ for some $n\}$ we have explicitly calculated the action of $U(t)$. Since

$$e^{it\sum_{K=1}^{n}(D_K-\lambda_K)} A\mu = e^{it\sum_{K=1}^{\infty}(D_K-\lambda_K)} A\mu$$

for $A \in \mathcal{B}_n(\mu)$, we see that our group $U(t)$, obtained from the GNS construction, is the same as the one obtained by clever guesswork in section III.

Concluding remark: The point of this example is to show why a "change of Hilbert space" is often necessary in quantum field theory and to give the general idea of how this is accomplished using the GNS construction.

Effective Lagrangians and Broken Symmetries

Bruno Zumino
CERN, European Organization for Nuclear Research
Geneva, Switzerland

CONTENTS

1. Introduction

The approximate symmetries of the strong interactions appear to be of two main types. An example of the first type is SU(3). If we imagine switching off the symmetry breaking, we find an ideal situation in which the elementary particles group themselves in multiplets belonging to linear representations of SU(3) and the amplitudes for the various physical processes are linearly related. Such a symmetry is sometimes called a true symmetry. In the presence of the symmetry breaking there are corrections which one hopes to estimate. An example of the second type is $SU(2) \times SU(2)$. Here if we switch off the symmetry breaking we find what is called a spontaneously broken symmetry, or sometimes a dynamical symmetry. Even in the ideal situation of no explicit symmetry breaking one does not find the linear multiplets of a true symmetry. One finds, instead, that certain particles must be massless, the so called Goldstone bosons. In the case of $SU(2) \times SU(2)$ the Goldstone boson is the pion and the fact that it is not massless is attributed to the presence of explicit breaking of the symmetry. Instead of the usual relations among amplitudes which are consequences of a true symmetry, a spontaneously broken symmetry implies relations among amplitudes which refer to processes involving different numbers of Goldstone bosons.

One of the techniques for the study of approximate spontaneously broken symmetries is that which makes use of non-linear group realizations and of non-linear effective Lagrangians. In these lectures we attempt to give a precise meaning to the effective Lagrangian or, more exactly, to the effective action, as being the functional which generates the one-particle irreducible vertices. We then proceed to apply the method of the effective action to the study of some spontaneously broken symmetries, in particular chiral and conformal symmetry.

From the point of view developed here a non linear effective Lagrangian may bear very little resemblance to the fundamental operator quantum field theory. Furthermore its being highly non-linear and having complicated derivative couplings does not exclude the possibility that the underlying operator theory may be renormalizable. As an example, let us imagine that the fundamental operator theory is one in which quark fields interact through a neutral vector field. Such an underlying theory is renormalizable. As a consequence of the interaction the quarks may bind together forming pions, nucleons, etc. As described in Section 2, we can associate with these particles phenomenological fields and an effective action to be used essentially in the tree approximation. Clearly the effective action which describes the low energy interactions between pions and nucleons bears no resemblance whatsoever to the fundamental quark Lagrangian. The symmetries of the fundamental Lagrangian will, of course, reflect themselves in special features of the effective action. However, the effective action could have symmetries, valid only at relatively low energies, which do not correspond to actual symmetries of the fundamental Lagrangian and are instead of dynamical origin. From this point of view it would be clearly incorrect to use the effective action at high energies of the (real or virtual) particles involved. Nevertheless it can be used for so-called hard-particle calculations at reasonably low energies.

2. Effective Action and Phenomenological Fields

Let $\Phi(x)$ be the renormalized Heisenberg field operator of a renormalizable theory. The τ-functions (time ordered vacuum expectation values)

$$\langle 0 | T (\Phi(x_1), \Phi(x_2), \cdots \Phi(x_n)) | 0 \rangle$$

can be generated by expanding the functional

$$Z[J] = \langle 0 | T e^{i \int d_4x \, J(x) \Phi(x)} | 0 \rangle$$

in a power series of the c-number source $J(x)$.

In perturbation theory these τ-functions correspond to sums of Feynman diagrams. It is not difficult to verify that the connected τ-functions, corresponding to connected Feynman diagrams, are generated in the same sense by a functional $W[J]$ related to $Z[J]$ by

$$Z[J] = e^{iW[J]} \; .$$

A further reduction can be achieved by introducing the one-particle irreducible τ -functions, also called simply vertices. These correspond to connected Feynman diagrams which cannot be separated into two parts by cutting a single line. Clearly all connected diagrams can be constructed by putting together (one-particle irreducible) vertices in structures having the topological character of trees, where the lines of the trees correspond to propagators which include higher order corrections. By definition the irreducible vertices do not have one-particle singularities. The one-particle singularities of a connected diagram are explicitly exhibited as poles of the propagators corresponding to the lines of the associated tree.

The one-particle irreducible vertices can be generated [1] by expanding a functional $A[\varphi]$, connected to the functional $W[J]$ by a functional Legendre transformation

$$W[J] = A[\varphi] + \int d_4x\, J\varphi \ ,$$

$$\frac{\delta W}{\delta J(x)} = \varphi(x) \quad , \quad \frac{\delta A}{\delta \varphi(x)} = -J(x) \ .$$

To show this, let us define the propagagator

$$G(x,x') = \frac{\delta^2 W}{\delta J(x)\,\delta J(x')} = \frac{\delta \varphi(x)}{\delta J(x')}$$

and the kinetic kernel

$$K(x,x') = \frac{\delta^2 A}{\delta \varphi(x)\,\delta \varphi(x')} = -\frac{\delta J(x)}{\delta \varphi(x')} \ .$$

Clearly $\quad \mathbb{G} = -\mathbb{K}^{-1} \ ,$

i.e., the associated integral operators are reciprocal. Using an obvious notation ($J'' \equiv J(x'')$, integrations over space-time not explicitly indicated) we have

$$\frac{\delta G}{\delta J''} = K^{-1}\,\frac{\delta K}{\delta J''}\,K^{-1} = G\,\frac{\delta K}{\delta J''}\,G \ .$$

On the other hand

$$\frac{\delta}{\delta J} = \frac{\delta \varphi}{\delta J}\,\frac{\delta}{\delta \varphi} = G\,\frac{\delta}{\delta \varphi} \ ,$$

so finally we obtain

$$\frac{\delta^3 W}{\delta J(x)\,\delta J(x')\,\delta J(x'')} = \int d_4z\, d_4z'\, d_4z''\, G(x,z)\,G(x',z')\,G(x'',z'')\,\frac{\delta^3 A}{\delta \varphi(z)\,\delta \varphi(z')\,\delta \varphi(z'')} \ ,$$

or, graphically,

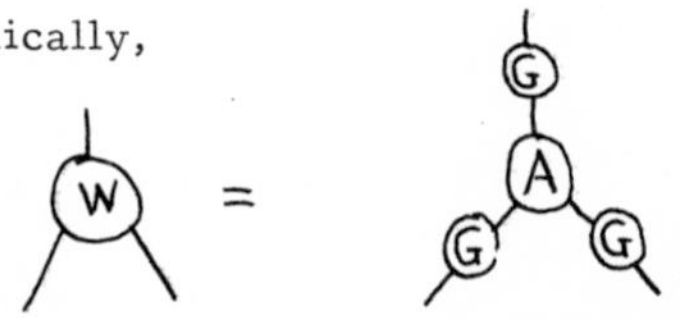

.

Differentiating once more one would find a relation which can be described graphically as

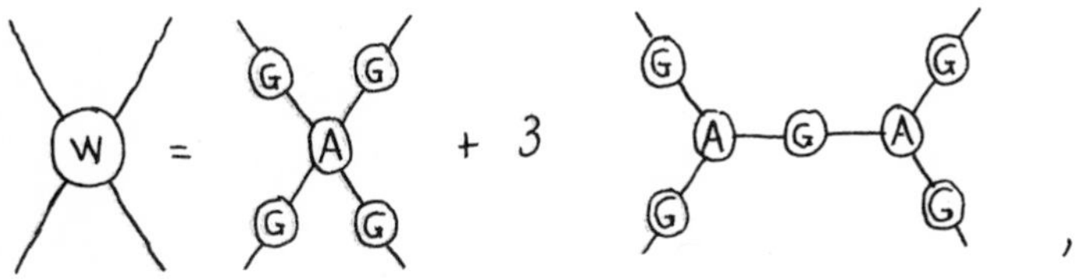

i.e., the connected four-point function can be constructed as a sum of the irreducible four-point vertex (with exact propagators in the external lines) plus three tree diagrams made up with irreducible three point vertices as indicated.

This procedure could be continued to give the one-particle structure of a connected τ -function with an arbitrary number of external lines. By comparison with the diagrams of perturbation theory, for instance, one may then convince oneself that the one particle structure has been fully analyzed, so that the resulting vertices are indeed irreducible.

If the exact functional $A[\varphi]$ were known, the exact τ-functions could be constructed by means of the relations given above. The relations between the connected τ-functions generated by $W[J]$ and the irreducible vertices generated by $A[\varphi]$ are exactly of the same type as those between the τ-functions in the tree approximation and the classical action integral of a given field theory. Because of this fact we shall call the exact $A[\varphi]$ the effective action and the c-number fields such as φ the phenomenological fields. It is clear that, if the dynamics of the theory generates bound states or resonances to which no fundamental quantum field is associated, it is always possible to introduce a local operator and a source to describe that bound state or resonance. By means of a functional Legendre transformation one can then introduce a phenomenological field. The purpose of this procedure is to obtain irreducible vertices which are free of the poles due to that bound state or resonance. These poles appear in the τ-functions through the propagators entering in the tree diagrams.

3. Ward Identities and the Effective Action

Let us consider a renormalizable field theory invariant under a group, which for definiteness we may take to be that of isospin rotations, although it will be clear that the formalism is quite general. In such a field theory there will be a conserved operator isospin current $V_i^{\,\mu}$, $(i=1,2,3)$,

$$\partial_\mu V_i^{\,\mu} = 0 \quad , \qquad \partial_\mu \equiv \frac{\partial}{\partial x^\mu} \quad ,$$

whose components satisfy the equal time commutation relations

$$\left[V_i^{\,0}(\underset{\sim}{x},t) \, , V_j^{\,0}(\underset{\sim}{x}',t) \right] = i\,\epsilon_{ijk}\, V_k^{\,0}(\underset{\sim}{x},t)\, \delta_3(\underset{\sim}{x}-\underset{\sim}{x}') \, ,$$

$$\left[V_i^{\,0}(\underset{\sim}{x},t) \, , V_j^{\,r}(\underset{\sim}{x}',t) \right] = i\,\epsilon_{ijk}\, V_k^{\,r}(\underset{\sim}{x},t)\, \delta_3(\underset{\sim}{x}-\underset{\sim}{x}') + \text{S.T.},$$

$$\epsilon_{123} = +1 \, , \; \epsilon_{321} = -1 \, , \text{ and cyclic permutations} .$$

Here S.T. denotes the so-called Schwinger terms. Denoting collectively with Ψ_a all fundamental field operators occurring in the theory (for instance the pion and nucleon fields) we have also

$$\left[V_i^{\,0}(\underset{\sim}{x},t) \, , \Psi_a(\underset{\sim}{x}',t) \right] = i\, T_{iab}\, \Psi_b(\underset{\sim}{x},t)\, \delta_3(\underset{\sim}{x}-\underset{\sim}{x}') \, ,$$

where the T_i are the matrices representing the generators of isospin transformations.

We are now interested in τ-functions which are vacuum expectation values of a number of fields and currents

$$\langle 0|\, T^*(\Psi_a(x),\ \Psi_b(x'), \ldots V_i^{\mu}(z), V_j^{\nu}(z'), \ldots)\,|0\rangle \ .$$

These can be generated by expanding the functional

$$Z[\eta_a, v_{\mu i}] = \langle 0|\, T^* e^{\, i\int d_4x\,(\eta_a \Psi_a + v_{\mu i} V_i^{\mu})}\, |0\rangle$$

in a power series of the sources η_a and the potentials $v_{\mu i}$, which are numerical functions (the sources for fermion fields should of course be totally anticommuting) . Observe that we have used the T^* product, instead of using the T product. The reason is that the T product is not a covariant concept when the operators involved, like the currents, do not commute at equal times. Without going here into a precise definition of the T^*product, let us only observe that it should be a covariant construction such that the τ-functions involving currents correspond to the actual physical amplitudes relevant to the electromagnetic and weak interactions. The covariant τ-functions satisfy Ward identities (which can be easily derived from the conservation equations and the equal time commutators) to which the Schwinger terms do not contribute.[2]

Let us introduce the functional $W[\eta_a, v_{\mu i}]$ which generates the connected τ-functions

$$Z = e^{iW} \ .$$

The Ward identities can be written compactly as follows (always summing over repeated indices):

$$\left(\partial_\mu \frac{\delta}{\delta v_{\mu i}} + \epsilon_{ijk}\, v_{\mu j} \frac{\delta}{\delta v_{\mu k}} + T_{iab}\, \eta_a \frac{\delta}{\delta \eta_b}\right) W = 0 \ .$$

We now make a functional Legendre transformation with respect to the sources η_a and introduce phenomenological fields ψ_a and an effective action $A[\psi_a, v_{\mu i}]$;

$$W = A + \int d_4x\, \eta_a \psi_a ,$$

$$\frac{\delta W}{\delta \eta_a} = \psi_a , \qquad \frac{\delta A}{\delta \psi_a} = -\eta_a .$$

The Ward identities for W can immediately be transformed into restrictions on A, taking the form

$$\left(\partial_\mu \frac{\delta}{\delta v_{\mu i}} + \epsilon_{ijk} v_{\mu j} \frac{\delta}{\delta v_{\mu k}} - T_{iab} \psi_b \frac{\delta}{\delta \psi_a} \right) A = 0 .$$

This equation gives compactly the Ward identities for the irreducible vertices; it simply states that the effective action is invariant

$$\delta A = 0$$

under the local (Yang-Mills) isospin transformations

$$\delta \psi_a(x) = \Lambda_i(x)\, T_{iab}\, \psi_b(x) ,$$

$$\delta v_{\mu k}(x) = \Lambda_i(x)\, \epsilon_{ikj}\, v_{\mu j}(x) + \partial_\mu \Lambda_k(x) .$$

In other words, the effective action has the invariance expected of a classical action which is the integral of a local Lagrangian invariant under isospin rotations and in which the fields are under the influence of external potentials $v_{\mu i}(x)$. This is generally the case: the correct Ward identities are most easily guessed by extending to the effective action the invariances of the classical theory, except of course when these invariances clash with basic principles such as unitarity, as is the case when anomalies are present.

The treatment given above can easily be extended to more

general groups (replacing ϵ_{ijk} by the structure constants of the group and T_i by the appropriate representation matrices) as well as to partially conserved currents (in which case the right hand sides of the Ward identities will not be zero) .

4. Goldstone's Theorem

As an example of an application of the above relations we prove Goldstone's theorem.[3] In the Ward identities for A set $\mathcal{V}_{\mu i} = 0$ and then integrate over all space time. The result is

$$\int d_4x \, T_{iab} \psi_b \frac{\delta}{\delta \psi_a} A = 0 \, ,$$

since the integral of a divergence vanishes. This equation just states, as expected, that $A[\psi_a] \equiv A[\psi_a, 0]$ is invariant under isospin transformations on ψ_a with constant (x -independent) parameters. Without changing the notation, let us think in terms of a more general invariance group, for which T_i are the matrices which represent the generators in the representation ψ_a . Differentiating the above equation with respect to $\psi_c(x')$ we obtain

$$\int d_4x \, T_{iab} \psi_b \frac{\delta^2 A}{\delta \psi_a \delta \psi_c'} + T_{iac} \frac{\delta A}{\delta \psi_a'} = 0 \, ,$$

or, since

$$K_{ca}(x', x) = \frac{\delta^2 A}{\delta \psi_a \delta \psi_c'} \, ,$$

$$\int d_4x \, K_{ca}(x', x) \, T_{iab} \psi_b(x) - \eta_a(x') \, T_{iac} = 0 \, .$$

Now set $\eta_a = 0$. If the resulting equation for ψ_a ,

$$\frac{\delta A}{\delta \psi_a} = 0 \quad ,$$

admits a non-vanishing solution

$$\psi_a = K_a$$

where the K_a are constants (by translational invariance) , we say that we are in presence of a spontaneously broken symmetry. For $\eta_c = 0$, K_{ca} is a function only of the difference $x' - x$. Going over to momentum space we obtain

$$\tilde{K}_{ca}(p=0)\, T_{iab} K_b = 0 \; .$$

The matrix $\tilde{K}_{ca}(p=0)$ has as many vanishing eigenvalues as there are non-vanishing independent eigenvectors $T_{iab} K_b$, the various eigenvectors being labelled by th index i while the index a gives the various components of each eigenvector. If we remember that the matrix K is the reciprocal of the propagator matrix

$$G_{ca}(x', x) = \frac{\delta^2 W}{\delta \eta_a \delta \eta_c'}$$

we can conclude that the vanishing eigenvalues of $\tilde{K}_{ca}(p=0)$ correspond to massless particles. We obtain in this way Goldstone's theorem: in the case of a broken symmetry there are as many massless particles as "broken" generators (i.e. generators which do not correspond to a symmetry of the solution) .

If in the effective action $A[\psi_a]$ we make the replacement $\psi_a = \mu_a$, where the μ_a are constants, we obtain an expression proportional to the infinite space-time volume

$$A[\mu_a] = F[\mu_a] \int d_4 x \; .$$

Nevertheless the function $F(\mu_a)$ is well defined and must be invariant under group transformations on μ_a . Clearly the equations

$$\frac{\partial F}{\partial \mu_a} = 0$$

admit the non-vanishing solution $\mu_a = K_a$. Furthermore,

$$\frac{\partial^2 F}{\partial \mu_a \partial \mu_b} = \int d_4x \, K_{ab}(x-x') = \tilde{K}_{ab}(p=0) \; .$$

In view of the physical interpretation of $\tilde{K}_{ab}$ one can say that the non-vanishing eigenvalues of $\tilde{K}_{ab}(p=0)$ must be positive. This condition often permits one to choose, among various possible solutions K_a , those which are physically acceptable as corresponding to a positive norm Hilbert space.

5. Non-linear Realizations

Non-linear group realizations arise naturally when a symmetry is spontaneously broken. As an example, consider a renormalizable field theory model, such as the σ -model [4] which involves a σ -field and a $\vec{\varphi}$ field and is invariant under the chiral $SU(2) \times SU(2)$ group ($\vec{\varphi}$ is the pseudoscalar isovector field of the pion and σ that of a scalar isoscalar π-π resonance) . By a chiral transformation of infinitesimal parameter $\vec{\alpha}$, the σ and $\vec{\varphi}$ fields transform into each other according to

$$\delta\sigma = -2\vec{\alpha}\cdot\vec{\varphi} \; ,$$

$$\delta\vec{\varphi} = 2\vec{\alpha}\,\sigma \; .$$

By an isospin transformation, of infinitesimal parameter $\vec{\beta}$, they transform of course as

$$\delta\sigma = 0 \quad ,$$
$$\delta\vec{\varphi} = 2\vec{\beta}\times\vec{\varphi} \quad .$$

The expression

$$\sigma^2 + \vec{\varphi}^{\,2}$$

is invariant. The above transformation laws give an irreducible linear representation of the chiral group extended by the parity operation $\sigma \rightarrow \sigma \; , \; \vec{\varphi} \rightarrow -\vec{\varphi}$.

Now let us consider the case in which the symmetry is spontaneously broken, i.e. the dynamics is such that, for vanishing sources, the fields do not all vanish. Without lack of generality we can take the solution

$$\vec{\varphi} = 0 \, , \quad \sigma = \kappa$$

where κ is a non-vanishing constant, the vacuum expectation value of the σ-field for vanishing sources. Any other constant solution can be transformed into this form by a chiral transformation. In this case it is convenient to define a new σ-field whose vacuum expectation value vanishes :

$$\sigma' = \sigma - \kappa \; .$$

Now, by choosing new fields, which are appropriate non-linear functions of the fields $\vec{\varphi}$ and σ' one may reduce the chiral transformations. For instance take the fields

$$R = \sqrt{(\sigma'+\kappa)^2 + \vec{\varphi}^{\,2}} - \kappa = \sigma' + \cdots \quad ,$$
$$\vec{\pi} = \vec{\varphi}\,\frac{\kappa}{\sqrt{(\sigma'+\kappa)^2 + \vec{\varphi}^{\,2}}} = \vec{\varphi} + \cdots \quad ,$$

where the dots indicate quadratic and higher expressions in the fields. In terms of these new fields the chiral transformation becomes

$$\delta R = 0 \ ,$$

$$\delta \vec{\pi} = 2\vec{\alpha}\sqrt{K^2 - \vec{\pi}^2} = 2\vec{\alpha}K + \cdots \ .$$

The R and $\vec{\pi}$ fields do not mix: we have succeded in reducing the chiral transformation at the price of having the $\vec{\pi}$ field transform non-linearly. The $\vec{\pi}$ field is the massless Goldstone boson of this spontaneously broken symmetry: only the transformation law of a massless field could contain an additive constant like $2\vec{\alpha}K$, since a mass term in the Lagrangian cannot be invariant (nor part of an invariant) under such a transformation. Observe that the isospin transformation law of the new fields is still linear

$$\delta R = 0 \ ,$$

$$\delta \vec{\pi} = 2\vec{\beta} \times \vec{\pi} \ ;$$

the generators which are not broken continue to be represented linearly.

Use of the fields R and $\vec{\pi}$ will give amplitudes which coincide on the mass shell with those calculated using σ' and $\vec{\varphi}$. One may sometime prefer to use still another pion field giving identical results on the mass shell.

For instance, the field

$$\vec{\pi}' = \frac{2\vec{\pi}}{1 + \sqrt{1 - \frac{\vec{\pi}^2}{K^2}}} = \vec{\pi} + \cdots \ , \quad \vec{\pi} = \frac{4\vec{\pi}'}{4 + \frac{\vec{\pi}'^2}{K^2}} = \vec{\pi}' + \cdots$$

transforms under a chiral transformation as

$$\delta \vec{\pi}' = \frac{1}{a}\left[\vec{\alpha}\,(1 - a^2\vec{\pi}'^2) + 2a^2\,\vec{\pi}'(\vec{\alpha}\cdot\vec{\pi}')\right], \quad a \equiv \frac{1}{2K} \ ,$$

and has been often used in the literature. [5]

As indicated by the above example, the use of a non-linear group realization allows one to work with a smaller number of fields. One may then proceed to the construction of invariant effective actions.

For instance, a chiral invariant Lagrangian involving only the pion field $\vec{\pi}'$ (and no σ-field) is

$$L = -\frac{1}{2} \frac{(\partial_\mu \vec{\pi}')^2}{(1 + a^2 \vec{\pi}'^2)^2}$$

For the solution of the general problem of the classification of non-linear realizations of internal symmetry groups and of the construction of invariant Lagrangians we refer to the existing literature. [6]

6. Massive Yang-Mills Fields as Phenomenological Fields.

Let us return to the effective action $A[\eta_a, v_{\mu i}]$ of Section 3. We assume now that the dynamics of the theory generates a vector resonance (the ρ meson) having the same quantum numbers as the currents V_i^μ. The τ-functions will then have poles corresponding to the mass m_ρ of this resonance and it is convenient to introduce effective vertices which are further reduced, by making a further Legendre transformation, this time with respect to the variables $v_{\mu i}$. In view of the difference in structure between a two-point function of currents and a vector propagator, we make the Legendre transformation in the form

$$A[\psi_a, v_{\mu i}] = B[\psi_a, \rho_{\mu i}] - \frac{1}{2} m_\rho^2 \int d_4x \left(\rho_{\mu i} - \frac{1}{g} v_{\mu i}\right)^2,$$

$$\frac{\delta A}{\delta v_{\mu i}} = \frac{m_\rho^2}{g}\left(\rho_i^\mu - \frac{1}{g} v_i^\mu\right), \quad \frac{\delta B}{\delta \rho_{\mu i}} = m_\rho^2\left(\rho_i^\mu - \frac{1}{g} v_i^\mu\right).$$

Clearly, the invariance of A under Yang-Mills transformations on ψ_a and $v_{\mu i}$ (stated in Section 3), implies now the invariance of B,

$$\delta B = 0 \quad ,$$

under the Yang-Mills transformations

$$\delta \psi_a = \Lambda_i T_{iab} \psi_b \quad ,$$

$$\delta \rho_{\mu k} = \Lambda_i \epsilon_{ikj} \rho_{\mu j} + \frac{1}{g} \partial_\mu \Lambda_k \quad .$$

Indeed, the difference

$$\hat{\rho}_{\mu k} = \rho_{\mu k} - \frac{1}{g} v_{\mu k}$$

transforms simply as

$$\delta \hat{\rho}_{\mu k} = \Lambda_i \epsilon_{ikj} \hat{\rho}_{\mu j} \quad ,$$

which shows that the various terms of the Legendre transformation formula are separately invariant. The matrix elements of the current can be obtained from $\frac{m_\rho^2}{g} \hat{\rho}_{\mu i} [\eta, v_\mu]$ by setting $v_\mu = 0$ and putting the external lines on the mass shell. They satisfy a conservation equation. We can also write

$$A[\psi_a, v_{\mu i}] + \frac{1}{2} \frac{m_\rho^2}{g^2} \int d_4x \, v_{\mu i}^2 = B[\psi_a, \rho_{\mu i}] - \frac{1}{2} m_\rho^2 \int d_4x \, \rho_{\mu i}^2 + \frac{m_\rho^2}{g} \int d_4x \, \rho_{\mu i} v^{\mu i} ,$$

from which we see that we can satisfy the Ward identities by using tree diagrams and an effective action which consists of the sum of a Yang-Mills invariant and a mass term which breaks the invariance. Observe, however, the importance of the term $\frac{1}{2} \frac{m_\rho^2}{g^2} \int d_4x \, v_{\mu i}^2$.

The simplest model for the gauge invariant part B of the effective action is obtained by assuming local vertices which are polynomials in the momenta of the fields ψ_a and $\rho_{\mu i}$. In this crude low-energy approximation B has the form of a Yang-Mills action

$$B = \int d_4x \, L_{YM} ,$$

$$L_{YM} = -\tfrac{1}{4} \rho_{\mu\nu i}^{2} + L_{INV}(\psi_a, \rho_{\mu i}) ,$$

$$\rho_{\mu\nu i} = \partial_\mu \rho_{\nu i} - \partial_\nu \rho_{\mu i} + g \epsilon_{ijk} \rho_{\mu j} \rho_{\nu k} ,$$

where L_{INV} is the invariant Lagrangian for the fields ψ_a, constructed in a standard way with the covariant Yang-Mills derivatives.

7. Broken Scale Invariance

We begin by studying a simple model, which was first considered by Nambu and Freund. [7] Let the action be

$$A = \int d_4x \, (L_{INV} + L_{SB}) ,$$

$$L_{INV} = -\tfrac{1}{2}(\partial_\mu \psi)^2 - \tfrac{1}{2} f^2 \psi^2 \varphi^2 - \tfrac{1}{2}(\partial_\mu \varphi)^2 ,$$

$$L_{SB} = \frac{\tau}{4}\left(\frac{\varphi^2}{g^2} - \frac{1}{2}\varphi^4 - \frac{1}{2g^4}\right) , \quad \tau \geq 0 .$$

Here ψ and φ are scalar fields. The parameters f and τ are pure numbers, while g has dimensions of a reciprocal mass, $(\hbar = c = 1)$.

A scale transformation on the fields,

$$\psi(x) \longrightarrow \lambda \psi(\lambda x) ,$$
$$\varphi(x) \longrightarrow \lambda \varphi(\lambda x) ,$$

induces the transformation

$$L_{INV}(x) \longrightarrow \lambda^4 L_{INV}(\lambda x) ,$$

so that the corresponding part of the action is scale invariant. The integral of L_{SB}, however, is not scale invariant. Setting

$$\lambda = 1 + \epsilon ,$$

with ϵ infinitesimal, we can write

$$\delta\psi = \epsilon (x\cdot\partial + 1)\psi , \qquad x\cdot\partial \equiv x^\lambda \frac{\partial}{\partial x^\lambda} ,$$
$$\delta\varphi = \epsilon (x\cdot\partial + 1)\varphi ,$$

and a simple calculation shows that

$$\delta L_{INV} = \epsilon \partial\cdot(x L_{INV})$$

while

$$\delta L_{SB} = \epsilon \partial\cdot(x L_{SB}) + \epsilon \frac{\tau}{2}\left(\frac{\varphi^2}{g^2} - \frac{1}{g^4}\right) .$$

The invariant part of the Lagrangian transforms by the addition of a divergence, but the symmetry breaking part has an extra term. We obtain $\delta A = \int d_4x \, \epsilon \frac{\tau}{2}\left(\frac{\varphi^2}{g^2} - \frac{1}{g^4}\right)$.

With the above action we have

$$\frac{\delta A}{\delta \psi} = \Box \psi - f^2 \varphi^2 \psi ,$$

$$\frac{\delta A}{\delta \varphi} = \Box \varphi - f^2 \psi^2 \varphi + \frac{\tau}{2}\left(\frac{\varphi}{g^2} - \varphi^3\right) .$$

If we set these expressions equal to zero and look for constant solutions we find either $\varphi = 0$ or $\varphi = \pm \frac{1}{g}$. A look at the second derivatives tells us that we must choose one of the latter solutions, for instance $\varphi = \frac{1}{g}$. We are now led to introduce the field

$$\varphi' = \varphi - \frac{1}{g}$$

which, under a scale transformation, transforms as

$$\delta \varphi' = \epsilon\,(x \cdot \partial + 1)\,\varphi' + \frac{\epsilon}{g} .$$

We recognize the additive constant term characteristic of the transformation law of a Goldstone boson.

We could now express the Lagrangian in terms of the field φ'. To establish contact with the work of Nambu and Freund we prefer to introduce the new field

$$\chi = \frac{g^2 \varphi^2 - 1}{2g} = \varphi' + \frac{g}{2}\,\varphi'^2$$

which transforms by the law

$$\delta \chi = \epsilon\,(x \cdot \partial + 2)\,\chi + \frac{\epsilon}{g} .$$

In terms of this field we have

$$L_{INV} = -\frac{1}{2}(\partial_\mu \psi)^2 - \frac{1}{2} m_\psi^2 \psi^2 (1 + 2g\chi) - \frac{1}{2}\frac{(\partial_\mu \chi)^2}{1 + 2g\chi}$$

and

$$L_{SB} = -\frac{1}{2} m_\chi^2 \chi^2$$

where

$$m_\psi^2 = \frac{f^2}{g^2} \quad , \quad m_\chi^2 = \frac{\tau}{g^2} \ .$$

We can now write

$$\delta A = \epsilon \frac{m_\chi^2}{g} \int d_4x \, \chi \ .$$

For $\tau \to 0$, g fixed, $m_\chi \to 0$, and the action becomes invariant. As we remove the symmetry breaking term from the action, the expectation value of the φ field remains equal to $\frac{1}{g}$ and we may take it to be equal to this value in the limit. Observe however that, if we actually set $\tau = 0$ in the action, the dynamical mechanism which fixes the value of the expectation value of the φ field is no longer present (the "potential" becomes a flat function with no minima) . So the situation occurring in the case of scale invariance is quite different from that which is usual in the case of spontaneously broken internal symmetries. There an explicit symmetry breaking may distort the form of the potential, but does not change its general qualitative behaviour. As the symmetry is switched off, the position of the minima will move, but they will not disappear.

In theories where scale invariance is a true symmetry, all masses (and parameters of positive mass dimensions) must vanish. When the symmetry is broken, the symmetry breaking is the sum of terms proportional to the various masses. Approximate scale invariance makes sense only if all the masses are small. On the other hand, from the model described in this section we see that the situation is different when scale invariance is treated as an approximate spontaneously broken symmetry. Only the mass of one scalar field (the Goldstone boson) must be small. The masses of all other fields (represented in the model by the field ψ) are arbitrary. The coupling of the Goldstone boson to any other field is given in terms of a universal constant and of the mass of that field. This universality of the coupling of the Goldstone boson is the most striking consequence of the idea of spontaneously broken scale invariance.

The model of Nambu and Freund has an interesting property, which is in fact the reason they introduced it. Construct the canonical energy momentum tensor associated with the total Lagrangian density

$$L = L_{INV} + L_{SB}$$

which is given by

$$T_{\mu\nu} = \partial_\mu \psi \, \partial_\nu \psi + \partial_\mu \varphi \, \partial_\nu \varphi + \delta_{\mu\nu} L ,$$

and take its trace. The result is

$$T_{\mu\mu} = (\partial_\mu \psi)^2 + (\partial_\mu \varphi)^2 + 4L =$$

$$= -(\partial_\mu \psi)^2 - (\partial_\mu \varphi)^2 - 2f^2\psi^2\varphi^2 + \tau\left(\frac{\varphi^2}{g^2} - \frac{1}{2}\varphi^4 - \frac{1}{2g^4}\right).$$

Now use the equations of motion. With a little algebra it follows that

$$T_{\mu\mu} + \tfrac{1}{2}\Box(\varphi^2 + \psi^2) = \frac{\tau}{2g^4}(g^2\varphi^2 - 1) = \frac{m_\chi^2}{g}\chi .$$

Defining the Callan-Coleman-Jackiw energy momentum tensor [8]

$$\Theta_{\mu\nu} = T_{\mu\nu} - \frac{1}{6}\left(\partial_\mu \partial_\nu - \delta_{\mu\nu} \Box\right)\left(\varphi^2 + \psi^2\right) ,$$

one obtains

$$g\, \Theta_{\mu\mu} = m_\chi^2\, \chi .$$

The Lagrangian of Nambu and Freund provides a simple model for scalar field dominance of the trace of the energy momentum tensor. As we shall see later, this result can be understood from the transformation properties of the action integral under scale transformations.

8. The Fifteen Parameter Conformal Group and the Weyl Transformations.

The Poincaré group (inhomogeneous proper Lorentz group) is a ten parameter group. If one enlarges it by the one parameter scale transformations

$$x^\mu \longrightarrow x^{\mu\prime} = \lambda x^\mu$$

and by the proper conformal transformations, depending upon four parameters a^μ,

$$x^\mu \longrightarrow x^{\mu\prime} = \frac{x^\mu + a^\mu x^2}{1 + 2a\cdot x + a^2 x^2} ,$$

one obtains a fifteen parameter group, the conformal group.[9] In infinitesimal form the scale transformations can be written as

$$\delta x^\mu = x^{\mu\prime} - x^\mu = \epsilon x^\mu \quad , \quad \lambda = 1+\epsilon$$

and the proper conformal transformations as

$$\delta x^\mu = x^{\mu\prime} - x^\mu = a^\mu x^2 - 2 x^\mu a\cdot x \ .$$

Observe that, for scale transformations,

$$\delta x^2 = 2\epsilon x^2$$

and, for proper conformal transformations,

$$\delta x^2 = -2 a\cdot x \, x^2 \ .$$

In either case δx^2 vanishes if x^2 does. This shows that a point on the light cone is transformed into a point on the light cone.

Finite conformal transformations mix up in a complicated way the topology of space time. For instance, take in the above transformation law $a^\mu = (a,0,0,0)$, $x^\mu = (t,0,0,0)$. One finds

$$t' = -\frac{1}{a}\cdot\frac{t}{t-\frac{1}{a}} \ .$$

Take $a>0$; we see that, as $t \to \pm\infty$, $t' \to \left(-\frac{1}{a}\right)_\mp$. Furthermore, for $t \to \left(\frac{1}{a}\right)_\mp$, $t' \to \pm\infty$.

These remarks show that it is not useful to interpret the proper conformal transformations as mappings of space-time. For us they will be instead symmetries of a part of the action, the invariant part, which are violated by the symmetry breaking term, and are used to generate the corresponding Ward identities. The conformal group must then be described as a group of transformations on fields. For scale transformations we have

$$\delta\psi_i = [(\delta x)\cdot\partial + d\epsilon]\psi_i$$

and for proper conformal transformations

$$\delta\psi_i = [(\delta x)\cdot\partial - 2d\,a\cdot x - 2a^\mu x^\nu \Sigma_{\mu\nu}]\psi_i \;.$$

Here ψ_i describes a field, possibly having various spin components (labelled by the index i), d is a number, called the dimension of the field, $\Sigma_{\mu\nu}$ is related to the spin operator and we have used the notation

$$\Sigma_{\mu\nu}\psi_i = (\Sigma_{\mu\nu})_i{}^j\psi_j \;.$$

For instance, for canonical fields, it is

	scalar	spinor	vector
d =	1	$3/2$	1
$(\Sigma_{\mu\nu})_i{}^j$ =	0	$-\frac{i}{2}(\sigma_{\mu\nu})_i{}^j$	$\eta_{\mu i}\delta_\nu^j - \eta_{\nu i}\delta_\mu^j$.

It is useful to study the Ward identities from a more general point of view. Let us introduce an external gravitational potential, described by a metric tensor $g_{\mu\nu}$, and let us write the action in a generally covariant form. It will then be invariant under general coordinate transformations. In infinitesimal form

$$\delta x^\mu = x^{\mu\prime} - x^\mu = \xi^\mu(x)$$

where the ξ^μ are arbitrary infinitesimal functions (which satisfy suitable boundary conditions at infinity) . The corresponding transformations on the external potentials and on the fields are

$$\delta g_{\mu\nu} = \xi^\lambda \partial_\lambda g_{\mu\nu} + \partial_\mu \xi^\lambda g_{\lambda\nu} + \partial_\nu \xi^\lambda g_{\lambda\mu}$$

and

$$\delta \psi = \xi^\lambda \partial_\lambda \psi ,$$

where, for simplicity, we have considered a scalar field only. We shall call these transformations Einstein transformations. The presently accepted theory of gravitation requires that the action be rigorously invariant under them.

Let us now introduce what we shall call Weyl transformations. For the metric tensor

$$\delta g_{\mu\nu}(x) = 2\Lambda(x)\, g_{\mu\nu}(x)$$

and, for a scalar field,

$$\delta \psi(x) = -\Lambda(x)\, \psi(x) .$$

It is easy to see that, if the action is invariant under Einstein and Weyl transformations, then that action, restricted to flat space $(g_{\mu\nu} = \eta_{\mu\nu})$, is invariant under the fifteen parameter conformal group. Let us verify this statement.

For scale transformations choose

$$\xi^\lambda = \epsilon x^\lambda$$

so that the Einstein transformations give

$$\delta g_{\mu\nu} = \epsilon\,(x^\lambda \partial_\lambda g_{\mu\nu} + 2 g_{\mu\nu})$$

$$\delta \psi = \epsilon x^\lambda \partial_\lambda \psi ,$$

and $\Lambda = -\epsilon$

so that the Weyl transformations give

$$\delta g_{\mu\nu} = -2\epsilon g_{\mu\nu}$$

$$\delta\psi = \epsilon\psi .$$

Combining these transformations we see that the action must be invariant under

$$\delta g_{\mu\nu} = \epsilon x^\lambda \partial_\lambda g_{\mu\nu}$$

$$\delta\psi = \epsilon(x^\lambda\partial_\lambda + 1)\psi .$$

Now, if $g_{\mu\nu} = \eta_{\mu\nu}$, then $\delta g_{\mu\nu} = 0$. Therefore the action restricted to flat space is invariant under the scale transformation on ψ. For proper conformal transformations the proof proceeds in the same way with the choice

$$\xi^\lambda = a^\lambda x^2 - 2x^\lambda a\cdot x \ , \quad \Lambda = 2a\cdot x \ .$$

Instead of working with the fifteen parameter conformal group, we can take an action which is invariant under Einstein transformation and study the restrictions imposed by (approximate) invariance under Weyl transformations. It is obvious that a mass term such as

$$-\frac{1}{2}\int d_4x \sqrt{g}\, m_\psi^2 \psi^2 \ , \quad g = -\det g_{\mu\nu}$$

is not invariant under Weyl transformations, since

$$\delta\sqrt{g} = 4\Lambda\sqrt{g} \ , \quad \delta\psi^2 = -2\Lambda\psi^2 .$$

We may, however, introduce a Goldstone boson χ. Its transformation law under Weyl transformations may be inferred from that of the field φ

$$\delta\varphi = -\Lambda\varphi$$

and from the relation (see Section 7. There the constant b was called g)

$$b^2\varphi^2 = 1 + 2b\chi$$

We obtain

$$\delta\chi = -2\Lambda\chi - \frac{\Lambda}{b} .$$

It is sometimes preferable to introduce a new field σ, related to χ by an allowed field transformation

$$1 + 2b\chi = e^{2b\sigma}$$

so that

$$b\varphi = e^{b\sigma} .$$

It transforms as

$$\delta\sigma = -\frac{\Lambda}{b} .$$

Using for instance the field σ one can write the expression

$$-\frac{1}{2}\int d_4x\sqrt{g}\; m_\psi^2\,\psi^2 e^{2b\sigma}$$

which is invariant under Weyl transformations. Expanding the exponential one obtains a mass term as well as interaction terms of which the simplest is the trilinear interaction

$$-b\,m_\psi^2\int d_4x\sqrt{g}\;\psi^2\sigma ,$$

with a coupling given in terms of a universal constant and of the mass square of the particle to which the Goldstone boson couples. This agrees with a result obtained in Section 7.

9. Conservation Identities and Trace Identities

The action must be invariant ,

$$0 = \delta A = \int d_4x \left(\frac{\delta A}{\delta g_{\mu\nu}} \delta g_{\mu\nu} + \frac{\delta A}{\delta \psi} \delta \psi \right) ,$$

under Einstein transformations

$$\delta g_{\mu\nu} = \xi^\lambda \partial_\lambda g_{\mu\nu} + \partial_\mu \xi^\lambda g_{\lambda\nu} + \partial_\nu \xi^\lambda g_{\mu\lambda}$$

$$\delta \psi = \xi^\lambda \partial_\lambda \psi .$$

Since ξ^λ is arbitrary, and with the definition

$$T^{\mu\nu} \sqrt{g} = 2 \frac{\delta A}{\delta g_{\mu\nu}} ,$$

one obtains

$$\partial_\mu \left(T^\mu{}_\lambda \sqrt{g} \right) - \frac{1}{2} \left(\partial_\lambda g_{\mu\nu} \right) T^{\mu\nu} \sqrt{g} - \partial_\lambda \psi \frac{\delta A}{\delta \psi} = 0.$$

Upon using the field equations for ψ ,

$$\frac{\delta A}{\delta \psi} = 0 ,$$

this relation becomes, as is well known, the covariant conservation equation for the energy momentum tensor, which is an ordinary conservation equation in flat space $(g_{\mu\nu} = \eta_{\mu\nu})$. On the other hand we may interpret A as the effective action which generates the one particle irreducible vertices with a certain number of external ψ lines and a certain number of external lines associated with the external gravitational potential. The above relation generates then, by differentiation with respect to ψ and to $g_{\mu\nu}$, an infinite set of Ward identities for the irreducible vertices. The analogy with the developments of Section 3 is obvious. Here we postulate the result, on physical grounds, rather than attempting to derive it from equal time commutators and operator conservation equations. The above Ward identities will be called conservation identities.

We consider now the change in the action

$$\delta A = \int d_4x \left(\frac{\delta A}{\delta g_{\mu\nu}} \delta g_{\mu\nu} + \frac{\delta A}{\delta \psi} \delta \psi \right)$$

under a Weyl transformation

$$\delta g_{\mu\nu} = 2 \Lambda g_{\mu\nu}$$

$$\delta \psi = - \Lambda \psi$$

(the $g^{\mu\nu}$ transform with the opposite sign to the $g_{\mu\nu}$ since the two are reciprocal matrices). Since Λ is arbitrary, we obtain

$$T^{\mu\nu} g_{\mu\nu} \sqrt{g} - \psi \frac{\delta A}{\delta \psi} = \frac{\delta A}{\delta \Lambda}$$

If we use in this relation the field equations for ψ, we obtain a connection between the trace of the energy momentum tensor and the change in the symmetry breaking part of the action. On the other hand, if we interpret again A as the effective action, the above relation gives an infinite set of identities among the irreducible vertices. We shall call them trace identities.

The trace identities are analogous to the Ward identities for the axial vector current. They correspond to a broken symmetry which, in the limit, becomes spontaneously broken. Furthermore if one attempts to verify them in specific renormalizable field theory models, one seems to find anomalies.[10] No anomalies should arise in the conservation identities.

10. Invariant Actions

The minimal action for a massless scalar field which is invariant under Einstein transformations

$$A = -\frac{1}{2}\int d_4x\sqrt{g}\; g^{\mu\nu}\partial_\mu\psi\,\partial_\nu\psi$$

is not invariant under Weyl transformations. Instead, as one can easily verify,

$$\delta A = \int d_4x\sqrt{g}\; g^{\mu\nu}\partial_\mu\psi\;\psi\partial_\nu\Lambda =$$

$$= -\frac{1}{2}\int d_4x\sqrt{g}\;\psi^2\,\Box\Lambda = -\frac{1}{2}\int d_4x\sqrt{g}\;\Lambda\,\Box\psi^2 ,$$

where

$$\Box\Lambda = \frac{1}{\sqrt{g}}\,\partial_\mu\left(g^{\mu\nu}\sqrt{g}\,\partial_\nu\Lambda\right) .$$

From the arguments of the previous section we see that, with

$$T^{\mu\nu}\sqrt{g} = 2\,\frac{\delta A}{\delta g_{\mu\nu}} ,$$

the field equations for ψ imply

$$T^{\mu\nu}g_{\mu\nu} = -\frac{1}{2}\,\Box\psi^2 .$$

On the other hand, if one adds to A a term

$$A' = \frac{1}{12}\int d_4x\sqrt{g}\;R\,\psi^2 ,$$

the total action $A + A'$ is invariant under Weyl transformations. This follows immediately from the transformation law

$$\delta R = -2\Lambda R + 6\Box\Lambda$$

which can be derived from the definition of the scalar curvature R. (We recall that

$$\Gamma^{\lambda}_{\mu\nu} = \frac{1}{2} g^{\lambda\sigma}(\partial_\mu g_{\sigma\nu} + \partial_\nu g_{\sigma\mu} - \partial_\sigma g_{\mu\nu}) = \Gamma^{\lambda}_{\nu\mu} ,$$

$$y^{\rho}_{\mu\nu\lambda} = \partial_\lambda \Gamma^{\rho}_{\mu\nu} + \Gamma^{\rho}_{\lambda\tau} \Gamma^{\tau}_{\mu\nu} ,$$

$$B^{\rho}_{\mu\nu\lambda} = y^{\rho}_{\mu\nu\lambda} - y^{\rho}_{\mu\lambda\nu} ,$$

$$R_{\mu\nu} = B^{\lambda}_{\mu\lambda\nu}$$

and

$$R = g^{\mu\nu} R_{\mu\nu} .)$$

If we define a new energy momentum tensor [8]

$$\theta^{\mu\nu}\sqrt{g} = 2\,\frac{\delta(A+A')}{\delta g_{\mu\nu}} ,$$

it follows that it is traceless

$$\theta^{\mu\nu} g_{\mu\nu} = 0 .$$

These considerations throw light on corresponding results of Section 7. Observe that, in flat space, $R = 0$. Therefore the action $A + A'$ agrees with the usual choice of the action in absence of a gravitational field.

The usual action for a massless vector field v_μ ,

$$A = -\frac{1}{4}\int d_4x\sqrt{g}\; v_{\mu\nu} v_{\lambda\rho}\, g^{\mu\lambda} g^{\nu\rho}\,, \quad v_{\mu\nu} = \partial_\mu v_\nu - \partial_\nu v_\mu\,,$$

is automatically invariant under Weyl transformations provided one attributes to the covariant vector field the transformation law

$$\delta v_\mu = 0\ .$$

In order to study the case of spin one-half we must learn to write the Dirac action in a curved space. This will be done in the next section. The equations for fields of spin higher than one in curved space present consistency difficulties and will not be discussed in these lectures.

Finally, an invariant action for the Goldstone field σ itself can be easily written if one remembers that the field

$$\varphi = \frac{1}{b} e^{b\sigma} = \frac{1}{b} + \sigma + \cdots$$

transforms like an ordinary field

$$\delta\varphi = -\Lambda\varphi\,.$$

The developments at the beginning of this section show that we may take

$$A = -\frac{1}{2}\int d_4x\sqrt{g}\; g^{\mu\nu}\partial_\mu\varphi\,\partial_\nu\varphi + \frac{1}{12}\int d_4x\sqrt{g}\; R\varphi^2$$

replacing for φ its expression in terms of σ By expanding the exponential we see that this action has the form

$$A = -\frac{1}{2}\int d_4x\sqrt{g}\; g^{\mu\nu}\partial_\mu\sigma\,\partial_\nu\sigma + \frac{1}{12b^2}\int d_4x\sqrt{g}\; R\left(1+2b\sigma+2b^2\sigma^2\right) + \cdots$$

where the dots denote terms of higher order in σ. The field σ is massless, unless the Weyl symmetry is explicitly broken.

It has been suggested incorrectly that the σ field can have a mass in the symmetric limit because a mass term is contained in the Weyl invariant action density :

$$-\frac{1}{16}\frac{m_\sigma^2}{b^2}\sqrt{g}\,e^{4b\sigma}$$

It is true that by expanding the exponential

$$e^{4b\sigma} = 1 + 4b\sigma + 8b^2\sigma^2 + \cdots ,$$

the quadratic term looks like a mass term for the field σ. This action density, however, is not acceptable since it does not vanish at infinity in space-time, where the field σ vanishes. If one modifies it by subtracting, for instance, its value at infinity, then it is no longer Weyl invariant. To generate a mass for the σ field it is necessary to add to the action an explicit symmetry breaking term, specified by its transformations. One could take, for instance, the symmetry breaking term of the Nambu Freund model which is, written in terms of the σ field,

$$-\frac{1}{2}m_\sigma^2\left(\frac{e^{2b\sigma}-1}{2b}\right)^2\sqrt{g} .$$

11. The Action for a Spin One-half Field

For a spin one-half field under the influence of a gravitational potential we follow H. Weyl's method.[11] The points of space-time are parametrized by introducing curvilinear coordinates x^μ. At each point of space-time a local Lorentz system is introduced, specified by four orthonormal vectors (a "vierbein") $e^a{}_\mu(x)$. The Lorentz index a(=0, 1, 2, 3) labels the four vectors, the index μ (=0, 1, 2, 3) their components along the curvilinear coordinates. A contravariant vector, such as the infinitesimal displacement dx^μ, can also be described by giving its components in the local Lorentz frame

$$dx^a = e^a{}_\mu\, dx^\mu \ .$$

The length of this vector is

$$ds^2 = dx^a \eta_{ab} dx^b$$

where η_{ab} is, as usual, the diagonal Lorentz metric tensor (-1, 1, 1, 1) . Substituting the expression for dx^a one can also write

$$ds^2 = dx^\mu g_{\mu\nu} dx^\nu$$

with

$$g_{\mu\nu} = e^a{}_\mu\, \eta_{ab}\, e^b{}_\nu \ .$$

So the vierbein field $e^a{}_\mu(x)$ is a kind of square root of the matrix $g_{\mu\nu}$. No symmetry property is assumed for $e^a{}_\mu$ however. Given $g_{\mu\nu}$, $e^a{}_\mu$ is not uniquely determined, since a Lorentz transformation with respect to the index a leaves $g_{\mu\nu}$ invariant.

We introduce the matrix $e_a{}^\mu$, which is the reciprocal transposed of $e^a{}_\mu$,

$$e_a{}^\mu e^b{}_\mu = \delta_a{}^b \ , \quad e^a{}_\mu e_a{}^\nu = \delta_\mu{}^\nu \ .$$

Clearly

$$g^{\mu\nu} = e_a{}^{\mu} \eta^{ab} e_a{}^{\nu} .$$

Lorentz indices can be raised or lowered by means of the Lorentz metric tensor

$$e^{a\mu} = \eta^{ab} e_b{}^{\mu} \quad , \quad e_{a\mu} = \eta_{ab} e^b{}_{\mu} .$$

It is easy to verify that this is consistent with the raising and lowering of greek indices by means of the metric tensor

$$e^{a\mu} = e^a{}_{\nu} g^{\nu\mu} \quad , \quad e_{a\mu} = e_a{}^{\nu} g_{\nu\mu} .$$

Observe that

$$-g = \det g_{\mu\nu} = -(\det e^a{}_{\mu})^2 ,$$

so that

$$\sqrt{g} = e \equiv \det e^a{}_{\mu} .$$

With respect to the index μ , $e^a{}_{\mu}$ transforms like a covariant vector

$$\delta e^a{}_{\mu} = \xi^{\lambda} \partial_{\lambda} e^a{}_{\mu} + \partial_{\mu} \xi^{\lambda} e^a{}_{\lambda} ,$$

and $e_a{}^{\mu}$ like a contravariant vector

$$\delta e_a{}^{\mu} = \xi^{\lambda} \partial_{\lambda} e_a{}^{\mu} - \partial_{\lambda} \xi^{\mu} e_a{}^{\lambda} .$$

With respect to the index a they transform like Lorentz vectors. So

$$\delta e^a{}_{\mu} = \theta^a{}_b e^b{}_{\mu} \quad , \quad \theta_{ab} = -\theta_{ba} ,$$

where $\theta_{ab}(x)$ are the infinitesimal parameters of a local Lorentz transformation.

The gravitational field (and the geometrical properties of space-time) can be described by means of the vierbein field. The action must be invariant with respect to general coordinate (Einstein) transformations as well as with respect to local Lorentz transformations. Let $A[e_{a\mu},\psi]$ be the action for the field ψ in an external field $e_{a\mu}$. Invariance under Einstein transformations requires that

$$0 = \delta A = \int d_4x \left[\frac{\delta A}{\delta e^a_{\ \mu}} \left(\xi^\lambda \partial_\lambda e^a_{\ \mu} + \partial_\mu \xi^\lambda e^a_{\ \lambda} \right) + \frac{\delta A}{\delta \psi} \delta\psi \right] .$$

If we use the field equations

$$\frac{\delta A}{\delta \psi} = 0$$

and the definitions

$$e\, T_a^{\ \mu} = \frac{\delta A}{\delta e^a_{\ \mu}} \quad , \quad T_\lambda^{\ \mu} = e^a_{\ \lambda} T_a^{\ \mu} \quad ,$$

we obtain the covariant conservation equation

$$\partial_\mu (e T_\lambda^{\ \mu}) - \partial_\lambda e^a_{\ \mu}\, e T_a^{\ \mu} = 0 .$$

On the other hand invariance under local Lorentz transformations requires

$$0 = \delta A = \int d_4x \left(\frac{\delta A}{\delta e^a_{\ \mu}} \theta^a_{\ b} e^b_{\ \mu} + \frac{\delta A}{\delta \psi} \delta\psi \right) .$$

If we use again the field equations for ψ, and remembering that $\theta_{ab}(x)$ is arbitrary (except for its antisymmetry), we obtain

$$T_a^{\ \mu} e^b_{\ \mu} = T_b^{\ \mu} e^a_{\ \mu} .$$

Therefore the Lorentz tensor

$$T_{ab} = T_a^{\ \mu} e_{b\mu}$$

is symmetric, and so is the Einstein tensor

$$T^{\nu\mu} = e^{\nu a} T_a{}^{\mu} \quad .$$

Using this symmetry one can verify that, for any infinitesimal variation

$$\delta g_{\rho\sigma} T^{\rho\sigma} = 2 \delta e^a{}_\rho T_a{}^\rho \quad .$$

This shows that the tensor $T^{\mu\nu}$ constructed above agrees with that defined in Section 9, and also (taking $\delta g_{\rho\sigma} \propto \partial_\lambda g_{\rho\sigma}$) that the covariant conservation equation derived for it agrees with that in Section 9. It is also clear that the above equations can be interpreted as Ward identities for the irreducible vertices, in analogy with the discussion given in Section 9.

We now proceed to construct the invariant action. Let ψ be a Dirac spinor defined with reference to the local Lorentz frame. Under Einstein transformations it behaves like a scalar

$$\delta\psi = \xi^\lambda \partial_\lambda \psi \,,$$

while under Lorentz transformations it behaves like a spinor

$$\delta\psi = \frac{i}{2} \theta_{ab} S^{ab} \psi \,, \quad S^{ab} = \frac{1}{4i}(\gamma^a\gamma^b - \gamma^b\gamma^a) \,.$$

Here γ^a are the usual constant gamma matrices which satisfy

$$\gamma^a\gamma^b + \gamma^b\gamma^a = 2\eta^{ab}$$

so that S^{ab} are the spin matrices (for instance $S^{12} = \frac{1}{2}\sigma^3$). We need a covariant derivative $D_\mu\psi$ having the property that, under Einstein transformations, it transforms as a covariant vector, while under local Lorentz transformations it transforms like the spinor ψ itself

$$\delta(D_\mu\psi) = \frac{i}{2}\theta_{ab}S^{ab}(D_\mu\psi)\,.$$

One can verify that it is given by

$$D_\mu\psi = (\partial_\mu - \frac{i}{2}\omega_{\mu ab}S^{ab})\psi,$$

where

$$\omega_{\mu ab} = e^c{}_\mu\,\omega_{cab}$$

$$\omega_{abc} = \frac{1}{2}(\Omega_{bca} + \Omega_{cab} - \Omega_{abc})$$

$$\Omega_{cab} = e_{c\nu}(e_a{}^\mu\partial_\mu e_b{}^\nu - e_b{}^\mu\partial_\mu e_a{}^\nu) = -\Omega_{cba}.$$

Using the covariant derivative the Dirac action can be written as

$$A = i\int d_4x\, e\left[\frac{1}{2}\psi^\dagger\gamma_0\gamma^\mu D_\mu\psi - \frac{1}{2}(D_\mu\psi)^\dagger\gamma_0\gamma^\mu\psi + m\psi^\dagger\gamma_0\psi\right]$$

where γ_0 equals γ_a for $a=0$ $(\gamma_0{}^2=-1)$, the $\dagger$ denotes complex conjugation, and the matrices $\gamma^\mu \equiv \gamma^a e_a{}^\mu$ satisfy

$$\gamma^\mu\gamma^\nu + \gamma^\nu\gamma^\mu = 2g^{\mu\nu}\,.$$

The above action is symmetrized so that the action density is real. Integrating by parts one can also write

$$A = i\int d_4x\, e\left[\psi^\dagger\gamma_0\gamma^\mu D_\mu\psi + m\psi^\dagger\gamma_0\psi\right].$$

To prove this, one must use

$$(S^{ab})^\dagger\gamma_0 = -\gamma_0 S^{ab},$$

where $\dagger$ denotes now hermitian conjugation, as well as the relations

$$D_\mu e \equiv \partial_\mu e = e\,\Gamma^{\rho}_{\rho\mu}\,,$$

and

$$D_\mu(\gamma^\lambda\psi) \equiv \left(\partial_\mu - \tfrac{i}{2}\,\omega_{\mu ab}\,S^{ab}\right)(\gamma^\lambda\psi) + \Gamma^{\lambda}_{\mu\nu}\,\gamma^\nu\psi = \gamma^\lambda D_\mu\psi\,.$$

Variation with respect to $\psi^\dagger$ gives immediately the Dirac equation

$$(\gamma^\mu D_\mu + m)\psi = 0$$

We wish now to study the behaviour of the Dirac action under Weyl transformations, which must be defined here as

$$\delta e^{a}{}_{\mu} = \Lambda e^{a}{}_{\mu}\,, \quad \delta e_{a}{}^{\mu} = -\Lambda e_{a}{}^{\mu}\,,$$

$$\delta\psi = -\tfrac{3}{2}\Lambda\psi\,.$$

We obtain successively

$$\delta\Omega_{abc} = -\Lambda\,\Omega_{abc} - (\eta_{ac}e_b{}^\mu - \eta_{ab}e_c{}^\mu)\partial_\mu\Lambda\,,$$

$$\delta\omega_{abc} = -\Lambda\,\omega_{abc} + (\eta_{ac}e_b{}^\mu - \eta_{ab}e_c{}^\mu)\partial_\mu\Lambda\,,$$

$$\delta\omega_{\mu bc} = (e_{c\mu}e_b{}^\rho - e_{b\mu}e_c{}^\rho)\partial_\rho\Lambda\,,$$

so that

$$\delta(D_\mu\psi) = -\tfrac{3}{2}\Lambda D_\mu\psi - \tfrac{1}{4}(\gamma^\rho\gamma_\mu - \gamma_\mu\gamma^\rho)\psi\,\partial_\rho\Lambda - \tfrac{3}{2}\psi\,\partial_\mu\Lambda\,.$$

If we observe that $\delta\gamma^\mu = -\Lambda\gamma^\mu$

and make use of the identity

$$\gamma^\mu(\gamma^\rho\gamma_\mu - \gamma_\mu\gamma^\rho) + 6\gamma^\rho = 0$$

we see that

$$\delta(\gamma^\mu D_\mu \psi) = -\tfrac{5}{2}\Lambda(\gamma^\mu D_\mu \psi).$$

It follows then immediately that the Dirac action is invariant under Weyl transformations, except for the mass term.[12]

We can, of course, use again the Goldstone field σ and generalize the mass term to the Weyl invariant form

$$im\int d_4x\, e\, \psi^+\gamma_0\psi\, e^{b\sigma}.$$

Expanding the exponential we see that there is now a trilinear coupling

$$imb\int d_4x\, e\, \psi^+\gamma_0\psi\,\sigma,$$

proportional to the mass of the spinor field.

12. Chiral and Conformal Invariant Actions

First a general remark. If $A[g_{\mu\nu}, \psi]$ is invariant under Einstein transformations, then

$$A[g_{\mu\nu}e^{2b\sigma}, \psi e^{-db\sigma}]$$

is clearly invariant under Einstein transformations as well as Weyl transformations. Here d is the weight of the field ψ

$$\delta\psi = -d\Lambda\psi$$

(The analogous statement when spinors and the vierbein field are present is obvious). Vice versa, any expression involving the fields $g_{\mu\nu}$, ψ, and σ invariant under

Einstein and Weyl transformations must have the above form. Indeed, let

$$A[g_{\mu\nu}, \psi, \sigma] = A[g_{\mu\nu} e^{2\Lambda}, \psi e^{-d\Lambda}, \sigma - \tfrac{1}{b}\Lambda]$$

Choosing $\Lambda = b\sigma$ we verify the statement.

A further simple observation is that every local Lorentz invariant action can be extended, by the introduction of the field σ, to an action invariant under the fifteen parameter conformal group. [13] All one needs to do is to extend it first to an action invariant under Einstein transformations, by the introduction of the $g_{\mu\nu}$ field (or the vierbein field); one then extends it to a Weyl invariant action, by the introduction of the σ field, and finally one restricts it to flat space. According to our results of Section 8 the final action is invariant under the fifteen parameter conformal group. Clearly this procedure is equivalent to working in a curved space of the special form

$$g_{\mu\nu} = \eta_{\mu\nu} e^{2b\sigma} \quad (\text{or} \quad e^{a}{}_{\mu} = \delta^{a}{}_{\mu} e^{b\sigma})$$

i.e. a conformally flat space. Incidentally, these remarks indicate that the consistency difficulties which beset the equations for fields of spin higher than one in an external gravitational field will also be encountered if one attempts to write for such fields equations which are invariant under the fifteen parameter conformal group. Conformally flat spaces are still general enough to cause trouble.

Using the procedure described above, a chiral invariant action can be easily made chiral and conformal invariant, provided the field σ is taken to be a chiral scalar.

Consider, for instance, the non linear chiral Lagrangian for SU(2) × SU(2)

$$- \frac{1}{2} (\partial_\mu \vec{\pi})^2 - \frac{1}{2} \left(\partial_\mu \sqrt{K^2 - \vec{\pi}^2} \right)^2 =$$

$$= - \frac{1}{2} (\partial_\mu \vec{\pi}^2) - \frac{1}{2} \frac{(\vec{\pi} \cdot \partial_\mu \vec{\pi})^2}{K^2 - \vec{\pi}^2}$$

which is invariant under the non-linear realization

$$\delta \vec{\pi} = 2\vec{\alpha} \sqrt{K^2 - \vec{\pi}^2}$$

(when no confusion can arise, we do not introduce explicitly the Lorentz matrix $\eta_{\mu\nu}$). The substitutions outlined above give a conformal invariant action having as Lagrangian density

$$L = - \frac{1}{2} \left(\partial_\mu \vec{\pi} - b \partial_\mu \sigma \vec{\pi} \right)^2 - \frac{1}{2} \frac{\left[\vec{\pi} \cdot (\partial_\mu \vec{\pi} - b \partial_\mu \sigma \vec{\pi}) \right]^2}{K^2 e^{2b\sigma} - \vec{\pi}^2} - \frac{1}{2b^2} \left(\partial_\mu e^{b\sigma} \right)^2 .$$

Here we have also added the Lagrangian for the field σ itself as described in Section 10.

We have taken the weight d of the field $\vec{\pi}$ to be equal to one. A simpler procedure, which is preferable in general, and gives identical results on the mass shell, is to use fields of weight zero. It would correspond here to the use of the field

$$\vec{\pi}' = \vec{\pi} e^{-b\sigma}$$

In the particular case under consideration we prefer to keep the field $\vec{\pi}$ of weight one, in order to exhibit a simple alternative form for the Lagrangian. It is easy to verify that it is identical to $L = -\frac{1}{2} (\partial_\mu \vec{\pi})^2 - \frac{1}{2} \left(\partial_\mu \sqrt{r^2 \varphi^2 - \vec{\pi}^2} \right)^2 - \frac{1}{2} (1 - r^2) (\partial_\mu \varphi)^2 ,$

where

$$r = Kb \quad , \quad \varphi = \frac{1}{b} e^{b\sigma} \quad .$$

This form of L could have been obtained directly from the original chiral Lagrangian. The simple replacement $K \to r\varphi$ would give an incorrectly normalized kinetic term for the φ field, unless $r = 1$. The last term

$$-\frac{1}{2}(1-r^2)(\partial_\mu \varphi)^2 \quad ,$$

which is invariant by itself, reestablishes the correct normalization of the kinetic term. [14] The value of the parameter r is not determined by invariance arguments.

We must now choose convenient symmetry breaking terms. From our experience with chiral symmetry breaking we are led to add to the Lagrangian L the expression

$$L' = -\frac{1}{4}\lambda_0 r^4 \varphi^4 - c(b\varphi)^{w-1}\sqrt{r^2\varphi^2 - \vec{\pi}^2} \quad + \text{const} \; .$$

Here the second term transforms like the fourth component of a chiral four vector and has conformal weight w. The first term is invariant under both groups. The constant breaks the conformal symmetry; it is needed because the field φ does not vanish at infinity in space time, to guarantee that the Lagrangian density does, but it has no observable consequences. Writing

$$\varphi = \frac{1}{b} + \varphi' \quad ,$$

and expanding in the field φ', we obtain to lowest order the old chiral Lagrangian for the field $\vec{\pi}$. We identify therefore

$$K = \frac{r}{b} = \frac{1}{2} F_\pi$$

where the pion decay constant is

$$F_\pi \approx 190 \text{ MeV} .$$

The coefficient of the term linear in φ' must vanish, to ensure that we have properly identified the vacuum expectation value of the field φ. This give the relation

$$\lambda_0 \frac{r^4}{b^3} + crw = 0 .$$

The terms quadratic in the fields determine the respective masses. We find

$$m_\pi^2 = -c\frac{b}{r} , \qquad m_{\varphi'}^2 = crb\,w(w-4)$$

which gives

$$m_{\varphi'}^2 = m_\pi^2 r^2 w(4-w) .$$

In particular, for $r = w = 1$,

$$m_{\varphi'}^2 = 3m_\pi^2 .$$

This result should not be unexpected. It is easy to see that, for $r = w = 1$, our Lagrangian is identical with that of the "linear" σ model with vanishing bare mass, which is known to have the above mass relation as a consequence. In order to make the identification one needs only to express the Lagrangian in terms of the fields $\vec{\pi}$ and $\sqrt{\varphi^2 - \vec{\pi}^2}$. However, for $r \neq 1$, or $w \neq 1$, our model is truly non-linear.

It is interesting to work out the trilinear $\sigma\pi\pi$ interaction. Expanding in φ' one finds

$$b\,\partial_\mu \varphi'\,\vec{\pi}\cdot\partial_\mu\vec{\pi} \quad - (w-2)\,\tfrac{1}{2}\,b\,m_\pi^2\,\varphi'\,\vec{\pi}^{\,2}$$

which, by partial integration, is equivalent to

$$-\tfrac{1}{2}\,b\,\Box\varphi'\,\vec{\pi}^{\,2} - (w-2)\,\tfrac{1}{2}\,b\,m_\pi^2\,\varphi'\,\vec{\pi}^{\,2} \;.$$

In the low energy limit the first term vanishes. However, if the particles are on the mass shell the above expression becomes

$$-\tfrac{1}{2}b\left[m_{\varphi'}^2 + (w-2)m_\pi^2\right]\varphi'\,\vec{\pi}^{\,2} \;.$$

The effective Lagrangian method allows one to do a little better than one does with simple low energy theorems. Actually the above result does not depend upon the particular SU(2) × SU(2) transformation property of the symmetry breaking. To see this, let us use the weight zero pion field $\vec{\pi}'$ defined earlier. We assume only that the term which breaks chiral symmetry has conformal weight w. Expanding in powers of the field $\vec{\pi}'$ we find

$$L = -\tfrac{1}{2}(\partial_\mu\vec{\pi}')^2 e^{2b\sigma} - \tfrac{1}{2}m_\pi^2\,\vec{\pi}'^{\,2}\,e^{wb\sigma} + \dots$$

$$= -\tfrac{1}{2}(\partial_\mu\vec{\pi}')^2 - \tfrac{1}{2}m_\pi^2\,\vec{\pi}'^{\,2} - b\sigma(\partial_\mu\vec{\pi}')^2 - \tfrac{1}{2}wbm_\pi^2\,\sigma\,\vec{\pi}'^{\,2} + \dots$$

On the mass shell the trilinear interaction terms agree with the result given before.

The simple SU(2) × SU(2) model studied in this section is probably not sufficiently realistic. The particle associated with the field σ (or φ') should be identified with the $\epsilon(700)$ resonance and its large mass forces a rather large value for r and consequently for b. At the SU(2) × SU(2) level it is probably necessary to introduce in the Lagrangian an additional term, proportional to φ'^2, which breaks only conformal invariance and contributes to the ϵ mass. There are, however, some qualitative aspects of the model which are of interest. In particular the derivative coupling in the $\epsilon\pi\pi$ interaction, which seems to be needed in a phenomenological description of pion-nucleon scattering [15] arises here in a very natural way. The main difficulty still seems to be the magnitude of the ϵ-nucleon coupling constant, which is predicted here to be bm_N, so that its ratio to the effective $\epsilon\pi\pi$ coupling constant should be

$$\frac{g_{\epsilon NN}}{g_{\epsilon\pi\pi}} = \frac{2m_N}{m_\epsilon^2 + (w-2)m_\pi^2} .$$

Most of the estimates of $g_{\epsilon NN}$ (from nucleon-nucleon scattering) are too low. However, the corresponding estimates of the ϵ mass are also too low (giving about 400 Mev) and more recent estimates seem to give larger values for the coupling constant, together with larger values (closer to 700 Mev) for the ϵ mass. [16] The situation changes, of course, if σ corresponds to a linear combination of several scalar particles.

13. SU(3) × SU(3) and Conformal Invariance.

The chiral group SU(3) × SU(3) can be realized by non-linear transformations on eight pseudoscalar fields ξ_ℓ $(\ell = 1, 2, \cdots 8)$, which correspond to eight Goldstone bosons. Let β_ℓ and α_ℓ be the sixteen parameters which specify a group element in terms of the SU(3) and chiral generators. We shall use Gell-Mann's SU(3) matrices λ_ℓ and later also $\lambda_0 = \sqrt{\frac{2}{3}}$. In terms of the three by three matrices

$$\xi = \sum_{\ell=1}^{8} \lambda_\ell \xi_\ell \,, \quad \beta = \sum_{\ell=1}^{8} \lambda_\ell \beta_\ell \,, \quad \alpha = \sum_{\ell=1}^{8} \lambda_\ell \alpha_\ell$$

one can write the non-linear realization as $\xi \to \xi'$ where

$$e^{-2i\xi'} = e^{-i(\alpha-\beta)} \, e^{-2i\xi} \, e^{-i(\alpha+\beta)} \,.$$

Clearly, for $\alpha = 0$, this gives a linear representation of the diagonal, parity conserving SU(3) subgroup

$$\xi' = e^{i\beta} \xi \, e^{-i\beta} \,.$$

Chiral transformations, however, are realized non-linearly. It turns out that the normalized pseudoscalar fields are given by

$$\xi_\ell = \frac{1}{a} \pi_\ell$$

where $a = F_\pi^{-1}$ up to SU(3) symmetry breaking effects.

The problem of constructing chiral invariant Lagrangians has been treated in the literature and we shall not go into it here. Symmetry breaking terms can be characterized by their transformation properties under the chiral group. This poses the problem of finding functions of the eight pseudoscalar fields which transform linearly under the group. [17] For instance the nine scalar fields S_0, S_ℓ and the nine pseudoscalars P_0, P_ℓ defined by

$$\sum_{a=0}^{8} (S_a + iP_a)\lambda_a = e^{-2i\xi} \,,$$

obviously transform linearly under the entire group. This is the so-called $(3,\bar{3})$ representation. One can write explicitly

$$4 S_a = \mathrm{Tr}\, \lambda_a \left(e^{-2i\xi} + e^{2i\xi} \right)$$

$$4 P_a = i\, \mathrm{Tr}\, \lambda_a \left(e^{-2i\xi} - e^{2i\xi} \right) .$$

The symmetry breaking can be taken as a linear combination of S_0 and S_8. This choice has been reasonably successful. [18]

As mentioned in the previous section, the formally simplest way to extend chiral invariant Lagrangians so as to have conformal invariance is to use pseudoscalar fields of weight zero, rather than one. The results are the same on the mass shell. We follow here this procedure, so the fields π_ℓ will be taken to have weight zero. We are therefore led to study the expression

$$-\tfrac{1}{4}\lambda_0 r^4 \varphi^4 - \left(a_0 S_0 + a_8 S_8 \right) (b\varphi)^{w} + \text{const}$$

as the simplest possible symmetry breaking ansatz. Actually the first term is a chiral and conformal invariant, while the second is characterized by its simple transformation properties under both groups. As before we expand in powers of the fields φ' and π_ℓ. We find first

$$S_0 = \sqrt{\tfrac{3}{2}} - 2\sqrt{\tfrac{2}{3}} \sum_{i=1}^{8} \xi_i^2 + \cdots = \sqrt{\tfrac{3}{2}} - 2\sqrt{\tfrac{2}{3}}\, a^2 \sum_{i=1}^{8} \pi_i^2 + \cdots =$$

$$= \sqrt{\tfrac{3}{2}} - 2\sqrt{\tfrac{2}{3}}\, a^2 \left(\vec{\pi}^2 + \bar{K}K + \eta^2 \right) + \cdots$$

$$S_8 = -2 \sum_{i,j=1}^{8} d_{8ij}\, \xi_i \xi_j = -2a^2 \left(\tfrac{1}{\sqrt{3}} \vec{\pi}^2 - \tfrac{1}{2\sqrt{3}} \bar{K}K - \tfrac{1}{\sqrt{3}} \eta^2 \right) + \cdots .$$

Setting equal to zero the coefficient of the term linear in φ' gives the relation

$$\lambda_0 r^4 + w a_0 b^4 \sqrt{\tfrac{3}{2}} = 0 .$$

For the masses one finds then

$$m_{\varphi'}^2 = a_0 \sqrt{\tfrac{3}{2}}\, b^2 w(w-4) ,$$

$$m_\pi^2 = -4a^2\left(a_0\sqrt{\tfrac{2}{3}} + a_8\sqrt{\tfrac{1}{3}}\right),$$

$$m_K^2 = -4a^2\left(a_0\sqrt{\tfrac{2}{3}} - a_8\tfrac{1}{2}\sqrt{\tfrac{1}{3}}\right),$$

$$m_\eta^2 = -4a^2\left(a_0\sqrt{\tfrac{2}{3}} - a_8\sqrt{\tfrac{1}{3}}\right).$$

Naturally, the pseudoscalar masses satisfy the Gell-Mann Okubo relation. For $a_8/a_0 = -\sqrt{2}$ one has $m_\pi = 0$ and chiral SU(2) × SU(2) is valid; from the actual masses one determines $a_8/a_0 \approx -1.25$. Finally one can write, as a consequence of the special way conformal invariance is broken, the relation

$$m_{\varphi'}^2 = w(4-w)\frac{r^2}{2}\left(m_\pi^2 + 2m_K^2\right) , \quad a = \frac{b}{2r} .$$

For $w = 1$ and identifying the field φ' with the $\epsilon(700)$ we would need $r \approx 0.8$.

14. Strong Gravitation

We have seen how the trace of the energy momentum tensor can be dominated by a scalar particle. We wish now to find a Lagrangian formulation which describes how the tensor itself can be dominated by a spin two particle. The lowest mass candidate for such a particle is the $f(1260)$ resonance. In view of its rather large mass, the extrapolations involved in applying the low energy theorems may seem to have little justification and the energy dependence of the form factors may be expected to alter considerably the results. Nevertheless, it is interesting that a very elegant Lagrangian formulation exists[19], and we feel that it deserves to be tested. In doing so one will also have to bear in mind the complication due to the mixing between the f and the $f'(1514)$, which carry the same quantum numbers.

We follow a line of thought completely analogous to that of Section 6, which provided us with a Lagrangian formulation of vector meson dominance. Here we satisfy the conservation identities of Section 9 or 11 by means of an effective action invariant under Einstein and local Lorentz transformations. The tensor dominance assumption requires that the gravitational field should not couple directly with the phenomenological fields of the hadrons, but should instead go through the intermediary of a spin two hadronic field to which it is coupled bilinearly. Finally we must ensure that, when all couplings are turned off, the spin two hadron describes a spin two particle without ghosts, i.e. satisfies the Fierz equations

$$f_{ab} = f_{ba} \qquad \partial_a f_{ab} = 0$$

$$(\Box - m_f^2)\, f_{ab} = 0 \qquad f_{aa} = 0$$

It turns out that these requirements are sufficient to determine the effective action in its general structure.

We describe the spin two hadron in terms of a vierbein field

$$u_a{}^{\mu} = \delta_{a\mu} - \lambda f_{a\mu}$$

where λ is a universal constant which turns out to characterize the strong interactions of the field $f_{a\mu}$. The gravitational field will also be described by a vierbein field

$$e_{a\mu} = \delta_{a\mu} + \mu h_{a\mu}$$

and the constant μ is related to the gravitational constant. Introducing for a moment full dimensions one has

$$\mu^2 c^2 = \frac{8\pi K}{c^2} = \frac{8\pi}{c^2} \, 6.7 \times 10^{-8} = 1.87 \times 10^{-27} \, \frac{\text{cm}}{\text{g}} \, .$$

If we denote collectively with ψ all other hadrons, the effective action can be written as

$$A = A_E [u_a{}^{\mu}, \psi] + \frac{m_f^2}{\lambda^2} \int d_4 x \, (3u - u_a{}^{\mu} u \, e_{a\mu} + e)$$

where $u = \det u_{a\mu}$ (in this section we write all local Lorentz indices as lower indices and do not explicitly use the Lorentz metric tensor) . Since the Ward identities require Einstein and Lorentz invariance of the total action, and the mass term is clearly invariant, the term A_E must also be invariant and we may approximate it at low energy by the Einstein action of general relativity. We can write then

$$A_E = - \frac{1}{2\lambda^2} \int d_4 x \, R u + \cdots$$

where the dots denote terms which refer to the other hadrons and to their interactions with the field $f_{a\mu}$. Since it is clear how one should write these terms in each case, we shall not make them explicit. The contracted curvature tensor is given by the formulas

$$R_{ab,\mu\nu} = \partial_\mu \omega_{\nu ab} - \partial_\nu \omega_{\mu ab} - \omega_{\mu ac}\omega_{\nu cb} + \omega_{\nu ac}\omega_{\mu cb} =$$
$$= -R_{ba,\mu\nu} = -R_{ab,\nu\mu} . \quad R_{a\mu} = \mu_b{}^\nu R_{ab,\mu\nu} . \quad R = \mu_a{}^\mu R_{a\mu} .$$

and the normalization $-\frac{1}{2\lambda^2}$ is chosen so as to obtain a correctly normalized kinetic energy for the field $f_{a\mu}$. The $\omega_{\nu ab}$ are given by the formulas of Section 11, but are expressed in terms of $\mu_a{}^\mu$ instead of $e_a{}^\mu$.

The above mass term consists of the sum of three terms. The second of them describes the "bilinear" interaction between the field $f_{a\mu}$ and the gravitational field. It could actually be written as a bilinear expression if we chose the field $\mu_a{}^\mu \mu$ instead of the field $\mu_a{}^\mu$ to describe the spin two hadron. The first and third term are needed in order to ensure that the action density vanishes at infinity in space time. The relative coefficients are chosen so that this happens and also so that there is no term linear in $f_{a\mu}$. Finally, the entire expression has the property that the terms quadratic in $f_{a\mu}$ give in flat space $(e_{a\mu} = \delta_{a\mu})$ simply a Pauli-Fierz mass term

$$-\frac{1}{2} m_f^2 \int d_4x \left(f_{ab} f_{ba} - f_{aa}^2 \right) .$$

This can be easily verified by using the expansion

$$\mu = \det \mu_{a\mu} = 1 + \lambda f_{aa} + \frac{1}{2}\lambda^2 \left(f_{ab} f_{ba} + f_{aa}^2 \right) + \dots .$$

It is interesting to observe that these requirements do not determine the mass term uniquely. In addition to the above there are exactly three other inequivalent choices, which we give here

$$\frac{m_f^2}{\lambda^2}\int d_4x\left(u - u_{a\mu}e_a{}^{\mu}e + 3e\right) ,$$

$$\frac{m_f^2}{\lambda^2}\int d_4x\left(u - u_a{}^{\mu}u^{1/2}e_{a\mu}e^{1/2} + 3e\right) ,$$

and

$$\frac{m_f^2}{\lambda^2}\int d_4x\left(3u - u_{a\mu}u^{1/2}e_a{}^{\mu}e^{1/2} + e\right) .$$

As is apparent, two of these four choices are obtained from the other two by switching the roles of the fields $u_a{}^{\mu}$ and $e_{a\mu}$. Depending on which mass term one chooses, it will be appropriate to use a different field for the spin two hadron. The first mass term we gave is preferable because the field $e_{a\mu}$ has a privileged role in describing the local properties of gravitation, as opposed say to the field $e_a{}^{\mu}e$. In the following we shall therefore use that first mass term.

The fact that the four possible mass terms, by construction, give the Pauli-Fierz mass term when expanded to second order already indicates that in absence of interaction no ghosts are present. To verify it explicitly let us derive the field equations, using the first form of the mass term. For arbitrary $\delta u_a{}^{\mu}$ one obtains, by standard methods,

$$\delta\left(-\frac{1}{2\lambda^2}\int d_4x\, Ru\right) = -\frac{1}{2\lambda^2}\int d_4x\; u\left(2R_{a\mu} - u_{a\mu}R\right)\delta u_a{}^{\mu}$$

On the other hand, since

$$\delta u = -u\, u_{a\mu}\, \delta u_a{}^{\mu} = u\, u_a{}^{\mu}\, \delta u_{a\mu} ,$$

we have

$$\delta \frac{m^2}{\lambda^2} \int d_4x \left(3u - u_a{}^a u + 1\right) d_4 x =$$

$$= \frac{m^2}{\lambda^2} \int d_4x \; u\left(-3u_{a\mu} + u_b{}^b u_{a\mu} - \delta_{a\mu}\right) \delta u_a{}^{\mu} .$$

Therefore, in flat space and in the absence of other hadrons, the field equations are

$$-u\left(R_{a\mu} - \tfrac{1}{2} u_{a\mu} R\right) + m^2 u\left(-3u_{a\mu} + u_b{}^b u_{a\mu} - \delta_{a\mu}\right) = 0 .$$

It is easy to see that, up to terms linear in $f_{a\mu}$,

$$\frac{2}{\lambda} R_{a\mu}^{LIN} = \partial_b^2 \left(f_{a\mu} + f_{\mu a}\right) + 2\partial_a \partial_\mu f_{bb} +$$

$$-\partial_a \partial_b \left(f_{\mu b} + f_{b\mu}\right) - \partial_\mu \partial_b \left(f_{ab} + f_{ba}\right) = \frac{2}{\lambda} R_{\mu a}^{LIN} ,$$

and

$$\frac{1}{\lambda} R^{LIN} = 2\partial_a \partial_b \left(-f_{ab} + \delta_{ab} f_{cc}\right) .$$

The linearized equations are, therefore,

$$-\frac{1}{\lambda}\left(R_{a\mu}^{LIN} - \tfrac{1}{2}\delta_{a\mu} R^{LIN}\right) + m^2 \left(f_{a\mu} - \delta_{a\mu} f_{bb}\right) = 0 .$$

It follows immediately that $f_{a\mu}$ is symmetric. Since

$$\partial_a \left(R_{a\mu}^{LIN} - \tfrac{1}{2} \delta_{a\mu} R^{LIN}\right) \equiv 0 ,$$

we see also that

$$\partial_a (f_{a\mu} - \delta_{a\mu} f_{bb}) = 0 ,$$

and this implies that

$$R^{LIN} = 0 .$$

Taking the trace of the linearized field equations we obtain then also

$$f_{aa} = 0 .$$

Finally, using these results, the field equations can be simplified to

$$(\Box - m_f^2) f_{a\mu} = 0 .$$

Therefore, in absence of interactions, $f_{a\mu}$ is a pure spin two Fierz field. Notice that it is not required a priori to be symmetric (the vierbein field is not symmetric) but it is so as a consequence of the field equations.

If we refer to our work of Section 11 we see that the energy momentum tensor is given by

$$e T_a{}^\mu = \frac{\delta A}{\delta e_{a\mu}} = -\frac{m_f^2}{\lambda^2} u_a^\mu u$$

We know from there that the field equations imply that the Lorentz tensor

$$T_{ab} = e_{b\mu} T_a{}^\mu$$

is symmetric and that the covariant conservation equation

$$\partial_\mu (e\, e_{a\lambda} T_a{}^\mu) - e\, \partial_\lambda e_{a\mu} T_a{}^\mu = 0$$

is valid. By going to flat space we see that the full non-linear field equations imply

$$\mu_a{}^b = \mu_b{}^a$$

or

$$f_{ab} = f_{ba}$$

and

$$\partial_\mu (\mu \mu_a{}^\mu) = 0 .$$

We understand in this way the origin of two results of the preceding paragraph.

Until now the gravitational field has been taken as a given external field. Let us now add its action to the terms already discussed. The total action becomes

$$-\frac{1}{2\lambda^2}\int d_4x \, \mu \, R[\mu_{a\mu}] - \frac{1}{2\mu^2}\int d_4x \, e \, R[e_{a\mu}] + \text{mass term} + \dots$$

where the dots denote terms describing the hadrons and their interactions with the field $\mu_{a\mu}$ and, if we wish, also the leptons and their interactions with the field $e_{a\mu}$. This total action presents a remarkable symmetry between the two fields $e_{a\mu}$ and $\mu_{a\mu}$. By expansion one verifies that the four mass terms coincide up to second order and have the form

$$\text{mass term} = -\frac{1}{2} m_f^2 \int d_4x \left\{ \left(f_{ab} - \frac{\mu}{\lambda} h_{ab}\right)\left(f_{ba} - \frac{\mu}{\lambda} h_{ba}\right) - \left(f_{aa} - \frac{\mu}{\lambda} h_{aa}\right)^2 \right\} + \\ + \dots$$

The fact that the two fields enter in the particular combination indicated is a consequence of the Einstein and local Lorentz invariance of the original full non-linear expression. Clearly the above mass matrix contains mixed bilinear terms and must be diagonalized. This can be easily achieved by introducing the new orthonormal fields

$$f'_{ab} = \left(1+\frac{\mu^2}{\lambda^2}\right)^{-1/2}\left(f_{ab} - \frac{\mu}{\lambda} h_{ab}\right)$$

$$h'_{ab} = \left(1+\frac{\mu^2}{\lambda^2}\right)^{-1/2}\left(h_{ab} + \frac{\mu}{\lambda} f_{ab}\right)$$

We see that the field f'_{ab} corresponds to a mass $\bar{m}_f$ slightly different from the parameter m_f :

$$\bar{m}_f^2 = m_f^2\left(1+\frac{\mu^2}{\lambda^2}\right).$$

The field h'_{ab} corresponds to a massless particle: this is a consequence of the invariances of the original mass term. The analogy with gauge invariant vector mixing [20] is apparent.

15. Concluding Remarks

We have tried to develop systematically the theory of effective actions as generating functionals of one-particle irreducible vertices. There is an alternative point of view in working with non-linear Lagrangians. One takes them seriously as Lagrangians to be quantized and attempts to calculate with diagrams containing arbitrary numbers of loops. One is then faced by the problem of divergences, since one is usually dealing with non-renormalizable field theories, and one attempts to solve the problem by means of partial summation of classes of diagrams. These partial sums can sometimes be shown to be more convergent than the individual terms of the sum after which the theory can be shown to be renormalizable or even finite. Apart from any consideration of physical plausibility, this approach is confronted by a technical difficulty because the partial summations produce ambiguous answers.[21] The ambiguity seems to be of the same order as that which one would encounter if one applied to the non-renormalizable field theory under consideration the usual method of renormalization theory, which would require the introduction of infinitely many counter terms in the Lagrangian and therefore of infinitely many arbitrary constants. The removal of the ambiguities should be based on some physical principle and not on ad hoc prescriptions, but until now no such principle appears to have been found. In view of this we have not followed this point of view in these lectures.

A number of fine points concerning conformal transformations have been completely ignored in these lectures. In particular we have mostly attributed to the fields their canonical dimensions, while it is known that interactions can change the dimension of a field.[22] Furthermore, we have used the trace identities in their simpler form, in spite of the fact that one knows from renormalizable field theory examples that they are not valid in that simple form.[23] Our only excuse is that in a beginning phenomenological analysis these modifications can probably be ignored. For a study of these questions we refer to the existing literature.

Acknowledgement

These lectures are based in part on work done in collaboration with J. Wess, to be published soon. The author acknowledges useful conversations with S. Deser, J. Ellis, and H. Schnitzer.

References

1. J. Schwinger, Proc. Nat. Acad. of Sciences 37, 452 and 455 (1951); G. Jona Lasinio, Nuovo Cimento 34, 1790 (1964). In this paper by Jona Lasinio one finds many of the ideas developed in Sections 2-4. See also G. Parisi and M. Testa, Nuovo Cimento 67, 13 (1970).
2. There may of course arise so called anomalous terms in the Ward identities. If so they can be included. For the definition of these terms see Adler's lectures in this series.
3. J. Goldstone, Nuovo Cimento 19, 154 (1961).
4. M. Gell-Mann and M. Levy, Nuovo Cimento 16, 53 (1960); B. W. Lee, Nucl. Phys. B9, 649 (1969).
5. See, e.g., J. Wess and B. Zumino, Phys. Rev. 163, 1727 (1967).
6. S. Coleman, J. Wess and B. Zumino, Phys. Rev. 177, 2239 (1969); ditto with C. G. Callan, Phys. Rev. 177, 2247 (1969). Earlier references can be found in the first of these papers.
7. P. G. O. Freund and Y. Nambu, Phys. Rev. 174, 1741 (1968).
8. C. G. Callan, S. Coleman and R. Jackiw, Annals of Physics (N.Y.) 59, 42 (1970).
9. See e.g. J. Wess, Nuovo Cimento 18, 1086 (1960); D. J. Gross and J. Wess, Phys. Rev. D2, 753 (1970), where other references can be found.
10. S. Coleman and R. Jackiw, MIT preprint (1970); C. G. Callan, Cal-Tech report CALT-68-257 (1970); K. Symanzik, DESY report 70/20 (1970).
11. H. Weyl, Z. Physik. 56, 330 (1929). See also J. Schwinger, Phys. Rev. 130, 800 and 1253 (1962), T. W. B. Kibble, Journal of Math. Phys. 4, 1433 (1963).

12. W. Pauli, H.P.A. 13, 204 (1940).

13. This was first pointed out, using a different method, by A. Salam and J. Strathdee, Phys. Rev. 184, 1750 and 1760 (1969). See also C. J. Isham, A. Salam and J. Strathdee, Phys. Letters 31B, 300 (1970) and ICTP reports, Trieste, IC/70/5 and IC/70/17. The work of this Section 12 is closely related to the work described in these papers. However, the point of view and some of the results are different.

14. J. Ellis, Cambridge Un. preprint AMTP 70/20 (1970).

15. P. Achutan and F. Steiner, University of Karlsruhe preprint (1970).

16. See e.g. Bugg, Nucl. Phys. B.5, 29 (1968).

17. See reference 6. The quantity ξ_ℓ of those papers is twice that used here.

18. The $(3,\bar{3})$ representation for $SU(3) \times SU(3)$ symmetry breaking has been used in particular by S. L. Glashow and S. Weinberg, Phys. Rev. Letters 20, 224 (1968) and by M. Gell-Mann, R. J. Oakes and B. Renner, Phys. Rev. 175, 2195 (1968). These authors do not use the method of effective actions.

19. J. Wess and B. Zumino, to be published. Lagrangians of the type considered in this section have been studied by V. I. Ogievetsky and I. V. Polubarinov, Annals of Physics 35, 167 (1965). Their motivation and point of view is rather different from ours. Furthermore our use of the vierbein field, which from our point of view is mandatory, allows a simpler and more satisfactory solution of the formal problems.

20. See e.g. N. Kroll, T. D. Lee and B. Zumino, Phys. Rev. 157, 1376 (1967).

21. See e.g. B. W. Lee and B. Zumino, Nucl. Phys. B13, 671 (1969). Reference to other work can be found in this paper.

22. K. G. Wilson, Phys. Rev. 179, 1499 (1969); also SLAC-PUB 737 and 741 (1970).

23. See the papers cited in ref. 10.

Additional References.

24. F. Gürsey, Ann. Phys. (N.Y.) 24, 211 (1963).

25. H. A. Kastrup, Phys. Rev. 150, 1183 (1966) and Nucl. Phys. B15, 179 (1969)

26. G. Mack, Nucl. Phys. B5, 499 (1968); Phys. Letters 26B, 575 (1968).

27. D. Boulware and S. Deser, J. Math. Phys. 8, 1468 (1968).

28. P. G. O. Freund, A. Maheshwari, E. Shonberg, Astrophys. J. 157, 857 (1969).

29. G. Mack and A. Salam, Ann. Phys. (N.Y.) 53, 174 (1969).

30. Raman, several Brown University preprints (1969).

31. M. Gell-Mann, report CALT-68-244 (1969).

32. B. Zumino, Proceedings of the Seminar on Vector Meson and Electro-magnetic Interactions, Dubna, Sep. 1969. At this seminar the author first described the point of view about effective actions developed in Sections 1 to 6 of the present lectures.

INDEX